Lecture Notes in Physics

Springer-Verlag Berlin Heidelberg GmbH

The Editorial Policy for Proceedings

The series Lecture Notes in Physics reports new developments in physical research and teaching – quickly, informally, and at a high level. The proceedings to be considered for publication in this series should be limited to only a few areas of research, and these should be closely related to each other. The contributions should be of a high standard and should avoid lengthy redraftings of papers already published or about to be published elsewhere. As a whole, the proceedings should aim for a balanced presentation of the theme of the conference including a description of the techniques used and enough motivation for a broad readership. It should not be assumed that the published proceedings must reflect the conference in its entirety. (A listing or abstracts of papers presented at the meeting but not included in the proceedings could be added as an appendix.)
When applying for publication in the series Lecture Notes in Physics the volume's editor(s) should submit sufficient material to enable the series editors and their referees to make a fairly accurate evaluation (e.g. a complete list of speakers and titles of papers to be presented and abstracts). If, based on this information, the proceedings are (tentatively) accepted, the volume's editor(s), whose name(s) will appear on the title pages, should select the papers suitable for publication and have them refereed (as for a journal) when appropriate. As a rule discussions will not be accepted. The series editors and Springer-Verlag will normally not interfere with the detailed editing except in fairly obvious cases or on technical matters.
Final acceptance is expressed by the series editor in charge, in consultation with Springer-Verlag only after receiving the complete manuscript. It might help to send a copy of the authors' manuscripts in advance to the editor in charge to discuss possible revisions with him. As a general rule, the series editor will confirm his tentative acceptance if the final manuscript corresponds to the original concept discussed, if the quality of the contribution meets the requirements of the series, and if the final size of the manuscript does not greatly exceed the number of pages originally agreed upon. The manuscript should be forwarded to Springer-Verlag shortly after the meeting. In cases of extreme delay (more than six months after the conference) the series editors will check once more the timeliness of the papers. Therefore, the volume's editor(s) should establish strict deadlines, or collect the articles during the conference and have them revised on the spot. If a delay is unavoidable, one should encourage the authors to update their contributions if appropriate. The editors of proceedings are strongly advised to inform contributors about these points at an early stage.
The final manuscript should contain a table of contents and an informative introduction accessible also to readers not particularly familiar with the topic of the conference. The contributions should be in English. The volume's editor(s) should check the contributions for the correct use of language. At Springer-Verlag only the prefaces will be checked by a copy-editor for language and style. Grave linguistic or technical shortcomings may lead to the rejection of contributions by the series editors. A conference report should not exceed a total of 500 pages. Keeping the size within this bound should be achieved by a stricter selection of articles and not by imposing an upper limit to the length of the individual papers. Editors receive jointly 30 complimentary copies of their book. They are entitled to purchase further copies of their book at a reduced rate. As a rule no reprints of individual contributions can be supplied. No royalty is paid on Lecture Notes in Physics volumes. Commitment to publish is made by letter of interest rather than by signing a formal contract. Springer-Verlag secures the copyright for each volume.

The Production Process

The books are hardbound, and the publisher will select quality paper appropriate to the needs of the author(s). Publication time is about ten weeks. More than twenty years of experience guarantee authors the best possible service. To reach the goal of rapid publication at a low price the technique of photographic reproduction from a camera-ready manuscript was chosen. This process shifts the main responsibility for the technical quality considerably from the publisher to the authors. We therefore urge all authors and editors of proceedings to observe very carefully the essentials for the preparation of camera-ready manuscripts, which we will supply on request. This applies especially to the quality of figures and halftones submitted for publication. In addition, it might be useful to look at some of the volumes already published. As a special service, we offer free of charge LaTeX and TeX macro packages to format the text according to Springer-Verlag's quality requirements. We strongly recommend that you make use of this offer, since the result will be a book of considerably improved technical quality. To avoid mistakes and time-consuming correspondence during the production period the conference editors should request special instructions from the publisher well before the beginning of the conference. Manuscripts not meeting the technical standard of the series will have to be returned for improvement.

For further information please contact Springer-Verlag, Physics Editorial Department II, Tiergartenstrasse 17, D-69121 Heidelberg, Germany

Aa. Sandqvist P. O. Lindblad (Eds.)

Barred Galaxies and Circumnuclear Activity

Proceedings of the NOBEL SYMPOSIUM 98
Held at Stockholm Observatory,
Saltsjöbaden, Sweden,
30 November – 3 December 1995

Springer

Editors

Aage Sandqvist
Per Olof Lindblad
Stockholm Observatory
S-13336 Saltsjöbaden, Sweden

Cataloging-in-Publication Data applied for.

Die Deutsche Bibliothek - CIP-Einheitsaufnahme

Barred galaxies and circumnuclear activity : proceedings of the Nobel Symposium 98, held at Stockholm Obeservatory, Saltsjöbaden, Sweden, 30 November - 3 December 1995 / Nobel Symposium 98. Aage Sandqvist ; Per Olof Lindblad (ed.). -
(Lecture notes in physics ; Vol. 474)
ISBN 978-3-662-14098-7   ISBN 978-3-540-70694-6 (eBook)
DOI 10.1007/978-3-540-70694-6

NE: Sandqvist, Aage [Hrsg.]; Nobel Symposium <98, 1995, Saltsjöbaden>; GT

ISSN 0075-8450
ISBN 978-3-662-14098-7

Originally published by Springer-Verlag Berlin Heidelberg New York in 1996
Softcover reprint of the hardcover 1st edition 1996

Typesetting: Camera-ready by the authors/editors
Cover design: *design & production* GmbH, Heidelberg
SPIN: 10520109   55/3142-543210 - Printed on acid-free paper

Preface

The year 1995 marked the centenary of two events: (i) 26 November, 1895 is the date of birth of Bertil Lindblad, and (ii) 27 November, 1895 is the date on which Alfred Nobel signed his testament that led to the Nobel Prizes. In order to celebrate these events a Nobel Symposium (No. 98) was held from 30 November to 3 December, 1995 in Saltsjöbaden, Sweden, near Stockholm Observatory where Bertil Lindblad was active as Director during most of his scientific life.

The topic of the Nobel Symposium, namely "Barred Galaxies and Circumnuclear Activity", was attuned to the activities of the research group at Stockholm Observatory, led by Per Olof Lindblad who had recently retired. With the successful refurbishing of the Hubble Space Telescope, the completion of the Keck Telescope and the Very Long Baseline Array, and the imminent construction of the Very Large Telescope, the time was considered ripe for bringing together the leading authorities on the observational, numerical, and theoretical aspects of the bars in barred galaxies and their interactions with the circumnuclear regions near the cores of these often active galaxies.

The Nobel Symposium was sponsored by the Nobel Foundation through its Nobel Symposium Fund. The invitations to attend were limited to about 30 scientists (by the symposium statutes of the Nobel Foundation). These included a son and two grandsons of Bertil Lindblad. There was an additional audience of about 40 other interested scientists and advanced students. The Scientific Organizing Committee consisted of Francoise Combes, Robert A.E. Fosbury, Steven Jörsäter, Per Olof Lindblad, and Aage Sandqvist, the latter three of whom also served as the Local Organizing Committee.

The Symposium was officially opened by Michael Sohlman, the Executive Director of the Nobel Foundation. After the opening, Alar Toomre gave some historical remarks on Bertil Lindblad's work on galactic dynamics. The symposium then got down to business and the papers are presented in these proceedings in the order in which they were given. Two of the invited participants were unfortunately unable to attend the symposium, but one had already prepared the manuscript and we are happy to include it in this volume.

The symposium banquet was held at the Stockholm Observatory, following aperitifs at the Observatory residence of Gunilla and Per Olof Lindblad (also formerly the residence of Bertil Lindblad). At this social event, George Contopoulos and Donald Lynden-Bell reminisced about their interactions with Bertil Lindblad. The following evening included a Stockholm Opera performance of Guiseppi Verdi's Requiem (which was given extra significance by being dedicated to the victims of the "Estonia") and a subsequent supper in the cellars of the eighteenth-century Old Stockholm Observatory.

As always, a symposium means a large amount of work for many people, and we should like to thank all of them. We especially wish to express our gratitude to Anne-Marie Tannenberg for functioning as secretary and hostess of this Nobel Symposium.

Saltsjöbaden, 31 May 1996 — Aage Sandqvist

Contents

List of Participants

Athanassoula, Evangelie lia@obmara.cnrs-mrs.fr
Observatoire de Marseille, 2 Place Le Verrier, 13248 Marseille Cedex 04, France

Axon, David axon@stsci.edu
Astrophysics Division of ESA, Space Telescope Science Institute, 3700 San Martin Drive, Baltimore MD 21218, USA

Bacon, Roland bacon@orion.univ-lyon1.fr
Centre de Recherche Astronomique de Lyon, Observatoire de Lyon, F-69561 St Genis-Laval cedex, France

Blitz, Leo blitz@astro.umd.edu
Astronomy Department, University of Maryland, College Park MD 20742, USA

Bosma, Albert bosma@batis.cnrs-mrs.fr
Observatoire de Marseille, 2 Place Le Verrier, F-13248 Marseille Cedex 4, France

Combes, Francoise bottaro@mesioa.obspm.fr
DEMIRM, Observatoire de Paris, 61 Av. de l'Observatoire, F-75014 Paris, France

Contopoulos, George gcontop@atlas.uoa.ariadne-t.gr
Department of Astronomy, University of Athens, Panepistimiopolis, GR-157 84 Athens, Greece

Ford, Holland ford@jhufos.pha.jhu.edu
Department of Physics and Astronomy, Johns Hopkins University, 3701 San Martin Drive, Baltimore MD 21218, USA

Fosbury, Robert rfosbury@eso.org
Space Telescope - European Coordinating Facility, D-85748 Garching bei München, Germany

Fridman, Alexei afridman@inasan.rssi.ru
Institute of Astronomy, Russian Academy of Sciences, 48 Pyatnitskaya St., Moscow 109017, Russia

Genzel, Reinhard genzel@mpe-garching.mpg.de
MPI für Extraterrestrische Physik, Giessenbachstrasse, D-85748 Garching, Germany

Heckman, Timothy heckman@rowland.pha.jhu.edu
Department of Physics and Astronomy, The Johns Hopkins University, Baltimore MD 21218, USA

Jörsäter, Steven steven@astro.su.se
Stockholm Observatory, S-133 36 Saltsjöbaden, Sweden

Kalnajs, Agris agris@merlin.anu.edu.au
Mount Stromlo and Siding Spring Observatories, Private Bag, Weston Creek PO, 2611, Australia

Knapen, Johan knapen@star.herts.ac.uk
Departement de Physique, Universite de Montreal, C.P. 6128, Succursale Centre-Ville, Montreal PQ H3C 3J7, Canada

Kristen, Helmuth helmuth@astro.su.se
Stockholm Observatory, S-133 36 Saltsjöbaden, Sweden

Lindblad, Per A.B. pabli@astro.su.se
Stockholm Observatory, S-133 36 Saltsjöbaden, Sweden

Lindblad, Per Olof po@astro.su.se
Stockholm Observatory, S-133 36 Saltsjöbaden, Sweden

Lynden-Bell, Donald asd@ast.cam.ac.uk
Institute of Astronomy, The Observatories, Cambridge CB2 0HA, UK

Nelson, Charles cnelson@stsci.edu
Space Telescope Science Institute, 3700 San Martin Drive, Baltimore MD 21218, USA

Pfenniger, Daniel pfennige@scsun.unige.ch
Geneva Observatory, University of Geneva, CH-1290 Sauverny, Switzerland

Sandqvist, Aage sandqvis@astro.su.se
Stockholm Observatory, S-133 36 Saltsjöbaden, Sweden

Sellwood, Jerry sellwood@physics.rutgers.edu
Rutgers University, Department of Physics and Astronomy, P.O. Box 849, Piscataway NJ 08855, USA

Shlosman, Isaac shlosman@asta.pa.uky.edu
Department of Physics and Astronomy, University of Kentucky, Lexington KY 40506-0055, USA

Sofue, Yoshiaki sofue@sof.mtk.ioa.s.u-tokyo.ac.jp
Institute of Astronomy, University of Tokyo, Mitaka, Tokyo 181, Japan

Teuben, Peter teuben@astro.umd.edu
Astronomy Department, University of Maryland, College Park MD 20742, USA

Toomre, Alar toomre@math.mit.edu
Massachusetts Institute of Technology, Cambridge MA 02139, USA

Tsvetanov, Zlatan zlatan@pha.jhu.edu
Department of Astronomy and Physics, Center for Astrophysical Sciences, Johns Hopkins University, Baltimore MD 21218, USA

Wilson, Andrew wilson@astro.umd.edu
Astronomy Department, University of Maryland, College Park MD 20742, USA

1. R. Genzel	2. R. Booth	3. L. Blitz	4. D. Lynden-Bell
5. H. Martinet	6. G. Contopoulos	7. S. Jörsäter	8. R. Bacon
9. J. Knapen	10. I. Shlosman	11. D. Pfenniger	12. P.O. Lindblad
13. A. Toomre	14. H. Ford	15. H. Kristen	16. Z. Tsvetanov
17. P. Teuben	18. R. Fosbury	19. A. Wilson	20. A. Kalnajs
21. Y. Sofue	22. D. Axon	23. C. Nelson	24. E. Athanassoula
25. F. Combes	26. P.A.B. Lindblad	27. J. Sellwood	28. A. Bosma
29. Aa. Sandqvist			

Some Historical Remarks on Bertil Lindblad's Work on Galactic Dynamics

Alar Toomre

Massachusetts Institute of Technology, Cambridge, MA 02139 USA

1 Introduction

When I was asked by Per Olof Lindblad to reflect here for a few minutes on the main contributions to galactic dynamics made by his famous father whose 100th birthday occurred on 26 November, I felt honored and delighted of course but at first also overwhelmed by the magnitude of this task. Then I remembered that I had felt even more overwhelmed two decades earlier when, after writing to Per Olof to request copies of just a few of Bertil Lindblad's papers to help me prepare an *Annual Reviews* article on theories of spiral structure, I received a hefty parcel containing about 55 reprints!

Somehow I survived that culture shock. In due course, I worked my way through that entire collection, and read perhaps two dozen of those papers quite thoroughly. To be sure, even then I had already been familiar with a few Lindblad items like his authoritative 1959 article in the *Handbuch der Physik* or his excellent Darwin Lecture from 1948. But it was this fresh round of self-inflicted reading that really filled me with admiration for this man's persistence and general wisdom, and even with considerable sympathy for his occasional foibles, most notably his long obsession with leading spiral arms which in retrospect was surely a mistake. Eventually I gathered my impressions under the headings "Some of Lindblad's ideas" and "Nearly kinematic waves" onto four of the frontmost pages of that review (Toomre 1977), "to start almost at the beginning".

Even today, I still recommend those four pages to anyone eager to follow up on this history. I also recommend the wonderful obituary that Oort (1966) composed soon after Lindblad's death following an operation less than a half-year before his 70th birthday, and likewise — for those who read Swedish — I recommend the warm biographical memoir that Lindblad's long-term colleague Öhman (1975) prepared for the Swedish Academy of Sciences about a decade later. Those two knew Bertil Lindblad very well. I myself, alas, never met him.

2 Major Contributions

Lindblad had essentially three triumphs in his long career on galaxy dynamics. The first occurred in the mid-1920s, when he was barely 30, whereas the other two came only when he was already in his 60s — a fact that should encourage others who are getting on in years. His first and last triumphs were actually rather shared discoveries, but the middle one was uniquely his own. All three dealt with disks in rotation.

2.1 Rotation of the Galaxy

One tends to forget these days that it wasn't always obvious that our Milky Way is a multi-component system supported against gravity largely by its rotation. There were real people involved in the long process of making this obvious, and those people were not only Kapteyn or Shapley or Strömberg, but especially Lindblad and Oort.

Those last two held a delightful tennis match of ideas on this subject during the years 1925-27, as reviewed in detail by P. O. Lindblad (1980) for a volume honoring Oort on his 80th birthday. Along with Kapteyn and Strömberg shortly before those three most glorious years, Oort himself had become very interested in high-velocity stars, and in how they seemed to avoid what we now know to be the forward direction; yet by his own admission it never occurred to him that either this avoidance or the asymmetric drift of moderately "hot" stars could arise plausibly from the orbiting of various parts of our system around a distant center roughly matching that for Shapley's globular clusters. It took Lindblad (1925, 1926) to point out how neatly both those riddles of star streaming could be solved by the hypothesis of rotation combined with different amounts of pressure support from random motions. Oort was initially very skeptical, but within a year or so he swung around completely, in large part thanks to his own brilliant supporting evidence (Oort 1927a,b) on the shear flow and spatial rotation of nearby disk stars, as symbolized nowadays by the Oort constants A and B.

Proofs of that first paper titled "Observational evidence confirming Lindblad's hypothesis of a rotation of the galactic system" were transmitted by young Oort to the man 5 years his senior in a famous letter which simultaneously expressed doubt that Lindblad would remember him from some brief past meeting, shame "that I had not been able to think of the explanation myself", and hope that here was "a paper in which you may be especially interested". Interested indeed, Lindblad (1927) soon swatted that ball back with extra spin from his side of the tennis court: It occurred to him in turn that even the curious ellipsoidal distribution of stellar random velocities parallel to the plane of our galaxy follows naturally from the idea that all these stars are in differential rotation, and he also pointed out that the ratio of those velocity axes agreed quite nicely with the ones implied by Oort's new A and B. With that, Lindblad introduced epicyclic motions into galaxy dynamics, though for years afterwards he avoided this Ptolemaic phrase and kept on referring to the epicycles only as "relative orbits".

2.2 Dispersion Orbits

Most remarkably after that fine beginning, it took Lindblad not three further months or years, but three whole decades, to connect this implied epicyclic frequency κ and the ordinary angular speed of rotation Ω into the kinematic wave speeds like $\Omega \pm \kappa/m$ which we very much associate with him nowadays, especially when muttering phrases like "Lindblad resonances".

It was not for a lack of trying. Indeed in several of his analytical papers from the 1940s and early 1950s Lindblad came fairly close to seizing upon such kinematic waves as something quite valuable in their own right. Of course, no one else did any better. For him, this proverbial penny dropped only when 21-cm rotation curves of our galaxy and M 31 began to indicate in the mid-1950s that the combination $\Omega - \kappa/2$ seems to remain reasonably constant over a surprisingly large range of interior radii.

This news, which he himself sleuthed out and reported in Lindblad (1955, 1956), prompted two excellent follow-on papers (Lindblad 1958, 1961) on the dynamics of what he called "dispersion" rings or orbits. He was hoping, it seems, that the interstellar gas and/or the colder stars would somehow aggregate on their own initiative into just a few such orbits in each galaxy — almost like some vastly expanded meteor streams. In retrospect, this dispersion theme never really panned out, but in the process Lindblad discovered something much more valuable: the essence of large-scale kinematic waves in a disk galaxy.

2.3 Density Waves

Even beyond such specific accomplishments, we remember Lindblad most of all for his tremendous lifelong struggle with trying to explain from Newtonian first principles the abundant spiral and bar structures of galaxies.

Much of this work, in retrospect, amounts to a magnificent failure, since even in his finest technical papers like Lindblad and Langebartel (1953) on bar-making the hard analysis always seemed to peter out before reaching any true climax, and it was usually followed only by sketches, speculations and/or — as still in that case — a total ambiguity as to whether the inferred spirals were leading or trailing! To be fair, such complex collective dynamics was perhaps too hard for anyone, no matter how talented, in those mid-20th-century decades before computers, plasma physics, or any inkling of massive halos.

And yet, once modern density-wave enthusiasm took hold from the mid-1960s onward, thanks especially to Lin and Shu but also to Kalnajs, Lynden-Bell and several others, it was possible for Dekker (1975) to reflect barely a decade later that "all problems that in later developments turned out to be important in the theory of spiral structure had, in one way or another, already been touched upon or even studied by Lindblad". Her words remain remarkably true even today.

For me, the best evidence on how close Lindblad actually came to grasping large chunks of the spiral problem resides in his two papers from 1963 and 1964. With rather vivid diagrams, he speculated there "on the possibility of a quasi-stationary spiral structure". For him — much like for Lin and Shu soon afterwards — this involved material travelling from arm to arm of a supposedly bisymmetric wave pattern in an overtaking sense at the smaller radii, and yet backward relative to the same revolving pattern at the larger radii. His details were unconvincing, but no one can accuse him of missing the big picture.

3 Farsighted Comments

Sprinkled amidst the Lindblad papers are various other comments and asides which attest that this man was often thinking far ahead of his time. Here are some of my favorites, to wrap up this summary using mostly his own words.

Regarding galaxy mergers (of all things!), Lindblad (1926) noted in passing that, although infrequent,

> sharp encounters between nebulae ... must probably be considered as highly 'unelastic' and must tend to convert translational into rotational kinetic energy. An encounter of this kind may even lead to a fusion of the respective bodies.

Similarly, though most of his struggles were with supposedly self-made spirals, Lindblad (1941) was open-minded enough to remark not only that for

> the classical spiral M 51 ... it is very probable that the tidal force of the well-known strong condensation at the end of one of the arms is to considerable extent responsible for the complete and beautiful spiral structure,

but also that in the case of M 81

> it is possible that the nebula M 82, situated 38' to the north of M 81, may have a certain influence on this system, if the two nebulae are actually neighbours in space, and not only form an optical pair.

Lindblad (1950) was astonishingly early into wondering about off-center $m = 1$ modes or instabilities, given that

> a wave of this kind means that ... the centre of the coordinate system ... will be subject to accelerations. In many nebulae, e.g. in the large spiral M101, there appears to be a deviation from rotational symmetry which may be due to this mode of variation.

And perhaps most aptly, given the heavy emphasis on bars at this conference, Lindblad (1951) also wrote that

> the most promising approach to a solution of the problem of spiral structure appears to be by way of the barred spiral nebulae.

He went on:

> In the ordinary spirals ... a plainly developed barred structure may not appear, but there is likely to be a characteristic density wave $s = 2$, which causes a departure from rotational symmetry and which incites the formation of spiral structure by its disturbing action on the internal motions in the system.

Translate that to $m = 2$, and he could have been writing yesterday!

References

Dekker, E. (1975): Spiral Structure and the Dynamics of Flat Stellar Systems, Ph. D. thesis, Leiden University

Lindblad, B. (1925): Arkiv Mat. Astron. Fysik **19A**, No. 21

Lindblad, B. (1926): Arkiv Mat. Astron. Fysik **19A**, No. 35

Lindblad, B. (1927): Arkiv Mat. Astron. Fysik **20A**, No. 17

Lindblad, B. (1941): Stockholm Obs. Ann. **13**, No. 10

Lindblad, B. (1948): MNRAS **108**, 214

Lindblad, B. (1950): Stockholm Obs. Ann. **16**, No. 1

Lindblad, B. (1951): Pubs. Obs. Univ. Michigan **10**, 59

Lindblad, B. (1955): Stockholm Obs. Ann. **18**, No. 6

Lindblad, B. (1956): Stockholm Obs. Ann. **19**, No. 7

Lindblad, B. (1958): Stockholm Obs. Ann. **20**, No. 6

Lindblad, B. (1959): in Handbuch der Physik **53**, Springer, Berlin, Heidelberg, p. 21

Lindblad, B. (1961): Stockholm Obs. Ann. **21**, No. 8

Lindblad, B. (1963): Stockholm Obs. Ann. **22**, No. 5

Lindblad, B. (1964): Astrophys. Norv. **9**, 103

Lindblad, B., Langebartel, R. G. (1953): Stockholm Obs. Ann. **17**, No. 6

Lindblad, P. O. (1980): in Oort and the Universe, ed. van Woerden, H., Brouw, W.N., van de Hulst, H.C., Reidel, Dordrecht, p. 59

Oort, J. H. (1927a): Bull. Astron. Inst. Netherlands **3**, 275

Oort, J. H. (1927b): Bull. Astron. Inst. Netherlands **4**, 79

Oort, J. H. (1966): QJRAS **7**, 329

Öhman, Y. (1975): Levnadsteckningar över Kungl. Svenska Vetenskapsakademiens ledamöter, No. 179 (in Swedish)

Toomre, A. (1977): ARA&A **15**, 437

Formation and Evolution Mechanisms of Barred Spiral Galaxies

D. Lynden-Bell

Institute of Astronomy, The Observatories, Cambridge CB2 0HA

Abstract. After reviewing the classical works on triaxial bodies with internal rotation six different proposals for the formation of bars are discussed. The way such bars respond to angular momentum loss to the spiral structure is also considered. Finally the Galactostrophic approximation is introduced and guiding centres are shown to follow Ergos curves. Lindblad-like epicyles about such curves yield a new integral for the stellar dynamics of barred galaxies.

1 Classical Considerations

On receiving the invitation to this Nobel Symposium I thought that preparing for it would spur me to a renewed interest in a beautiful subject and allow me to finish work in hand with J. Barot on the Galactostrophic approximation for stellar orbits in rotating systems. In practice the giving of a new lecture course left me too little time, so this review will be an historical one putting the theory of the barred spirals into the context of the much older subject of rotating gravitating bodies. Nevertheless, some results on the Galactostrophic approximation are included since they give both a new integral of the equations of motion and a useful generalisation of Bertil Lindblad's epicyclic theory.

One of the early works devoted to explaining the barred galaxies was published in the Stockholm Observatoriums Annaler for 1953 by Bertil Lindblad, whose birth we commemorate in this symposium, and R.G. Langebartel. Their Fig. 9 reproduced here as our Fig. 1 shows the $m = 2$ density wave to which the uniformly rotating stellar disc is unstable. It shows the four whorls noted by the authors.

Following Fridman (1994) these can be interpreted as cyclones top and bottom and anticyclones to the sides. Such an interpretation follows kinematically if one imagines the lines of vorticity drawn together with the mass concentrations and expanded apart at the density minima. However, in the interpretation in terms of cyclones it is the gravitational well caused by the matter concentration that dominates over any excess pressure, so the *high density* regions are cyclonic with the Coriolis force balancing the excess gravity of the matter there.

This perturbed configuration not only gave rise to the first mention of density waves in the literature and one of the early indications of a bar-like instability, but gave also a pretty example for the Galactostrophic approximation to galactic orbits which we develop in Sect. 3.

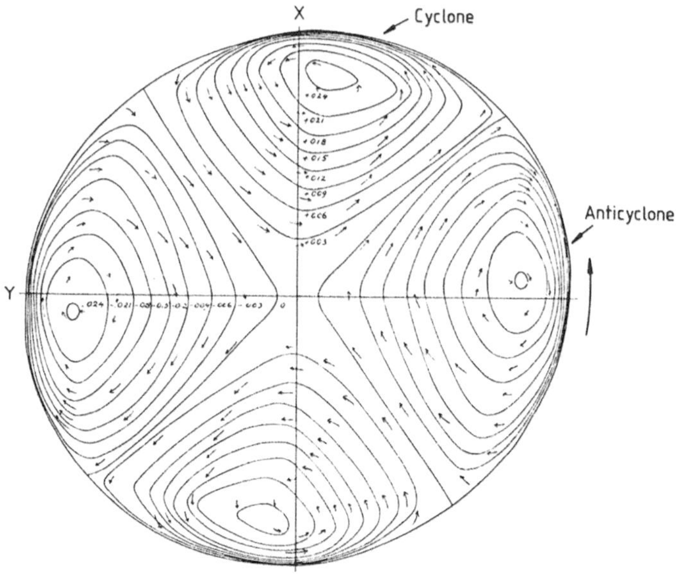

Fig. 1. The four whorls of the density wave $m = 2$. The direction of rotation of the x, y-system is counter-clockwise

The classical theory of rotating bodies started with Newton's prediction of the oblate shape of the slowing rotating Earth. That was followed by Maclaurin's discovery that a homogeneous fluid in rapid uniform rotation, under its own gravity, could be exactly spheroidal. Then Jacobi made the remarkable discovery that $a = b$ was not the only solution for an homogeneous ellipsoidal body in rapid rotation. Indeed at each given angular momentum the Jacobi ellipsoid has lower energy and lower angular velocity whenever it exists.

It is interesting to consider not only the angular momentum $\mathbf{L}$ but also the equatorial circulation C of such bodies following Lynden-Bell (1965).

$$\mathbf{L} = \tfrac{1}{5} M(a^2 + b^2)\Omega \tag{1.1}$$

$$C = \int \mathbf{u} \cdot \mathrm{d}\mathbf{l} = \int \boldsymbol{\omega} \cdot \mathrm{d}\mathbf{S} = \omega \pi ab = 2\Omega \pi ab \tag{1.2}$$

where $\boldsymbol{\omega} = \operatorname{curl} \mathbf{u}$ and the final equality is only true provided the system rotates uniformly.

Notice that

$$\frac{|L|}{MC} = \frac{1}{10\pi}\left(\frac{a}{b} + \frac{b}{a}\right) \tag{1.3}$$

is a function of $\frac{b}{a}$ only, and minimises when $a = b$.

Now consider a homogeneous rotating body whose density is increasing as it shrinks in size. L, C and M are all conserved, so if the system continues to have a uniform angular velocity (spatially) $\frac{b}{a}$ must remain constant.

Thus shrinking Maclaurin spheroids will keep $a = b$, they cannot enter the lower energy $a \neq b$ Jacobi sequence without violating conservation of circulation. Of course a magnetic field or viscosity may allow such a transition but only on the timescale determined by the field strength or the viscosity and the latter can be impressively long (Lynden-Bell 1964).

Of greater interest is the conundrum posed by a shrinking Jacobi ellipsoid because $\frac{b}{a}$ changes along the Jacobi sequence. For an inviscid fluid L, M and C are all conserved yet the dimensionless 'control' parameter $\mu = (\mathbf{L}/M)^2(GM)^{-4/3}(G\rho)^{1/3}$ is being increased with time as ρ increases. μ also increases along the Jacobi sequence but along that $\frac{b}{a}$ decreases whereas we have shown by (1.3) that $\frac{b}{a}$ must remain constant. So what happens as a Jacobi ellipsoid starts to shrink?

The escape route from this conundrum lies through another remarkable classical discovery opened by Dirichlet and Dedekind and explored more fully by Riemann and Chandrasekhar (1969). Dedekind showed that bodies of precisely the same shapes as the Jacobi ellipsoids could have fixed orientations in space with their non-spheroidal shapes maintained by an internal flow bounded by that ellipsoidal surface. It was Riemann who first explored the more general configurations of homogeneous bodies with any ellipsoidal shapes rotating with some pattern speed Ω_{p} (not necessarily along the angular momentum) within which there was an internal fluid flow. Indeed there is a beautiful theorem that any uniform inviscid fluid whose surface is initially ellipsoidal and whose internal velocities are initially linear functions of positions will maintain both these properties for all time. This holds for liquid, shrinking liquids and zero pressure gas falling under gravity and rotation.

Once we start shrinking an inviscid Jacobi ellipsoid it leaves that uniformly rotating sequence and follows a constant C, constant L trajectory through the Riemann ellipsoids. Thus its surface develops a pattern speed Ω_{p} different from a half of its internal vorticity and though its vorticity remains uniform it develops an internal flow which is no longer just uniform rotation. This difference between the pattern speed of the bar-like elongation and the circulation of the stars along their orbits within that bar is the basis of all modern theories of bar formation.

Here I will pay a tribute to a great figure of twentieth century astrophysics who died earlier this year.

"If you go to work with Chandra take with you a problem closely related to what he is working on NOW. It is no good taking one concerning a subject he has already finished". Following these wise words of Guido Munch I took just such a problem to Yerkes in 1962, and my reward was Chandrasekhar's enthusiasm and delight at seeing how his new methods could be successful in fully time-dependent problems.

Chandrasekhar and Norman Lebovitz had already been working on ellipsoidal configurations of equilibrium and their stability for several years. When I left CalTech the well known ELS paper on the collapse of the Galaxy was barely completed. Modelling that collapse as a pressureless spinning spheroid I had already shown that it fell through a sequence of spinning spheroids (Lynden-Bell

1962) but I wanted to know whether any small departures from that shape would be exaggerated during the collapse. Chandra's virial tensor theorem proved a wonderful tool for discussing large scale perturbations into ellipsoidal shapes and I gained greatly from his enormous enthusiasm for such research.

I found that a spinning collapsing spheroid was indeed unstable to becoming a Riemann ellipsoid with the ellipsoidal bulge rotating with a pattern speed $\frac{1}{2}\Omega$ where the internal vorticity was 2Ω. In such a configuration the fluid streamlines have two lobes and one turn relative to the axes turning with the pattern but there are no inner or outer (or corotation) Lindblad resonances. Thus the turning rate of the two-lobe-one-turn orbits in the perturbed flow is significantly faster than the (zero) pattern speed of unperturbed inner Lindblad orbits.

My 1964 paper on this subject was followed up by Lin, Mestel and Shu (1965) who also discussed the non-linear development of such instabilities in the non-rotating case subsequently applied to cosmology by Zeldovich (1970). Work on the classical ellipsoids is far from over, see Christodoulou et al. (1995).

Thus as a product of galaxy formation we may expect an elongated rotating body whose pattern speed is below its internal rotation rate.

Whereas such elementary considerations might apply to the central regions of a galaxy the continued infall championed by Gunn will certainly add an extensive outer halo and outer parts with more angular momentum which will orbit around the central body at greater radii.

There are close analogies between the circulation invariants around stream lines in fluid flows and the actions around corresponding orbits in stellar dynamics.

$$C = \int \mathbf{u} \cdot \mathrm{d}l \rightarrow \int \mathbf{p} \cdot \mathrm{d}\mathbf{q} = 4\pi J_f \tag{1.4}$$

For orbits with two lobes and one turn (inner Lindblad resonance type of closure) we have the *fast* action

$$2J_f = J_\phi + 2J_R = h + 2J_R \tag{1.5}$$

[At corotation closure we likewise have $J_f = J_R$]

The invariant eccentricity of an orbit is defined by

$$\left(\frac{b}{a}\right) \sim (1-e^2)^{\frac{1}{2}} = (1 + J_R/h)^{-1} = h/(J_f + \tfrac{1}{2}H) \tag{1.6}$$

Now torques on a bar due to the trailing spiral structure decrease h, the angular momentum, while leaving the fast adiabatic invariant J_f (the circulation) unchanged. Thus from (1.6) such a torque will make bars thinner. It also makes them slow down because the pattern speed of the orbital lobes $\Omega_s = \left(\frac{\mathrm{d}H}{\mathrm{d}h}\right)_{Jf}$ (Collett and Lynden-Bell 1987) decreases with h in the region where the orbits cooperate to form bars.

2 The Mechanism of Bar Formation

I shall list six mechanisms that have been discussed separately in the literature although I regard several of these as redescriptions of the same process with different emphasis.

1. Maclaurin-Jacobi-Riemann-Chandrasekhar. While Jacobi ellipsoids are certainly elongated the fluid in them rotates uniformly with the pattern whereas real bars have internal streaming motions. Thus the true analogues of bars in the classical works are the Riemann ellipsoids and, as we have seen, the simplest of all galaxy formation pictures - rotating collapsing material leads to such a configuration. It is natural to add Freeman's flat ellipsoidal stellar dynamical bars here as they have internal circulation and a significant pattern speed (Freeman 1966).

2. Density waves reflected through the centre. This concept of bars has been advocated by Lindblad, Lin and Toomre among others. While it is certainly true that $m = 2$ standing waves have some bar-like pattern in the middle this can hardly be the explanation of the early type barred spirals where the bars can be strong with very little external spiral wave. It does appear to be true that a galaxy with a core and an inner Lindblad resonance does not absorb all the wave so that a significant returning wave is present. I believe Toomre will address the problems of such a view of bars.

3. Interaction - Interaction with another galaxy that passes by in the direct sense can certainly promote bar formation. Even at closest approach the interloper's angular velocity about the galactic centre is not normally large. There can be a tide that interacts strongly at inner Lindblad resonance aligning the orbital lobes and as the interloper retreats taking some angular momentum from the oval so formed and so thinning it from an oval to a bar. Noguchi has seen bars generated in both galaxies of an interacting pair and I get the impression that pairs of strongly barred spirals may be more common than chance would predict. The above mechanism which lowers h while keeping J_f fixed may be responsible. Toomre has been a prominent advocate of a tidal origin for many of the more prominent normal spirals. Whether the mechanism described as 3 will promote strong lasting bars remains more problematical.

4. Contopoulos (1980) has long advocated that the distortion of circular orbits into rotating ovals due to a bar-like disturbance in the gravity field is the prime bar making mechanism. The correct form of distortion occurs within corotation and would end at inner Lindblad resonance if such resonances are encountered. Bars do rotate so that they end before corotation which is usually a factor 1.2 or so further out. Whereas this description of what happens is correct it does not give us a clear reason as to why the bars and the associated corotations should be at the radii at which they are observed.

5. I suggested that the region of bar formation was the region where orbital lobes could cooperate with one another (Lynden-Bell 1979). When an orbit with a lobe is subject to a gravitational torque urging it forwards it will gain angular momentum (at constant J_f) but this can lead to the pattern speed of the orbital

lobe decelerating as it does in the *outer* parts of the isochrone. However, in the regions that form bars, gain in h leads to increase in lobe rotation rate Ω_s. The mechanism proposed was for orbits with inner-Lindblad-type closure - i.e. two lobes and one turn. Such orbits not only attract one another gravitationally but react to their mutual attraction by congregating with their lobes aligned at the pattern speed (or vibrating somewhat about that rate). In the opposite case lobes will set perpendicularly to one another and so can not cooperate to make bars. Thanks to fine work analysing orbits in barred galaxy simulations (Sparke and Sellwood 1987) we now know that bars are dominated by aligned orbits with two lobes and one turn just as this mechanism suggests. However, the mechanism was originally advocated as operating at or close to the inner-Lindblad-resonance of the underlying galaxy so the orbits with two lobes and one turn could exist at about the observed pattern speeds. While fine work on bar pattern speeds using gas shocks, as a diagnostic (Sanders and Tubbs 1980; Athanassoula 1990) have found most barred spirals have inner Lindblad resonances, nevertheless strong bars often rotate so fast that no inner Lindblad resonances exist within the bars themselves. This led Sellwood and Wilkinson (1993) to dismiss the mechanism as irrelevant to the observed strong bars. Here I need your help - On the one side it seems that all the requirements of the mechanism i.e. 2 lobes and one turn and Ω_s increasing with h are in fact met by the observed bars, nevertheless the Ω_p of the bars are such that without the bar the orbits with 2-lobe-one-turn form will only close at a significantly slower pattern speed. There are two natural escape routes. One is to suggest that the presence of the bar naturally accelerates the pattern speed as in the case of collapsing spheroids discussed earlier. The other is to say that the considerably elongated orbits seen in bars are circular orbits violently deformed by the bar's forcing. Such an interpretation of strong bars would be a combination of Contopoulos' idea with the mechanism of lobe attraction.

6. Almost all the above approaches start with a round galaxy and then look for reasons why it should become barred. I first heard of a totally different approach from Collett. He suggests that one might look to statistical mechanics for a reason behind the shapes seen. Thus an assembly of centrally pivotted gravitating rods with some net angular momentum will fall into a bar like configuration if the energy is low but disperse into a disk when the energy exceeds some critical value at which a phase transition occurs. While it may be difficult to formulate the theory of bars in precisely this way I believe that there is sufficient truth in the idea that one should try such formulations.

3 The Galactostrophic Approximation and Ergos Curves

Analytical stellar dynamics is greatly aided by a knowledge of the integrals of the motion and it has been a long-standing frustration to me that no integral other than the energy (Jacobi constant) can be written down for realistic barred configurations. There are two well-known special cases which though very special do give considerable insight. The elliptical disc configurations beautifully constructed and explored by Freeman while he was briefly my student, and the exactly separable rotating potential discovered by Vandervoort (1979) and later more fully explored with Contopoulos (Contopoulos and Vandervoort 1992). The idea behind our development was given initial development by Prendergast even before Freeman's exploration of the exact elliptical discs.

The equation of planar motion of a star relative to rotating axis in which the bar or more generally the potential is fixed is

$$\ddot{\mathbf{R}} + 2\Omega \times \dot{\mathbf{R}} = \nabla\psi + \Omega^2\mathbf{R} = \nabla\Phi \tag{3.1}$$

where

$$\Phi = \psi + \tfrac{1}{2}\Omega^2 R^2 \tag{3.2}$$

The lines of constant Φ for a barred potential are shown in Fig. 2. One first integral is the energy in the rotating axes often called the Jacobi constant

$$E_R = \tfrac{1}{2}\dot{\mathbf{R}}^2 - \Phi \tag{3.3}$$

Following Lindblad's treatment of nearly circular orbits we suppose that the motion of a guiding centre is considerably smoother than a typical orbit which contains also a gyration about the guiding centre motion. This gyration is typically elliptical with its long axis pointing along the local equipotential.

At lowest order the galactostrophic approximation looks for the guiding centre motion and neglects their smooth acceleration in comparison with the Coriolis force. The motion of the guiding centres is at zero order determined by the balance between Coriolis force and $\nabla\Phi$. Thus solving for $\dot{\mathbf{R}}$ at zero order

$$\dot{\mathbf{R}} = \frac{1}{2}(\nabla\Phi \times \Omega)\Omega^{-2} \tag{3.4}$$

Since $\dot{\mathbf{R}}$ is perpendicular to $\nabla\Phi$ the velocity is clearly along the lines of constant Φ, the equipotentials, and the sense of motion is that given by (3.4). Unfortunately, we can already see that (3.4) is inconsistent with the exact conservation of the energy unless $|\nabla\Phi| = f(\Phi)$. The point is that the star moves around an equipotential so conservation of $\frac{1}{2}\dot{\mathbf{R}}^2 - \Phi$ implies $|\dot{\mathbf{R}}|$ = constant whereas (3.4) implies that $|\dot{\mathbf{R}}|$ varies proportionally to $|\nabla\Phi|$ there. The equipotentials of Fig. 2 do not keep a constant spacing around each equipotential so $|\nabla\Phi|$ is not a function of Φ. Thus we need a higher order approximation to replace (3.4) by something consistent with energy conservation.

Let $\hat{\mathbf{n}}$ be the unit normal to an equipotential so $\hat{\mathbf{n}} = \nabla\Phi/|\nabla\Phi|$. Let $\hat{\mathbf{s}}$ be $\hat{\mathbf{n}} \times \hat{\Omega}$ where $\hat{\Omega}$ is the unit vector along the rotation axis. Then a guiding centre's velocity can be written

$$\hat{\mathbf{v}} = S\hat{\mathbf{s}} + N\hat{\mathbf{n}} \tag{3.5}$$

where we expect the component N across the equipotentials to be significantly smaller than the component S along the equipotentials. The equation of motion (3.1) can be rewritten

$$\dot{S}\hat{\mathbf{s}} + \dot{N}\hat{\mathbf{n}} + N(\mathbf{v}\cdot\nabla)\hat{\mathbf{n}} + S(\mathbf{v}\cdot\nabla)\hat{\mathbf{s}} - 2\Omega N\hat{\mathbf{s}} + 2\Omega S\hat{\mathbf{n}} = \nabla\Phi \tag{3.6}$$

Now since $\hat{\mathbf{n}}$ and $\hat{\mathbf{s}}$ are unit vectors the two $\mathbf{v}\cdot\nabla$ terms are perpendicular to $\hat{\mathbf{n}}$ and $\hat{\mathbf{s}}$ respectively. Further more since $\hat{\mathbf{n}}\cdot\hat{\mathbf{s}} = 0$

$$\hat{\mathbf{n}}\cdot[(\mathbf{v}\cdot\nabla)\hat{\mathbf{s}}] + \mathbf{s}\cdot[(\mathbf{v}\cdot\nabla)\hat{\mathbf{n}}] = 0 \tag{3.7}$$

Thus we may write

$$(\mathbf{v}\cdot\nabla)\hat{\mathbf{s}} = KS\hat{\mathbf{n}} - K'N\hat{\mathbf{n}} \qquad (\mathbf{v}\cdot\nabla)\hat{\mathbf{n}} = K'N\hat{\mathbf{s}} - KS\hat{\mathbf{s}} \tag{3.8}$$

where the curvatures of the equipotentials and their normals are

$$K = [(\hat{\mathbf{s}}\cdot\nabla)\hat{\mathbf{s}}]\cdot\hat{\mathbf{n}}$$

and

$$K' = [(\hat{\mathbf{n}}\cdot\nabla)\hat{\mathbf{n}}]\cdot\hat{\mathbf{s}} \tag{3.9}$$

The components of equation (3.6) now take the form

$$\dot{S} + K'N^2 - KNS - 2\Omega N = 0 \tag{3.10}$$

$$\dot{N} + KS^2 - K'NS + 2\Omega S = |\nabla\Phi| \tag{3.11}$$

So far the equations are exact and indeed multiplying (3.10) by S and (3.11) by N we recover the energy integral exactly.

$$\tfrac{1}{2}(S^2 + N^2) - \Phi = E_R \tag{3.12}$$

Before exploring the general case we see how the exact case in which the equipotentials have a fixed centre of curvature fits into our formalism. Evidently then $K' \equiv 0, K = 1/R$ and $N = \dot{R}$, so equation (3.10) reduces to $\dot{h} = 0$ where

$$h = R(S + \Omega R) = \frac{1}{K}\left(S + \frac{\Omega}{K}\right) = \text{const} \tag{3.13}$$

Thus the angular momentum about the curvature centre is exactly constant. This is true of all orbits but the guiding centre orbits have $N = 0$ and K constant along each of them.

We now make the key approximation – for guiding centre motions $N \ll S$ so the velocity is *nearly* along equipotentials. Writing S_0 for the lowest approximation to S we have $\dot{S}_0$ is small $O(N)$ from (3.10) and from (3.11)

$$KS_0^2 + 2\Omega S_0 = |\nabla\Phi| \tag{3.14}$$

This expression is readily interpreted, the speed S_0 around an equipotential should be chosen so that the centripetal acceleration around the curvature (of the equipotential) together with the Coriolis acceleration are exactly the net effective gravity field in the rotating axes.

Solving for the speed S_0 we find

$$S_0 = \frac{1}{K}\left[-\Omega + \sqrt{K|\nabla\Phi| + \Omega^2}\right] = \frac{|\nabla\Phi|}{\Omega + \sqrt{K|\nabla\Phi| + \Omega^2}} \quad (3.15)$$

Notice that when K is neglected this reduces to the Galactostrophic expression (3.4). Neglecting N^2 we now use this approximation to the speed to find the orbits of the guiding centres from the energy equation so the 'Ergos' curve of energy E_R is g iven by

$$\mathcal{E}(x,y) \equiv \tfrac{1}{2}\frac{|\nabla\Phi|^2}{[\Omega + (K|\nabla\Phi| + \Omega^2)^{1/2}]^2} - \Phi(x,y) = E_R \quad (3.16)$$

Thus contours of $\mathcal{E}(x,y)$ give the guiding centre orbits in this approximation. It is simple to go back to (3.10) and (3.11) with the zeroth approximation S_0 for S and so find better approximations to any chosen order but J. Barot found (3.16) gave such good results anywhere near corotation that we could not distinguish computed guiding centre orbits from the Ergos curves.

We show examples of the equipotentials Fig. 2 and the Ergos curves for the same potential as Fig. 3, on the same scale.

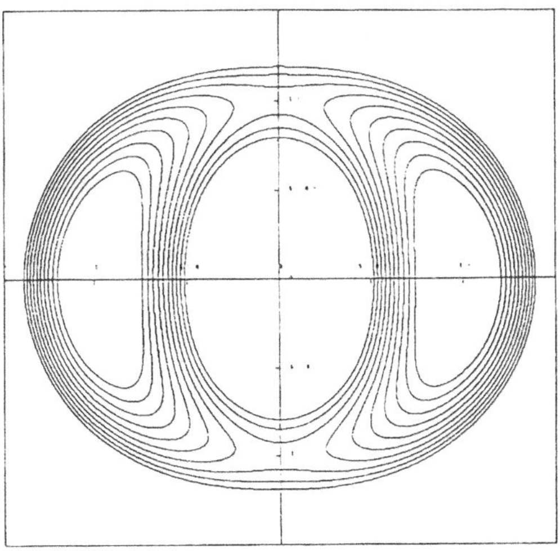

Fig. 2. Equipotentials near corotation. The bar is vertical

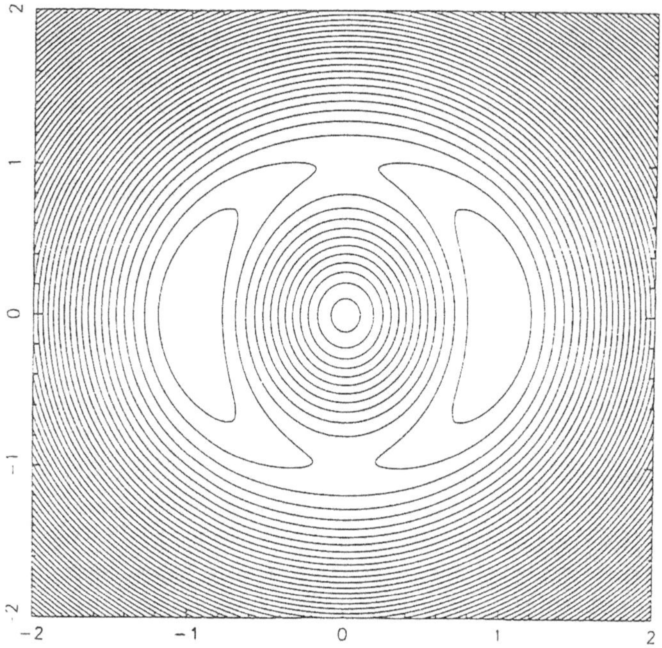

Fig. 3. Ergos curves on the scale of Fig.2

4 Adiabatic Invariants and Integrals for Epicycles

Small oscillations about the guiding centre orbits give us generalised Lindblad elliptic epicycles. They are oriented with their long axes along the equipotentials and their short axes across them as in Lindblad's nearly circular orbits but here the ratio of the axes of the elliptic epicyclic gyration slowly changes along the orbit. Although our method is perfectly general the principles behind it are best learned from the much simplified example in which the effective total potential Φ takes the form $\Phi = \frac{1}{2}K_1^2x^2$ where K_1 varies weakly with y. The equations of motion take the form

$$\ddot{x} - 2\Omega\dot{y} = K_1^2 x \tag{4.1}$$

$$\ddot{y} + 2\Omega\dot{x} = \frac{1}{2}\frac{\mathrm{d}K_1^2}{\mathrm{d}y}x^2 \simeq 0 \tag{4.2}$$

where

$$\dot{y} + 2\Omega x = C \tag{4.3}$$

where $\dot{C}$ is small, hence

$$\ddot{x} - 2\Omega(C - 2\Omega x) = K_1^2 x \tag{4.4}$$

so

$$\ddot{X} + \kappa^2 X = 0 \tag{4.5}$$

where

$$\kappa^2 = 4\Omega^2 - K_1^2 \tag{4.6}$$

and

$$X = x - \frac{2\Omega C}{\kappa^2} \tag{4.7}$$

Now K^2 and C vary little over one oscillation so we have an adiabatic invariant

$$J = \kappa^{-1} E_x = \kappa^{-1} \tfrac{1}{2}(\dot{X}^2 + \kappa^2 X) = \kappa^{-1} \left(E_R + \frac{1}{2} \frac{K_1^2}{\kappa^2} C^2 \right) \tag{4.8}$$

In this expression κ and K_1 are known slowly varying functions of y, the distance along the equipotential $x = 0$, while C varies slowly. Normally one uses adiabatic invariants when the energy varies slowly. Here E_R is an exact invariant and J is adiabatically invariant so expression (4.8) tells how C slowly varies with y to compensate for the changes in κ and K_1. Inserting the expression for C our integral for this case is

$$J = \kappa^{-1}[E_R + \tfrac{1}{2} \frac{K_1^2}{\kappa^2} (\dot{y} + 2\Omega x)^2]$$

With a little extra complication due to the curvatures of the equipotentials this same method can be applied to all cases in which K, K' and $|\nabla \Phi|$ vary slowly along the Ergos curve that approximates the guiding centre motions. A fuller account of the method will be given in Lynden-Bell and Barot (1995).

References

Athanassoula, E. (1991): in Dynamics of Disc Galaxies, ed. B. Sundelius, Göteborg University, Sweden

Chandrasekhar, S. (1969): Ellipsoidal Figures of Equilibrium, Yale University Press, republished Dover 1987

Christodoulou, D.M., Kuzanas, D., Schlosman, I., Tohline, J.E. (1995): ApJ **446**, 472

Collett, J., Lynden-Bell, D. (1987): MNRAS **224**, 489

Contopoulos, G. (1980): A&A **81**, 198

Contopoulos, G., Vandervoort, P.O. (1992): ApJ **389**, 118

Freeman, K.C. (1966): MNRAS **132**, 1

Fridman, A.M. (1994): in Physics of the Gaseous and Stellar Disks of the Galaxy, ed. I.R. King, PASPC **66**, p. 15

Lin, C.C., Mestel, L., Shu, F.H. (1965): ApJ **142**, 1431

Lindblad, B., Langebartel, R.G. (1953): Stockholms Observatoriums Annaler, Band **17**, No. 6.

Lynden-Bell, D. (1962): Proc. Camb. Phil. Soc. **58**, 709
Lynden-Bell, D. (1964): ApJ **139**, 1195
Lynden-Bell, D. (1965): ApJ **142**, 1648
Lynden-Bell, D. (1979): MNRAS **187**, 101
Lynden-Bell, D., Barot, J. (1995): in preparation
Sanders, R.H., Tubbs, A.D. (1980): ApJ **235**, 803
Sellwood, J.E., Wilkinson, A. (1993): Rep. Prog. Phys. **56**, 173
Sparke, S., Sellwood, J.E. (1987): MNRAS **225** 653
Toomre, A. (1981): in The Structure and Evolution of Normal Galaxies, eds. S.M. Fall, D. Lynden-Bell, Cambridge University Press, p. 111
Vandervoort, P.O. (1979): ApJ **232**, 91
Zeldovich, Ya.B. (1970): A&A **5**, 84

Orbits in Barred Galaxies

G. Contopoulos, N. Voglis and C. Efthymiopoulos

Department of Astronomy, University of Athens,
Panepistimiopolis, GR 157 84-Athens, Greece

Abstract. We study ordered and chaotic orbits in barred galaxies. Both types of orbits are important in constructing self-consistent models. We have developed two new criteria for characterizing ordered and chaotic orbits, the stretching numbers (or short-time Lyapunov characteristic numbers) and the helicity angles (the angles between the current deviations from a given orbit and a fixed direction). The distributions of successive stretching numbers and helicity angles are the spectra of an orbit. These are invariant with respect to initial conditions in a chaotic domain. The helicity angles give the fastest method up to now for separating ordered and chaotic orbits. They are more efficient than rotation angles, which cannot always be defined. A comparison of our method with Laskar's frequency analysis method is made. In 3-D systems we define one spectrum for stretching numbers and three spectra for helicity angles. A clear distinction between Arnold diffusion and resonance overlap diffusion is made.

1 Introduction

This symposium is appropriately devoted to the leading galactic astronomer of our time, Bertil Lindblad. Lindblad was a pioneer and had an exceptional foresight. Three of his most important contributions, namely the theory of galactic rotation, the theory of epicyclic motions, and the density wave theory of spiral structure, are the basic elements of present-day galactic dynamics. I (G.C.) was fortunate to work under Professor Lindblad in 1956. Following his suggestions I worked on two related subjects. The first was a generalisation of the epicyclic theory, to include higher order terms (Contopoulos 1957) and the second was a first study of the orbits in 3 dimensions (Contopoulos 1958). With the help of P.O. Lindblad (who was one of the first to calculate galactic orbits with a computer) I calculated two orbits numerically. These orbits were so unexpected, that they led, later, to the theory of the third integral. I am deeply grateful to Bertil Lindblad for his inspiration and I am happy to devote the present study of orbits in barred galaxies to his memory.

2 Ordered and Chaotic Orbits

A barred galaxy contains both ordered and chaotic orbits (Contopoulos et al. 1989; Contopoulos and Grosbøl 1989; Hasan and Norman 1990; Martinet and Udry 1990) [1]. If the bar is not very strong the orbits along the bar are mostly

[1] The references are only indicative. Further references can be found in the papers cited.

ordered (Figs 1 and 2). Close to the center, the main family of periodic orbits is the family x_1, consisting of nearly elliptical orbits along the bar. In many models the x_1 orbits develop a cusp along the x-axis close and beyond the 4/1 resonance (Fig. 1) but further out these orbits are unstable. Beyond the 4/1 resonance the dominant (stable) family of periodic orbits is 4/1, which contains elongated orbits near parallelograms [2]. Both families, x_1 and 4/1, trap quasiperiodic orbits around them (Fig. 2), that support a self-consistent bar. When a bar is weak and tends to a circle, both families x_1 and 4/1 tend to circular orbits. Further out

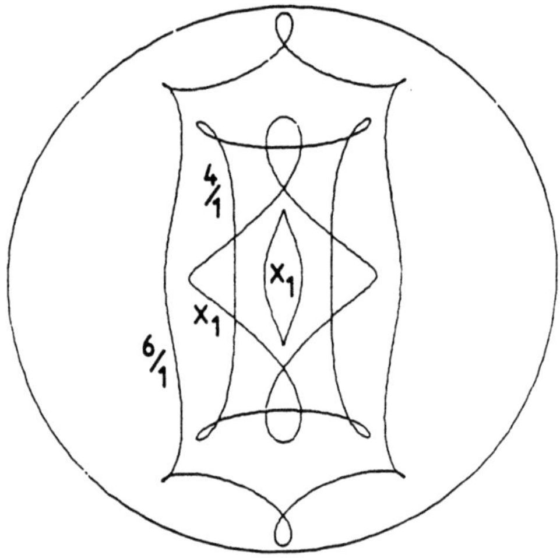

Fig. 1. Periodic orbits of the families x_1, 4/1 in the Florida bar model (Contopoulos et al. 1989). The outer circle represents corotation

along the bar, and close to corotation, we have families of periodic orbits like 6/1 (Fig.1), 8/1, etc. There are also other families of periodic orbits, like 1/1, 3/1, etc, that may be important in some models (Contopoulos and Grosbøl 1989). But the most important families, as regards self-consistency of the bar, are the families x_1 and 4/1.

Close to the Lagrangian points L_4, L_5 near corotation there are two more families of stable periodic orbits, the short periodic orbits, like ovals, and the long periodic orbits, like bananas (Fig. 3).

Beyond corotation, up to the outer Lindblad resonance, the main family of periodic orbits, x_1, is elongated perpendicularly to the bar. Considering the

---

[2] In other models (Athanassoula et al. 1983) the elliptic orbits x_1 become less elongated outwards and form near parallelograms, while the new family, 4/1, contains orbits with loops along the bar axis. The difference between the two types of models depends on the 4θ component of the bar. We have shown (Contopoulos 1988) that for strong enough bars the arrangement of the families of periodic orbits is of the form of Fig. 1 (loops for x_1, parallelograms for 4/1).

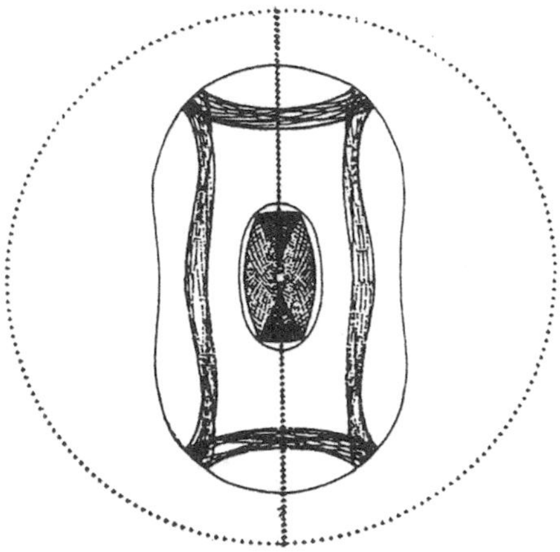

Fig. 2. Quasiperiodic orbits trapped around the periodic orbits x_1 and 4/1. These orbits are tangent to the corresponding equipotentials (curves of zero velocity, CZV). The outer circle is corotation

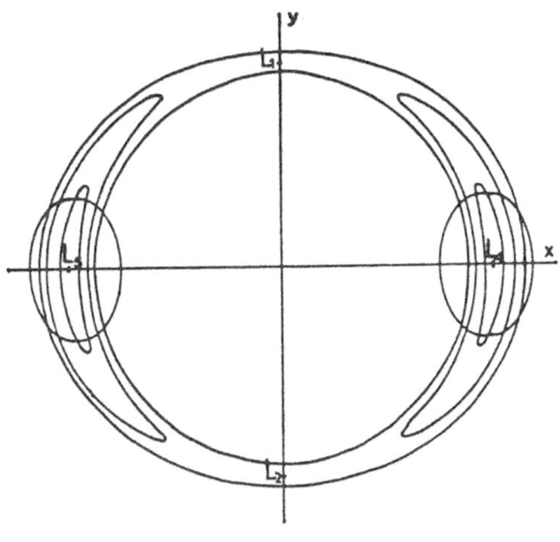

Fig. 3. Elongated banana orbits around L_4 and L_5 (long period orbits), and short period orbits that appear for larger energiés

various effects that support or tend to destroy the bar we found that bars can extend up to corotation, but not beyond it in general (Contopoulos 1980).

However, beyond corotation we may have a spiral, starting roughly at the end of the bar. In such a model the family x_1 consists of deformed circular orbits that support the spiral from the outer $-4/1$ resonance all the way to the outer Lindblad resonance (Fig. 4).

Close to corotation the orbits are mostly chaotic. Such orbits are of 3 types (i) completely inside corotation (Fig.5), and along the bar (ii) completely outside

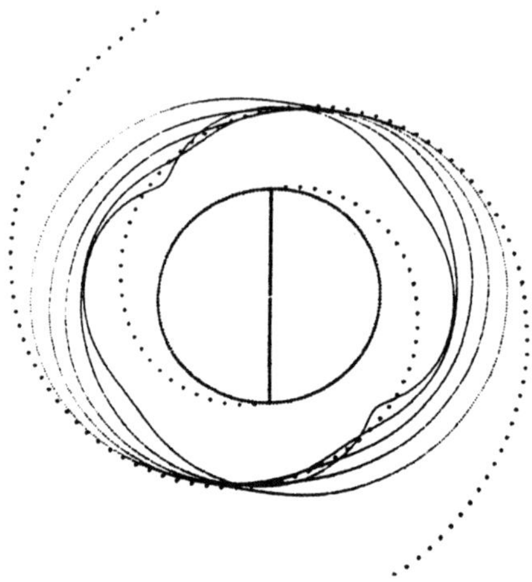

Fig. 4. A set of stable periodic orbits in a spiral outside corotation, between the $-4/1$ resonance and the $-2/1$ resonance (outer Lindblad resonance) that support the spiral (Model of Fig. 1)

corotation forming rings (Fig. 6), and (iii) partly inside corotation, along the bar and partly outside corotation, along the spiral (Fig. 7) (Kaufmann and Contopoulos 1993). The area covered by such orbits is larger when the bar and spiral perturbation is larger.

A strong bar is thin, while the equipotentials are broader. But chaotic orbits fill the whole area inside the equipotential corresponding to its energy, therefore they do not support a thin bar. This is probably the reason why the observed bars have an axis ratio larger than about 0.2. Thinner bars cannot be self-consistent (Martinet 1984). Furthermore thin bars are easily destroyed by the addition of an extra mass at the center of the galaxy (Hasan and Norman 1990).

The chaotic orbits well outside corotation do not support the spiral or an extended bar. But the orbits close to corotation partly support the spiral (Fig. 7) for very long times, of the order of 10^{10} years (Kaufmann and Contopoulos 1995).

Another mechanism that supports the spirals outside corotation is based on the behaviour of gas. Even in a model consisting of a bar plus a spiral outside the $-4/1$ resonance (up to the outer Lindblad resonance), the gas bridges the region between corotation and the $-4/1$ resonance (Contopoulos et al. 1989). The partial support of the spiral in this region by stochastic orbits allows the construction of long lived self-consistent bar-spiral models. Of course the spirals cannot last for ever, due to the existence of torques (Lynden-Bell and Kalnajs 1972). But the torques allow the survival of the spirals outside bars for a sufficiently long time of the order of several 10^9 years (Gnedin et al. 1995).

The role of chaotic orbits in galaxies has been overlooked in previous years, because of the success of integrable models, esp. Stäckel-type models (Lynden-

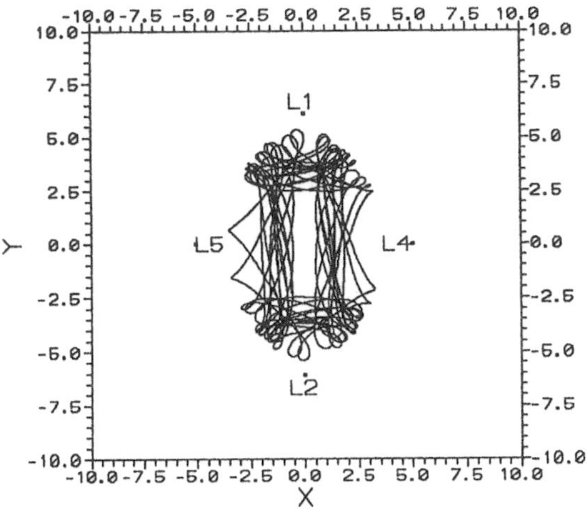

Fig. 5. A chaotic orbit completely inside corotation (Model of Kaufmann and Contopoulos 1995; initial conditions $x = 2$, $\dot{x} = 50$ and Jacobi constant $E_J = -191550$)

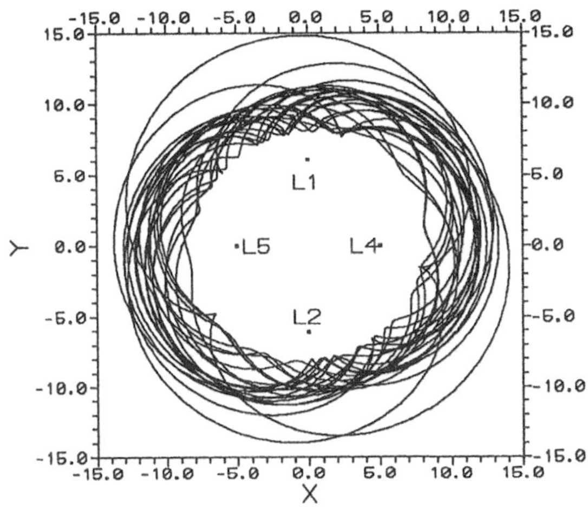

Fig. 6. A chaotic orbit completely outside corotation (Model of Fig. 5; initial conditions $x = -12.0$, $\dot{x} = 0$ and Jacobi constant $E_J = -198084$)

Bell 1962; Vandervoort 1984; de Zeeuw and Lynden-Bell 1985; de Zeeuw 1985; de Zeeuw et al. 1987; Hunter 1988) for elliptical galaxies. However these models are not rotating.

The only known rotating Stäckel model (Contopoulos and Vandervoort 1992), besides the trivial homogeneous case, is not realistic. In general, rotating models are nonintegrable and they contain a large degree of chaos, especially near corotation. Furthermore recent studies of models with a density cusp at the center contain many chaotic orbits, and it seems impossible to construct such models

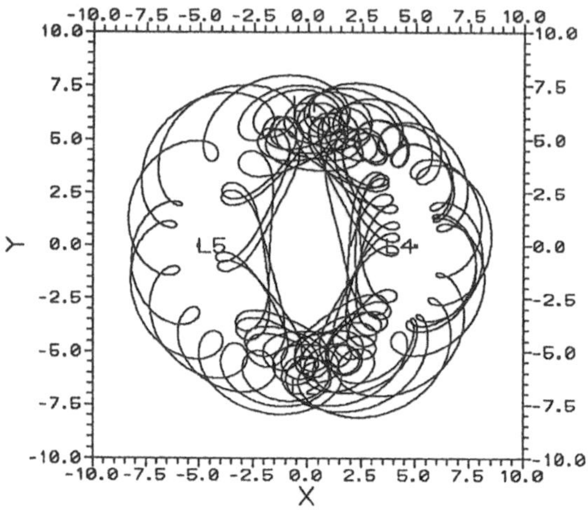

Fig. 7. A chaotic orbit near corotation that enters also in the the bar region (Model of Fig. 5; initial conditions $x = 2, \dot{x} = 50$ and Jacobi constant $E_J = -188500$). This orbit partially supports the bar and the spiral outside corotation

with only regular orbits. On the other hand Merritt and Fridman (1995) have constructed self-consistent models of elliptical galaxies (non rotating) with moderate cusps and a large proportion of chaotic orbits. But a very strong cusp may be inconsistent with a strong triaxiality.

This result is consistent with studies of orbits in bars with a central mass concentration, indicating the onset of chaos when the central mass is large, leading to the destruction of bars (Hasan and Norman 1990; Hasan et al. 1993; Norman et al. 1995).

The study of nonplanar orbits in systems of 3 degrees of freedom is a subject of great current interest (Contopoulos et al. 1982; Pfenniger 1984; Contopoulos 1986; Patsis and Grosbøl 1995; Contopoulos et al. 1995). We will come back to this problem further on, when dealing with Arnold diffusion in 3-D systems.

3 Stretching Numbers and Helicity Angles

The traditional methods to distinguish between ordered and chaotic orbits are (a) the distribution of the consequents on a Poincaré surface of section, and (b) the Lyapunov characteristic number (LCN), which is positive for chaotic orbits and zero for ordered orbits. A more recent method (c) based on a frequency analysis, was developed by Laskar (1990, 1993) and Laskar et al (1992). We will discuss later a fourth method (d), which seems to be the fastest of all, and is based on the "helicity angles".

We consider a 2D-mapping on a Poincaré surface of section, defined by the successive consequents (intersections of an orbit), and a small (linear) deviation $\underline{\xi}_i$, from every consequent $\underline{x}_i$.

We define the "helicity angle", ϕ_i, as the angle between the direction of $\underline{\xi}_i$ and a given fixed direction, say the axis x. At the same time we define a "stretching number" which is the logarithm of the ratio between two successive deviations

$$a_i = \ln \mid \frac{\underline{\xi}_{i+1}}{\underline{\xi}_i} \mid . \tag{1}$$

The average value of a_i, as the number of consequents tends to infinity, is the usual Lyapunov characteristic number (LCN).

However, the LCN, being an average value, does not give all the essential information about a given dynamical system. Much more complete information is provided by the spectra of the stretching numbers and the helicity angles, i.e. the distribution of the values of a_i and ϕ_i (Voglis and Contopoulos 1994).

Namely we calculate the quantities $S(a)$ and $S(\phi)$:

$$S(a) = \frac{\mathrm{d}N(a)}{N\mathrm{d}a}, \qquad S(\phi) = \frac{\mathrm{d}N(\phi)}{N\mathrm{d}\phi}, \tag{2}$$

where $\mathrm{d}N(a)$ is the number of values of a in the interval $(a, a+\mathrm{d}a)$, and $\mathrm{d}N(\phi)$ the number of values of ϕ in the interval $(\phi, \phi + \mathrm{d}\phi)$ after N iterations. The values of $S(a)$ and $S(\phi)$ can be provided in the course of calculating the Lyapunov characteristic number without much extra effort.

The spectra $S(a)$ and $S(\phi)$ are invariant (a) with respect to initial conditions along an orbit, (b) with respect to the initial direction of $\underline{\xi}$ (within certain limits in the directions; Contopoulos and Voglis 1996), and (c) with respect to different initial conditions in the same chaotic domain (or (c') with respect to initial conditions along the same invariant curve in the case of ordered orbits). These spectra provide much more information about a dynamical system than the LCN alone. A striking example is provided in Fig. 8. We consider two different maps, namely the standard map

$$\begin{aligned} x_{i+1} &= x_i + y_{i+1} \\ & \qquad\qquad\qquad (mod 1) \\ y_{i+1} &= y_i + \tfrac{K}{2\pi} \sin 2\pi x_i \end{aligned} \tag{3}$$

and the conservative Hénon map

$$\begin{aligned} x_{i+1} &= 1 - K' x_i^2 - y_i \\ & \qquad\qquad\qquad (mod 1) \\ y_{i+1} &= b x_i \end{aligned} \tag{4}$$

with $b = 1$.

Taking the same initial conditions and appropriate values of K and K' we find a completely chaotic distribution of consequents in both cases with no conspicuous islands of stability. The two figures look quite the same (Figs 8a,d). Furthermore the values of K and K' were chosen is such a way that the Lyapunov characteristic numbers are equal, LCN = 1.276. However, the spectra of

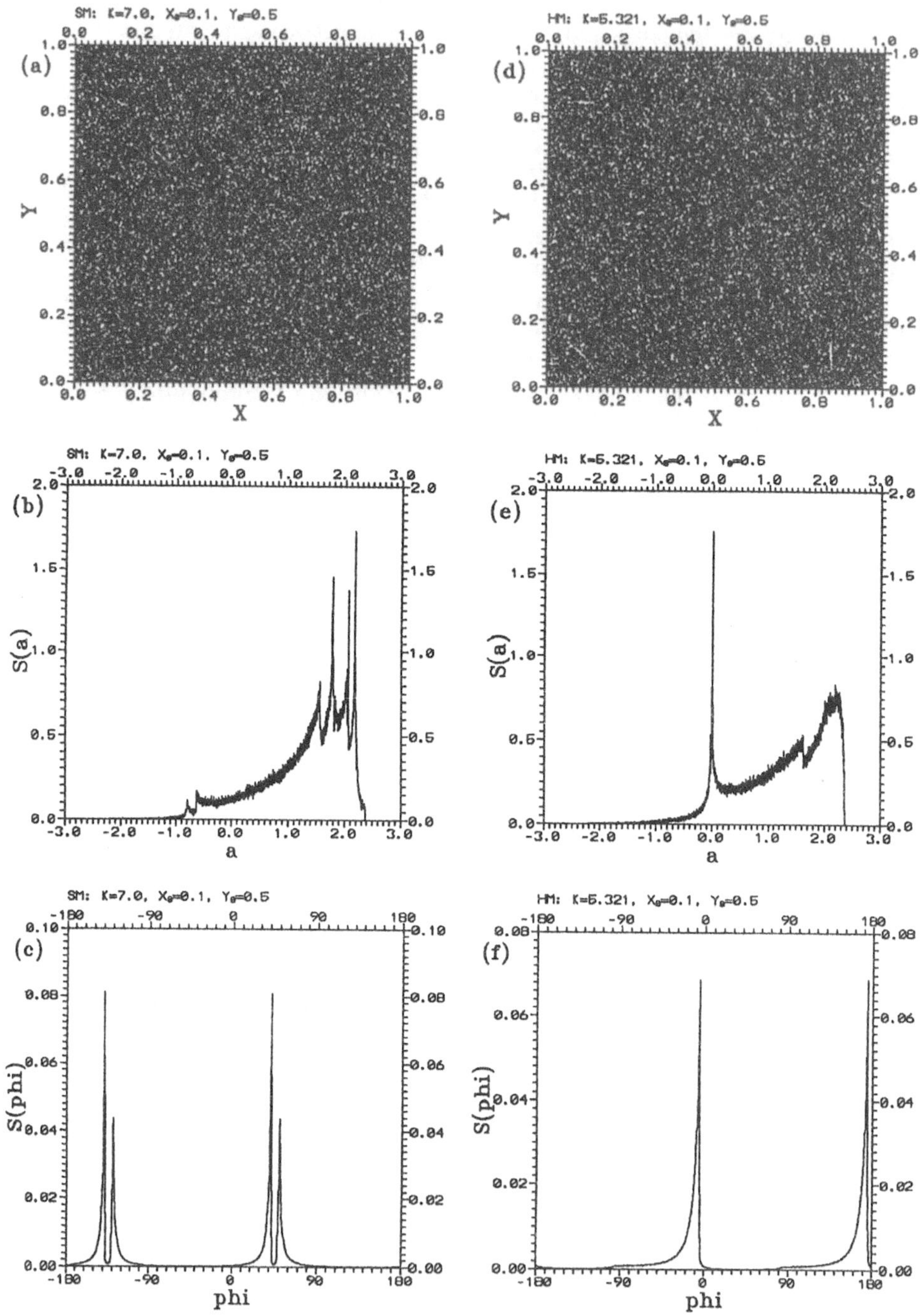

Fig. 8. The distribution of 8×10^4 consequents in the standard map ($K = 7$, $x_o = 0.1$, $y_o = 0.5$, $\xi_{x_o} = 1$, $\xi_{y_o} = 0$) (**a**), and in the Hénon map ($K' = 5.321$ and the same initial conditions) (**d**), and the corresponding spectra of stretching numbers (**b**,**e**), and helicity angles (**c**,**f**)

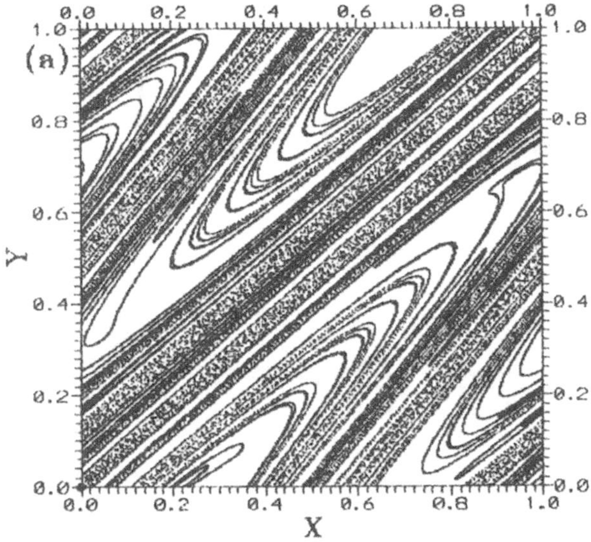

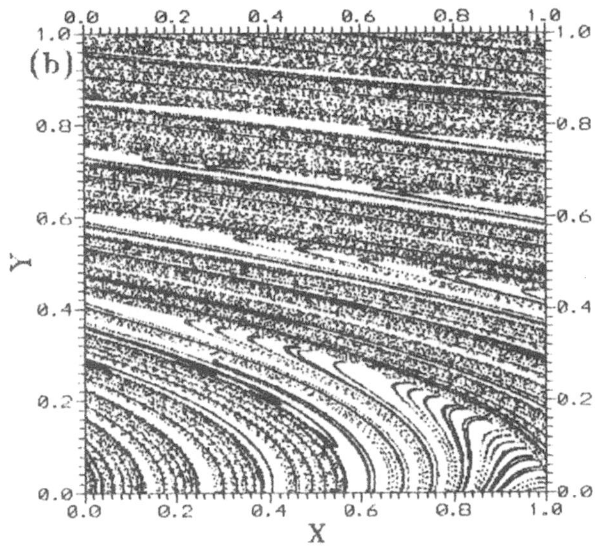

Fig. 9. One unstable asymptotic curve from the main periodic orbit (of period 1) in the cases: **a** in the standard map ($K = 7$) and **b** in the Hénon map ($K' = 5.321, b = 1$). The periodic orbit is marked by a large dot

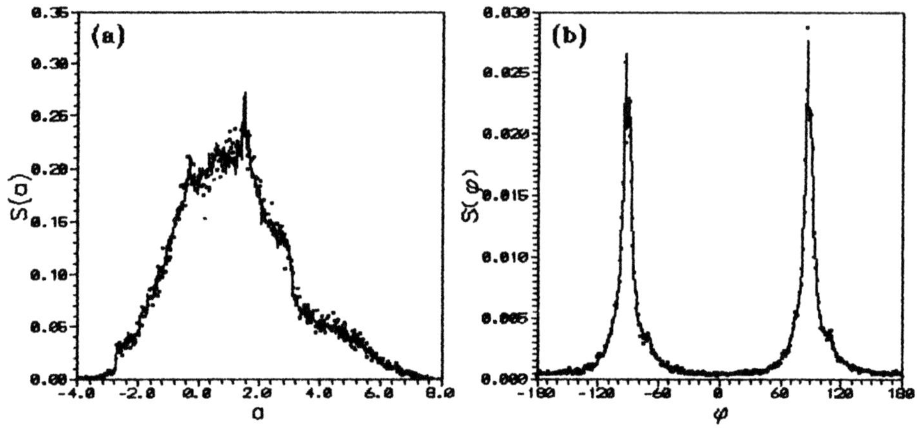

Fig. 10. Two spectra of orbits, for the orbit of Fig. 5 and another orbit with the same Jacobi constant, calculated for 10^5 periods each (intersections with a Poincaré surface of section). Stretching numbers (**a**) and helicity angles (**b**)

stretching numbers (Figs. 8b,e) and helicity angles (Figs. 8c,f) are quite different. Thus the two chaotic systems are, in fact, quite different.

The main difference between the two systems is the arrangement of the asymptotic curves of the unstable periodic orbits. In Fig. 9 we give one asymptotic curve in each of the cases of Fig. 8. The asymptotic curves of Figs. 9a and 9b are quite different. Thus, although the distributions of the consequents (Figs. 8a,d) look similar, the asymptotic curves provide a very different underlying order in the phase space. The same order is provided by the asymptotic curves of all other periodic orbits, because the unstable asymptotic curves of various orbits cannot be intersected.

The spectra of stretching numbers and helicity angles can be applied to Hamiltonian systems, and in particular to barred galaxies.

We have calculated many spectra in a model of a barred galaxy (Kaufmann and Contopoulos 1995) consisting of a disk, a bar, a spiral and a halo.

In Fig. 10a we show two chaotic spectra of stretching numbers, and in Fig. 10b the corresponding spectra of helicity angles, for the orbit of Fig. 5 and another orbit with the same value of the Jacobi constant, inside corotation. We notice that the spectra are practically the same, i.e. they are invariant, although the initial conditions are quite different. Similar results for the orbit of Fig. 6 and another orbit outside corotation (for the same Jacobi constant) are given in Fig. 11a,b. Finally in Fig. 12a,b we give the spectra of Fig. 7 and of another orbit partly inside and partly outside corotation (with the same Jacobi constant). In all cases the spectra are invariant with respect to the initial conditions for the same Jacobi constant. But the spectra for different Jacobi constants are different.

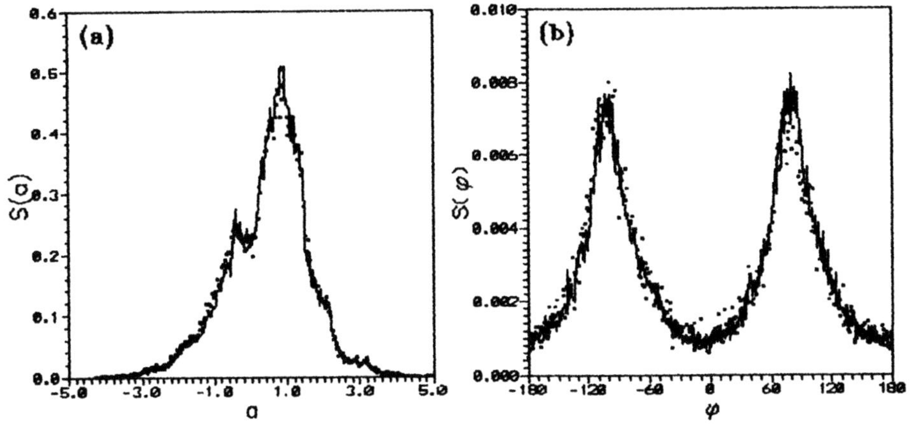

Fig. 11. Spectra, as in Fig. 10, for the chaotic orbit of Fig. 6, and another orbit with the same Jacobi constant

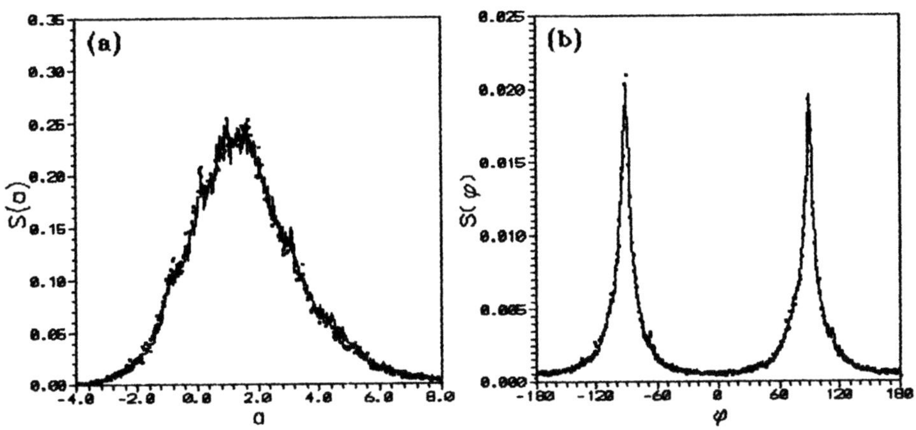

Fig. 12. Spectra, as in Fig. 10, for the chaotic orbit of Fig. 7, and another orbit with the same Jacobi constant

4 Rotation Numbers and Helicity Angles

One way to find the ordered and chaotic domains around a stable invariant point of a map (a stable periodic orbit) is by calculating the rotation numbers of orbits starting at different distances from the invariant point. The rotation number is defined as the average angle (with the circle as unit between the vectors from the invariant point to successive consequents) of an orbit in the Poincaré surface of section (Contopoulos 1966). If the orbit is on an invariant curve (KAM curve) the rotation number is well defined. In an integrable system the rotation number (rot) is always defined (Fig. 13a). The rotation curve (rot vs. x) has a constant value at an island, with rot=2/3, around a stable triple periodic orbit, while it has a vertical tangent at an unstable triple periodic orbit and at the boundaries of the island.

But when the system is nonintegrable there are regions where the rotation number is not defined. In Fig. 13b these regions are close to the unstable periodic orbit 2/3 (marked by a cross) and at the boundaries of the islands. These are the regions where chaos appears. In reality there are infinite small intervals where the rotation number is not defined, close to all the periodic orbits of the system. But as the set of KAM curves has a positive measure we have the appearance of a seemingly continuous rotation curve in Fig. 13b. In Fig. 13c, that corresponds to a larger nonlinearity, only small intervals give the appearance of continuous rotation curves. In general the rotation number cannot be defined except for some periodic orbits and some small islands of stability in the chaotic domain.

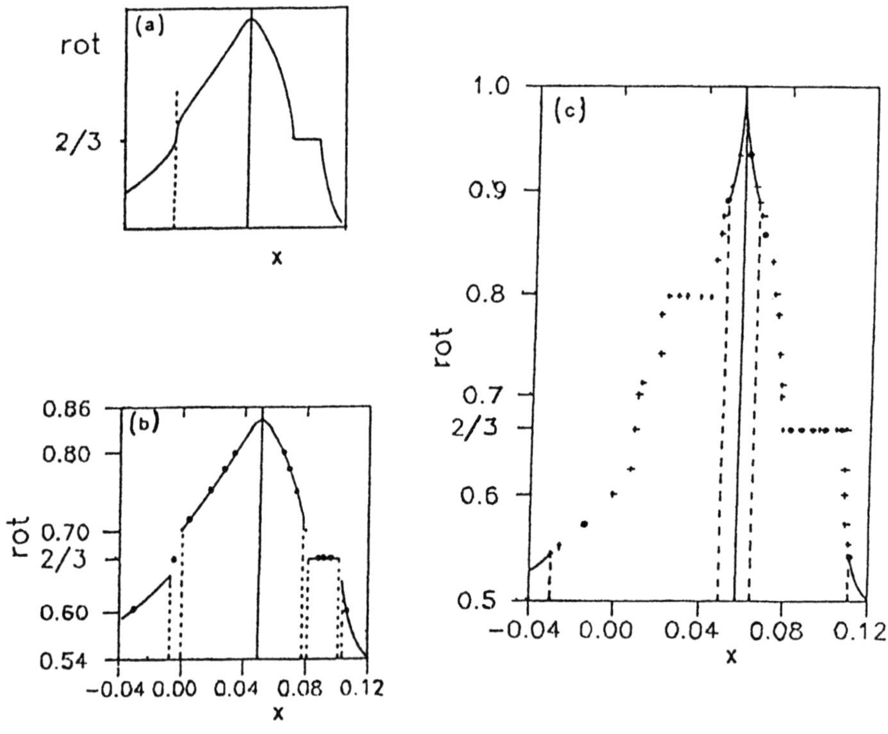

Fig. 13. The rotation number (rot) as a function of x (Contopoulos 1966). **a** An integrable case (schematically), **b** a slightly nonintegrable case, and **c** a strongly nonintegrable case; some stable (•) and unstable (+) periodic orbits are marked. The vertical lines mark the unstable periodic orbit 2/3 in (a) and the ends of the (apparently) continuous rotation curves in (b) and (c)

Laskar et al. (1992) have overcome this difficulty by defining "a rotation number over a fixed number of iterations" (or a fixed time interval, $2T$, in a continuous system). Such a "rotation number" always exists. But in the chaotic regions this "rotation number" varies wildly and is not a monotonic function of the distance $(x - x_o)$. Furthermore successive intervals $2T$ give different rotation numbers. These characteristics of the "rotation number" (or "fundamental frequency") have been used by Laskar et al. (1992) to define the chaotic domains. This method is much faster than establishing the nonexistence of the rotation number by taking large numbers of consecutive consequents, or calculating the Lyapunov characteristic number and finding that it is positive. In one example the method of Laskar et al. (1992) requires 2×10^4 iterations versus 5×10^6 iterations needed to calculate a reliable Lyapunov characteristic number.

Laskar et al. (1992) developed a fast method, called numerical analysis of fundamental frequencies (NAFF), to derive the fundamental frequencies of Hamiltonian systems and similar dynamical systems, like maps, of 2 or more degrees of freedom. This method is more accurate than similar methods giving the frequency spectrum of a system by using Fourier transforms. Spectral methods were introduced in stellar dynamics by Binney and Spergel (1982) but they were well known to physicists and chemists long ago (e.g. Noid et al. 1977).

One can define rotation numbers of higher order around any stable periodic orbit, e.g. in the island around the stable periodic orbit 2/3 of Figs 13 b,c there are closed invariant curves and periodic orbits with multiplicities that are multiples of 3. But it is sometimes impossible to define a rotation number, even for a periodic orbit. This applies to some families of periodic orbits, called "irregular families" (Contopoulos 1970). These families are not generated by bifurcation from a central family, but they appear at a "tangent bifurcation" as a couple of a stable and an unstable periodic orbit, at a finite value of the nonlinearity parameter and they do not exist for smaller values of this parameter.

In some cases such families appear in the region of an island, when the invariant curves surrounding the center of the island have been destroyed, e.g. in the region with constant rotation number 2/3 of Fig. 13c appear periodic orbits with a very different rotation number.

In many cases the successive consequents of an irregular periodic orbit cannot be arranged around an obvious "center" in order to define the average rotation number. In such cases the average (rotation number) depends on the choice of the center, which is arbitrary (Contopoulos 1980).

The existence of irregular families of periodic orbits is not an exception, but a common phenomenon in chaotic systems. Such families appear near heteroclinic points, which are the basic elements of the "interactions of resonances" and chaos. The existence of infinite periodic orbits near heteroclinic points was proven by Birkhoff. These families do not exist for small perturbations, when the heteroclinic orbits disappear. They are generated at a finite perturbation and are not produced by bifurcation from other preexisting families. In fact the heteroclinic orbits are asymptotic to different periodic orbits as $t \to \infty$ and as $t \to -\infty$, therefore orbits in their neighbourhood cannot be multiples of the one or the other type of limiting orbits.

The number of irregular families is extremely large. It grows exponentially with multiplicity (like e^n), while the regular bifurcations grow like a power (like n^2) (Guckenheimer and Holmes 1983).

The methods of the frequency analysis are not well suited for the irregular families of periodic orbits and their neighbourhoods. These methods give correctly the chaotic domains, but the structure of the chaotic domains requires new methods of analysis. One such method is the study of the spectra of stretching numbers and helicity angles.

The helicity angles are always defined, whether an orbit is ordered or chaotic, nonperiodic or periodic, regular or irregular.

An important property of the helicity angles is that they allow the fast separation of the ordered and chaotic domains, even faster than the frequency analysis method.

The average value of the helicity angle $<\phi>$ is the limit of the average $<\phi>_n$, after n iterations, when n tends to infinity. The limit $<\phi>=<\phi>_\infty$ always exists, and it is the same in a connected chaotic domain, in the same way as the average stretching number $<a>=<a>_\infty$, which is equal to the Lyapunov characteristic number. In the ordered domains the value of $<\phi>$ changes from one invariant curve to the next, and the variation of $<\phi>$ is smooth, while we move inside an ordered domain (e.g. an island).

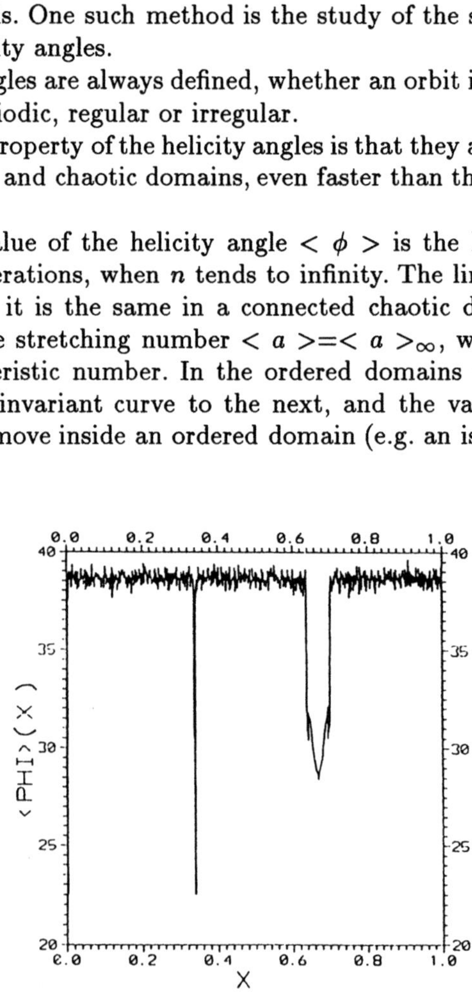

Fig. 14. The average values of the helicity angles $<\phi>_n$ for $n = 10^4$ in the standard map (K=5), for constant $y = 0.34$ and successive values of x with a step $\Delta x = 0.001$ from $x = 0$ to $x = 1$

If n is not very large the values $<\phi>_n$ have a dispersion around the average $<\phi>$. In Fig.14 we have calculated the values of $<\phi>_n$ for $n = 10^5$ along a line $y =$ const. at successive values of x from $x = 0$ to $x = 1$ with a step

$\Delta x = 0.001$. This line goes both through a chaotic domain and an ordered island, composed (mostly) of invariant curves. In the chaotic domain the dispersion of the individual values of $< \phi >_n$ appears as noise around a well defined mean value $< \phi >$. In the ordered domain, on the other hand, the value of $< \phi >_n$ changes smoothly from one point to the next. But the most important fact is that the values of $< \phi >_n$ in the ordered domain differ considerably from the average value $< \phi >$ of the chaotic domain and they change abruptly at the boundary of the island. Furthermore the dispersion in the ordered domain is smaller than in the chaotic domain.

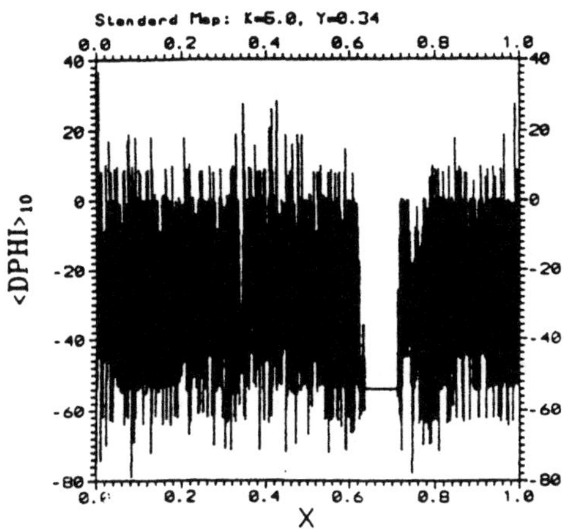

Fig. 15. The helicity angles $< \Delta\phi >$ for $n = 10$ (after the first 10 transient points) along the line of Fig. 14

One can see some details in the ordered region of Fig. 14 that look like local noise. These are due to secondary resonances inside the main island, namely islands, or localised chaotic zones. These features are symmetric with respect to the center of the islands, thus they are not due to random noise. A more detailed scanning of the ordered region shows the details of the secondary islands and chaotic zones inside the main island. If we are only interested in distinguishing between the main ordered and chaotic domains we can take n much smaller than $n = 10^4$, namely n of order 10 only. In these cases the noise is larger, but we can still distinguish between the chaotic and the ordered domains. This is shown in Fig. 15 which gives the average values of the differences $\Delta\phi$ of successive helicity angles for $n = 10$ iterations (excluding the first 10 iterations which are just transients). Despite the noise we can distinguish clearly the region of the island from the large chaotic domain.

We conclude that the method of the helicity angles gives a very fast separation of the chaotic and ordered domains, much faster than the frequency analysis method.

5 3-D Systems: Arnold Diffusion

Spectra of stretching numbers and helicity angles can be defined also for systems of 3 degrees of freedom. However there are some differences due to the larger dimensionality. We notice, first, that the 6-D phase space is reduced to 5-dimensions if the potential is time-independent. Thus a Poincaré surface of section is 4-dimensional. As a consequence the deviation $\underline{\xi}$ is a vector in 4-dimensions, therefore it defines 3 helicity angles. Thus we have one spectrum for the stretching numbers and 3 spectra for the helicity angles. The spectra are always invariant with respect to the initial conditions along an orbit. Furthermore the chaotic spectra are invariant both with respect to the direction of the initial deviation $\underline{\xi}$, and the initial conditions in the same chaotic domain. But the spectra of ordered orbits are invariant only if the initial conditions are on the same invariant surface and the deviation $\underline{\xi}$ is on the same integral surface of the variational equations. Thus different initial deviations $\underline{\xi}$ may give different spectra in the ordered cases.

The most important difference between systems of 2 and 3 degrees of freedom is Arnold diffusion. It is well known that in nonintegrable systems there is some degree of chaos at every resonance, namely near every unstable periodic orbit. However, in systems of 2 degrees of freedom, that are close to integrable, most resonances are separated from each other by closed KAM curves and chaos is limited. Only for large perturbations, when most resonances interact with each other, chaos is important. On the other hand, in systems of 3 (or more) degrees of freedom the KAM surfaces do not separate the various resonances and there is always some resonance interaction that leads to a slow diffusion of the orbits, called Arnold diffusion. Thus in 3-D systems there is no separation of the various chaotic domains and orbits that are not exactly on a KAM surface gradually visit all parts of phase space.

But the time needed for Arnold diffusion is usually so extremely long that it may not have any practical consequence. Some authors (Chirikov 1979; Laskar 1993) make a distinction between Arnold diffusion and resonance overlap diffusion which is similar to the diffusion in 2-D systems.

This distinction is not quite clear, because Arnold diffusion itself is due to interaction of many resonances. However, we will present evidence that although there is a continuous transition from Arnold diffusion to resonance overlap diffusion the two cases can be clearly distinguished.

As an example we consider a 4-D mapping

$$\begin{aligned} x_1' &= x_1 + y_1', \qquad y_1' = y_1 + \frac{K}{2\pi}\sin 2\pi x_1 - \frac{\beta}{\pi}\sin 2\pi(x_2 - x_1), \\ & \qquad\qquad\qquad\qquad\qquad\qquad\qquad (mod 1) \qquad\qquad (5) \\ x_2' &= x_2 + y_2', \qquad y_2' = y_2 + \frac{K}{2\pi}\sin 2\pi x_2 - \frac{\beta}{\pi}\sin 2\pi(x_1 - x_2), \end{aligned}$$

which is similar to the mapping on a Poincaré surface of section of a 3-D system.

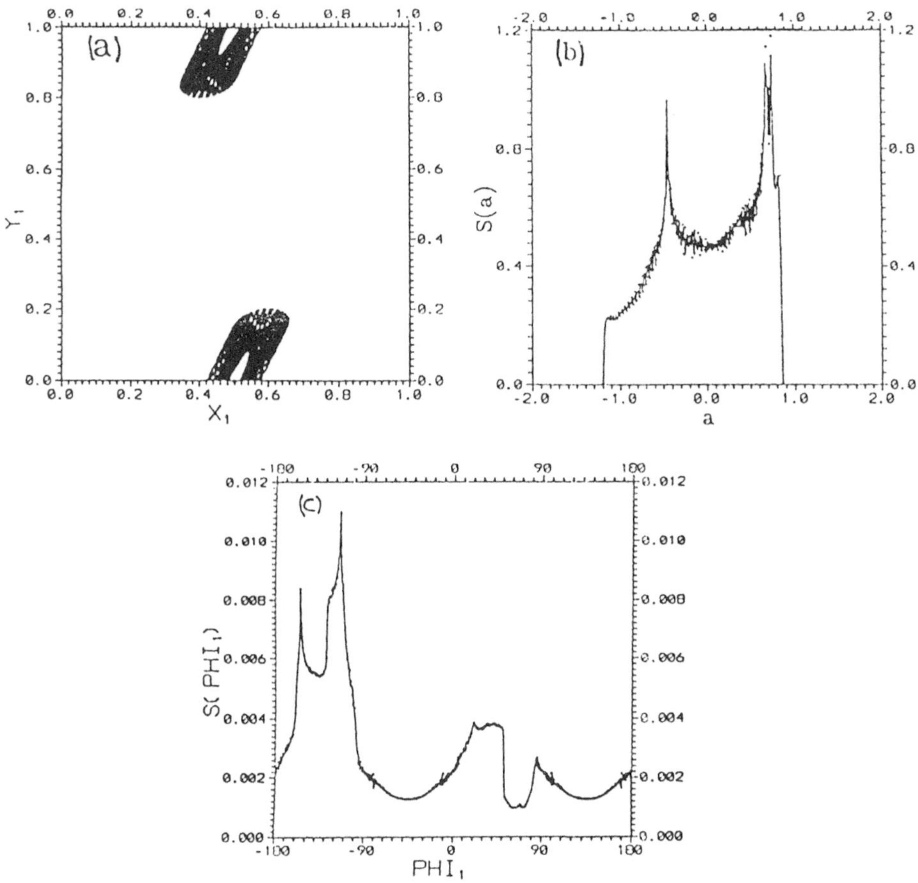

Fig. 16. The distribution of the projections of the consequents on the plane (x_1, y_1) (**a**) in an ordered case (initial conditions $x_1 = 0.55$, $y_1 = 0.1$, $x_2 = 0.62$, $y_2 = 0.2$, $\xi_{x_1} = \xi_{y_1} = \xi_{y_2} = 0$, $\xi_{x_2} = 1$, for $K = 3$ and $\beta = 0.3$) and the corresponding spectra of the stretching numbers (**b**), and of the helicity angle ϕ_1 (**c**), for 10^6 iterations

If $\beta = 0$ $(K > 0)$ this mapping is separated into 2 independent standard maps, which have both an ordered and a chaotic domain. If the coupling constant β is small we have again an ordered domain, if the initial conditions (x_1, y_1) and (x_2, y_2) are in the corresponding ordered domains of the case $\beta = 0$, and a chaotic domain if one pair, at least, of the initial conditions (say (x_1, y_1)) is in the corresponding chaotic domain. The distinction between order and chaos is seen in Figs. 16 and 17, together with the corresponding spectra of stretching numbers and of one of the three helicity angles, i.e. the angle of the projection of the deviation ξ on the plane (x_1, x_2) with the x_2 axis. Each spectrum is drawn for 10^6 iterations with a continuous line and the dots represent the corresponding spectrum for the next 10^6 iterations. The two spectra coincide exactly, demonstrating their invariance. The distribution of the consequents and the spectra in

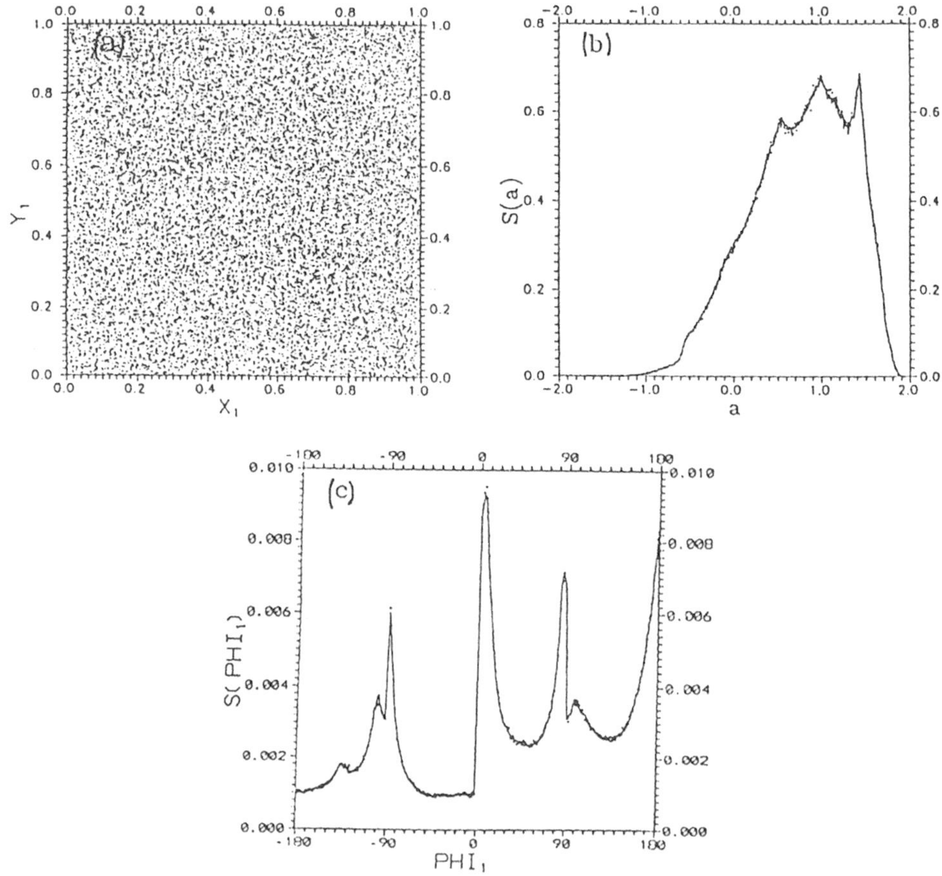

Fig. 17. As in Fig. 16 for a chaotic case (initial conditions $x_1 = 0.1$, $y_1 = 0.5$, $x_2 = 0.2$, $y_2 = 0.6$, and the other constants the same)

Figs. 16 and 17 are quite different from each other. But one expects that the distribution and the spectra of Fig. 16 are transient and after a long time they should tend to the distribution and spectra of Fig. 17. We found indeed such a transition for somewhat larger β, e.g. in Fig. 18 we find a transient spectrum (line a),which is almost invariant for $N < 10^5$ periods. But later the spectrum changes and tends to a truly invariant spectrum (line b) calculated for $N = 10^7$ periods.

The transition time is a function of β, decreasing for larger β. In Fig. 19

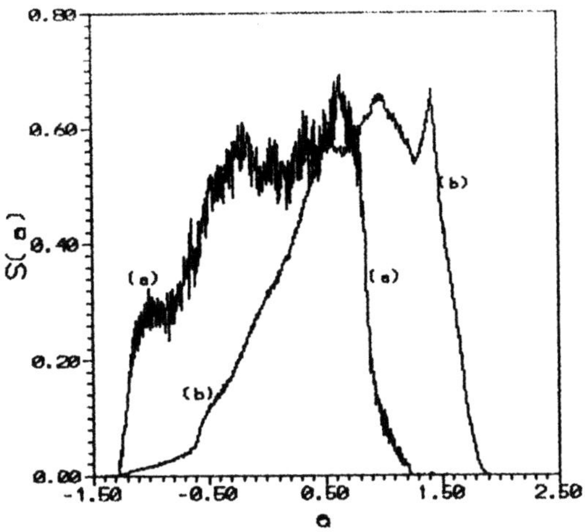

Fig. 18. The spectrum of stretching numbers for $K = 3, \beta = 0.30513$ and the same initial conditions as in Fig. 17a, which are located in an ordered domain. Line (a) is a transient spectrum for $N = 0.5 \times 10^5$ periods and line (b) is the final spectrum, calculated for $N = 10^7$ periods

we give the logarithm of the transition time for the same initial conditions and many values of β. We see that for $\beta > \beta_c = 0.305124$ the transition time can be represented by an exponential law

$$\log_{10} T = 4.94 - 4160(\beta - \beta_c). \tag{6}$$

However, for $\beta < \beta_c$ the transition time is much larger than expected by Eq. (6) and for $\beta < 0.30510015$ it is longer that 10^{10} periods.

Fig. 19 has a lot of noise indicating that the law (6) is not exact, and small variations of β (or of the initial conditions) change the exact value of T. Nevertheless the order of magnitude of T remains the same for each value of β, and increases as β decreases. This increase is exponential if $\beta > \beta_c$, but is much more dramatic if $\beta < \beta_c$. To give an idea of this increase, if we draw another straight line for the values of T when $\beta < \beta_c$ in Fig. 19 and extrapolate we find T of the order of $T = 10^{400}$ for $\beta = 0.3$. (But there is no way to check if this estimate is correct).

Thus we have clearly two quite different regimes of diffusion, that we identify with Arnold diffusion when $\beta < \beta_c$, and resonance overlap diffusion when $\beta > \beta_c$. The change from one regime to the other is continuous, but the derivative $\mathrm{d}T/\mathrm{d}\beta$ changes abruptly and discontinuously when β goes beyond the critical value β_c.

As a consequence, in most cases we can safely ignore Arnold diffusion and make a clear distinction between ordered and chaotic domains in 3 degrees of freedom. Also the spectrum of Fig. 16 is practically invariant for extremely long times. This applies both to the spectra of stretching numbers and helicity angles. Thus we can use the helicity angles in the same way as in systems of 2 degrees of freedom, to separate ordered and chaotic domains in phase space.

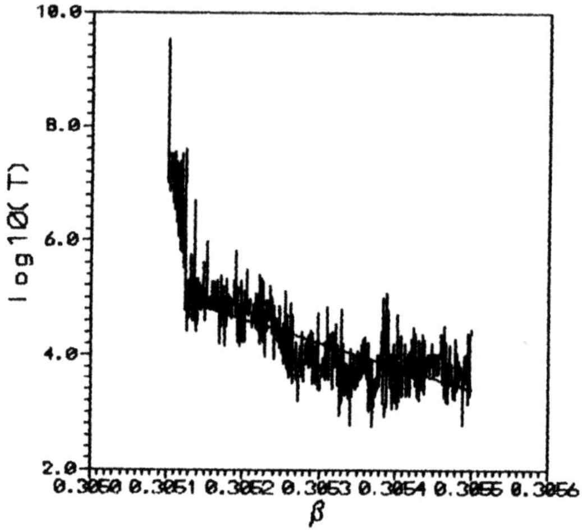

Fig. 19. The logarithm of the transition time from the ordered to the chaotic domain as a function of the coupling parameter

In galactic dynamics, where the age of the Universe is of the order of $10^2 - 10^3$ orbital periods the phenomenon of Arnold diffusion is insignificant. All diffusion effects can be attributed to resonance overlap. But in plasma physics and accelerator physics there may be cases in which Arnold diffusion is important.

6 Conclusions

1) Barred galaxies contain both ordered and chaotic orbits. Chaotic orbits are mainly around corotation, but appear also near the center if the density is large there. Both types of orbits are important in self-consistent models, e.g. chaotic orbits near corotation support the spirals emanating from the ends of the bar for long times, of the order of some 10^9 years. Chaotic orbits in systems with a cusp at the center play an important role in the evolution of the bars.

2) Two new methods for characterizing the orbits of a dynamical system, like a barred galaxy, are provided by the spectra of stretching numbers and helicity angles. The stretching number (or short-time Lyapunov characteristic number) is the logarithm of the ratio of successive deviations of two nearby orbits.The helicity angle is the angle between the deviation from a given orbit and a fixed direction. The distributions of the stretching numbers and helicity angles are the corresponding spectra. Their property is that they are invariant with respect to initial conditions in the same chaotic domain. (In the case of ordered orbits the spectra are invariant only with respect to initial conditions along the same invariant curve).

The spectra give much more information than the usual Lyapunov characteristic numbers. We give one example of two systems with the same Lyapunov characteristic number and the same appearance of the distribution of their consequents on a Poincaré surface of section, which have quite different spectra. Both systems are chaotic but their intrinsic properties are different, e.g. the asymptotic curves of their periodic orbits are very different.

The spectra of chaotic galactic orbits are invariant for the same Jacobi constant, but they are different if the orbits are inside corotation, outside corotation, or both inside and outside corotation.

3) We compare the helicity angles with the rotation angles around a central point. The average rotation angle after n iterations (if $n \to \infty$) is the rotation number. This limit does not exist for chaotic orbits, but, following Laskar, one can define an average rotation angle for fixed n. This varies considerably in chaotic domains and in this way one can separate the chaotic from the ordered orbits. But in strongly chaotic systems the rotation angle may not be defined because there is no obvious center of the system (e.g. some irregular periodic orbits have no well defined rotation number). On the other hand the helicity angles are always defined and can provide the fastest method to distinguish between order and chaos. The average value of the helicity angle is constant in the chaotic domain, but varies smoothly in an ordered domain (island). Short time averages show noise around the average in the chaotic case, but less noise in the case of islands. Thus even $n = 10$ iterations are sufficient to distinguish between chaotic and ordered domains.

4) In the case of three degrees of freedom we define one stretching number and three helicity angles. The spectra are again invariant in the chaotic domain. On the other hand the ordered domains (regions containing many KAM tori in nonintegrable systems) contain small chaotic domains that are not separated from the main chaotic domain but there is a slow diffusion (Arnold diffusion) that joins all chaotic domains. Thus the spectra in the ordered domains are only transient.

We studied two coupled 2D-maps, that represent the a surface of section of a system of three degrees of freedom. If the coupling is relatively large we found transient spectra that lead to chaotic invariant spectra after a long time. But for a small coupling constant the diffusion time is so extremely long that the transient spectra are invariant for all practical purposes. Thus we can again use helicity angles to distinguish between chaotic and ordered domains.

5) We found that there are two types of diffusion in systems of three degrees of freedom that are clearly distinguished. One is the resonance overlap diffusion, which gives a diffusion time that increases exponentially as the coupling parameter decreases. The other is Arnold diffusion, which gives a much larger diffusion time for small couplings. The transition between the two regimes is sharp (Fig. 19). As a conclusion we can disregard Arnold diffusion in galactic dynamics, but resonance overlap diffusion may be important in some barred galaxies.

Acknowledgements. This research was supported in part by the EEC Human Capital and Mobility Program (grant ERB 4050 PL930312). C.E. received support by the Greek Foundation of State Scholarships.

References

Athanassoula, E., Bienaymé, O., Martinet, L., Pfenniger, D. (1983): A&A **127**, 349
Binney, J., Spergel, D. (1982): ApJ **252**, 308
Chirikov, B.V. (1979): Phys. Rep. **52**, 263
Contopoulos, G. (1957): Stockholms Obs. Ann. **19**, No 10
Contopoulos, G. (1958): Stockholms Obs. Ann. **20**, No 5
Contopoulos, G. (1966): in Les Nouvelles Méthodes de la Dynamique Stellaire, eds F. Nahon, M. Hénon, CNRS, Paris Bull. Astron. Ser. 3, 2, Fasc. 1, p. 223
Contopoulos, G. (1971): AJ **75**, 96
Contopoulos, G. (1980): A&A **81**, 198
Contopoulos, G. (1986): A&A **161**, 244
Contopoulos, G. (1988): A&A **201**, 44
Contopoulos, G. (1993): Physica **D64**, 310
Contopoulos, G., Gottesman, S.T., Hunter, J.H., England, M.N. (1989): ApJ **343**, 608
Contopoulos, G., Grosbøl, P. (1989): A&AR **1**, 261
Contopoulos, G., Magnenat, P., Martinet, L. (1982): Physica **D6**, 123
Contopoulos, G., Vandervoort, P. (1992): ApJ **389**, 118
Contopoulos, G., Voglis, N. (1996): Celest. Mech. Dyn. Astron., in press
Contopoulos, G., Voglis, N., Efthymiopoulos, C. (1995): in Hamiltonian Systems with 3 or more degrees of freedom, ed. C. Simo, Plenum Press, in press
de Zeeuw, P.T. (1985): MNRAS **216**, 273
de Zeeuw, P.T., Hunter, C., Schwarzschild, M. (1987): ApJ **317**, 607
de Zeeuw, P.T., Lynden-Bell, D. (1985): MNRAS **215**, 713
Gnedin, O.Y., Goodman, J., Frei, Z. (1995): AJ **110**, 1105
Guckenheimer, J., Holmes, P. (1983): Nonlinear Oscillations, Dynamical Systems,, Bifurcations of Vector Fields, Springer Verlag, New York
Hasan, H., Norman, C. (1990): ApJ **361**, 69
Hasan, H., Pfenniger, D., Norman, C. (1993): ApJ **409**, 91
Hunter, C. (1988): Ann. New York Acad. Sci. **536**, 25
Kaufmann, D.E., Contopoulos, G. (1995): A&A, in press
Laskar, J. (1990): Icarus **88**, 266
Laskar, J. (1993): Physica **D67**, 257
Laskar, J., Froeschlé, C., Celetti, A. (1992): Physica **D56**, 253
Lynden-Bell, D. (1962): MNRAS **124**, 95
Lynden-Bell, D., Kalnajs, A.J. (1972): MNRAS **157**, 1
Martinet, L. (1984): A&A **132**, 381
Martinet, L., Udry, S. (1990): A&A **235**, 69
Merritt, D., Fridman, T. (1995): ApJ, in press
Noid, D.W., Koszykowski, M. L., Marcus, R.A. (1977): J. Chem. Phys. **67**, 404
Noid, D.W., Koszykowski, M. L., Marcus, R.A. (1981): Ann. Rev. Phys. Chem. **32**, 267
Norman, C.A., Hasan, H., Sellwood, J.A. (1995): preprint

Patsis, P.A., Grosbøl, P. (1996): A&A, in press
Pfenniger, D. (1984): A&A **134**, 373
Powell, G.E., Percival, I.C. (1979): J.Phys. **A12**, 2053
Vandervoort, P. (1984): ApJ **287**, 475
Voglis, N., Contopoulos, G. (1994): J. Phys. **A27**, 5357

Secular Evolution in Barred Galaxies

J. A. Sellwood and Victor P. Debattista

Rutgers University, Department of Physics & Astronomy, P O Box 849, Piscataway, NJ 08855, USA

Abstract. A strong bar rotating within a massive halo should lose angular momentum to the halo through dynamical friction, as predicted by Weinberg. We have conducted fully self-consistent, numerical simulations of barred galaxy models with a live halo population and find that bars are indeed braked very rapidly. Specifically, we find that the bar slows sufficiently within a few rotation periods that the distance from the centre to co-rotation is more than twice the semi-major axis of the bar. Observational evidence (meagre) for bar pattern speeds seems to suggest that this ratio typically lies between 1.2 to 1.5 in real galaxies. We consider a number of possible explanations for this discrepancy between theoretical prediction and observation, and conclude that no conventional alternative seems able to account for it.

1 Introduction

Chandrasekhar (1943) showed that a massive object moving though a background "sea" of light particles would experience a drag. The force is the gravitational attraction by the wake produced by the motion of the massive object (see e.g., Binney and Tremaine 1987, §7.1). When the mass, M, of the perturber is much larger than that of the individual background particles, the acceleration takes the form

$$\frac{dv_M}{dt} \propto -\rho \, M \, f\left(\frac{v_M}{\sigma}\right), \tag{1}$$

where ρ is the background density and f is a function of the ratio of the perturber's velocity, v_M, to the (assumed isotropic) velocity dispersion, σ, of the background particles. For a Maxwellian distribution of velocities, this function is a maximum when $v_M \simeq 1.37\sigma$.

A similar process must occur as a massive bar rotates inside a halo. In this case, the bar creates a wake in the halo which lags the bar and the gravitational attraction between the bar and the wake produces a torque which removes angular momentum from the bar and adds it to the halo. Weinberg (1985), adapting the perturbation theory approach of Tremaine and Weinberg (1984a), estimated the magnitude of the frictional drag force. Assuming a massive, rigid bar rotating in an isothermal halo, he concluded that the spin-down time for the bar would be as little as five bar rotations, for reasonable parameters!

Early low-quality, but fully self-consistent disc-halo simulations (Sellwood 1980) had previously revealed a rapid loss of angular momentum to the halo once a bar formed in the disc, at a rate roughly consistent with Weinberg's prediction. To our knowledge, no other such simulations have been conducted in the past 15 years; Combes et al. (1990) had only a bulge, not an extensive halo, of live particles, Raha et al. (1991) did not evolve their models for long enough, the "halo" particles of Little and Carlberg (1991) were confined to a plane, and the bar used by Hernquist and Weinberg (1992) was rigid and their model had no disc. We here report new fully self-consistent, simulations superior in many respects to those of Sellwood (1980), which were designed to reproduce the expected dynamical friction and to determine the secular changes to the bar, disc and halo in the long term. Athanassoula (work in progress) is conducting similar experiments using a direct N-body code on a GRAPE device.

2 New Simulations

2.1 Initial Set-up and Numerical Details

We begin by setting up a disc-halo equilibrium. The disc particles represent 30% of the total mass and are laid down with the Kuz'min-Toomre surface density distribution

$$\Sigma(R) = \frac{Mq}{2\pi a^2}\left(1+\frac{R^2}{a^2}\right)^{-3/2}. \tag{2}$$

We truncate this profile sharply at the rather small radius of $R = 4a$ in order to limit particle loss from the grid at later times. We also disperse the disc particles about the mid-plane in a Gaussian fashion having a uniform rms thickness of $0.4a$.

The remaining 70% of the mass is represented by the "halo" particles, a dynamically uniform population which could be thought of as comprising both a luminous bulge and a more extended dark halo. The halo particles are set in equilibrium in the manner first used by Raha et al. (1991). They are selected from an isotropic DF having a (lowered) polytropic form $f = f[(-E)^m]$, with the limiting energy $E = \Phi_m$, being the potential at some limiting radius. The combined disc and halo gravitational potential distribution to be used in this DF is determined iteratively in the manner adopted by Prendergast and Tomer (1970) and Jarvis and Freeman (1985). The polytropic index $m = 1.5$ in our case; n.b. this corresponds to a standard $n = 3$ polytrope, where $m = n - \frac{3}{2}$ (Binney and Tremaine 1987). The resulting halo mass distribution is not far from spherical and, when combined with the disc, gives rise to the circular velocity curve in the mid-plane shown in Fig. 1.

Having determined the potential of our initial mass distribution, we set the disc particles in motion. Their initial orbits are almost circular, but have enough random motion to maintain the vertical thickness and to set Toomre's $Q = 0.1$, in the case we focus on here. We have run other models in which $Q \geq 1$ at the start and find that the essential results we describe here are independent of the initial Q value.

Our simulations are performed on a 3-D Cartesian grid having 129^3 cubic cells; the code used was described by Sellwood and Merritt (1994). We set the

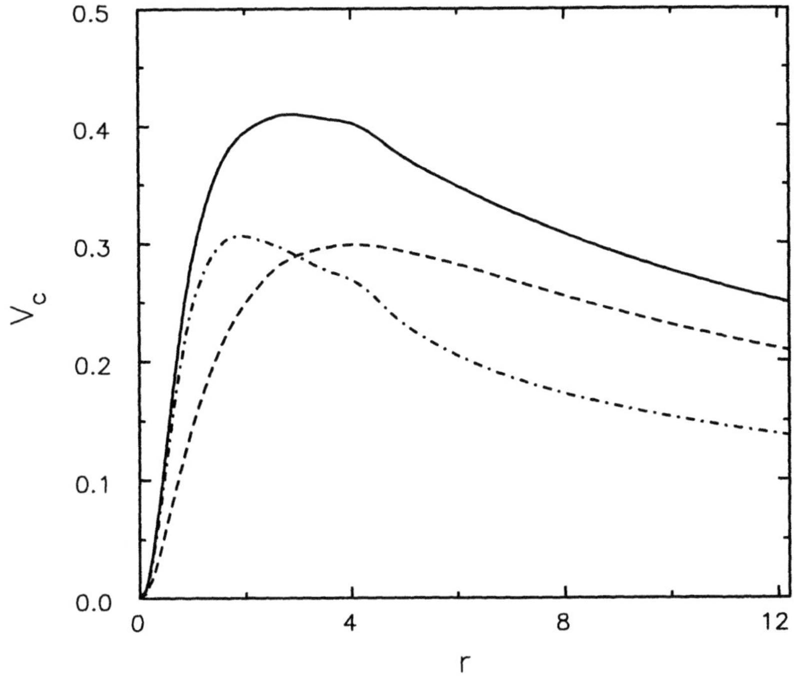

Fig. 1. The circular velocity curve at the start of the simulation (solid curve). The separate contributions from the disc (dot-dashed curve) and bulge + halo (dashed curve) are also shown

length scale $a = 5$ mesh spaces, and chose a time step for the leap-frog integration of 0.05 times the dynamical time $\sqrt{a^3/GM}$. We employ 300 K equal mass particles, of which 90 K represent the disc. We adopt units such that $G = M = a = 1$; the rotation period of a particle at the disc half-mass radius ($R \sim 2.3$) is about 35 in these units.

2.2 Evolution

This model is deliberately designed to be unstable and forms a strong, rapidly rotating bar within the first 100 dynamical times. As soon as the bar forms, a strong torque develops that begins to reduce the total angular momentum of the disc and to set the halo into rotation, as shown in Fig. 2(a). Total angular momentum is conserved, of course; almost all that which is lost from the disc goes into the halo, the tiny remainder being carried away by escaping particles.

A bi-symmetric distortion is readily detectable in the distribution of halo particles which initially lags the bar by $\sim 45°$. As the evolution proceeds, both the torque and the lag angle gradually decrease, until the rate of angular momentum transfer ceases almost entirely by $t \simeq 1600$, at which point the distortion in the halo has become aligned with the bar.

As usual, the bar suffers a bending instability in the early stages (e.g. Raha et al. 1991). It is first detectable at $t \simeq 250$ and is over by $t \simeq 450$. In this model, the bar amplitude is not greatly affected by the buckling instability and the torque on the halo is only slightly reduced by this event.

The pattern speed of the bar also begins to decrease after its formation, as shown in Fig. 2(b). Fig. 2(c) shows that the angular momentum remaining in the inner part of the disc (where the bar resides) decreases in a similar fashion. The similar shapes of these two curves indicates that the bar has a positive moment of inertia which is approximately, though not exactly, constant, justifying Weinberg's original assumption.

It is interesting that the secular changes seem to end before the halo was brought to co-rotate with the bar. Late in the simulation, the mean angular rotation rate of the halo particles in the very centre is about half that of the bar, but the rotation of the outer halo is characterised more by a constant mean orbital speed rather than by uniform rotation. The alignment of the halo distortion with the bar appears to indicate that a large fraction of the halo particles are trapped into resonances with the bar.

As evolution appeared to have almost ceased, we stopped the calculation at $t = 2000$, which corresponds to 40 rotation periods at the initial rotation rate of the bar.

Our principal result is displayed in Fig. 3, which shows estimates of the bar length and co-rotation radius at many times during the run. The distance, $D_{\rm L}$, is that from the centre to the Lagrange point on the bar major axis, and is determined from the potential and pattern speed at each instant. The semi-major axis of the bar, $a_{\rm B}$, is estimated to lie where the $m = 2$ coefficients of a Fourier expansion of the particle distribution depart from the constant phase and linear fall-off in amplitude characteristic of the outer bar region. This definition is consistent with those adopted by many observers.

The ratio $D_{\rm L}/a_{\rm B}$ increases steadily from a value of ~ 1.3 at $t = 200$, reaching ~ 3 by $t = 1200$. A large value of this ratio is quite unlike the values believed to pertain in real barred galaxies, as we review next.

3 Pattern Speeds of Bars in Galaxies

The only known technique to estimate the pattern speed directly from observations was proposed by Tremaine and Weinberg (1984b). It has been successfully applied in just one case, the SB0 galaxy NGC 936: Merrifield and Kuijken (1995) considerably improved Kent's (1987) original measurement for this galaxy. It is to be hoped that this technique will soon be applied to more galaxies, though it is unlikely to be successful for later Hubble types.

The new spectroscopic and photometric data on NGC 936 yield an estimate of $\Omega_{\rm p} \sin i = 3.1 \pm 0.75$ km s^{-1} arcsec^{-1}, which when combined with Kormendy's (1983) estimated inclination $i = 41°$ and rotation curve, places co-rotation at a distance of 69 ± 15 arcsec from the centre of the galaxy. This is somewhat outside

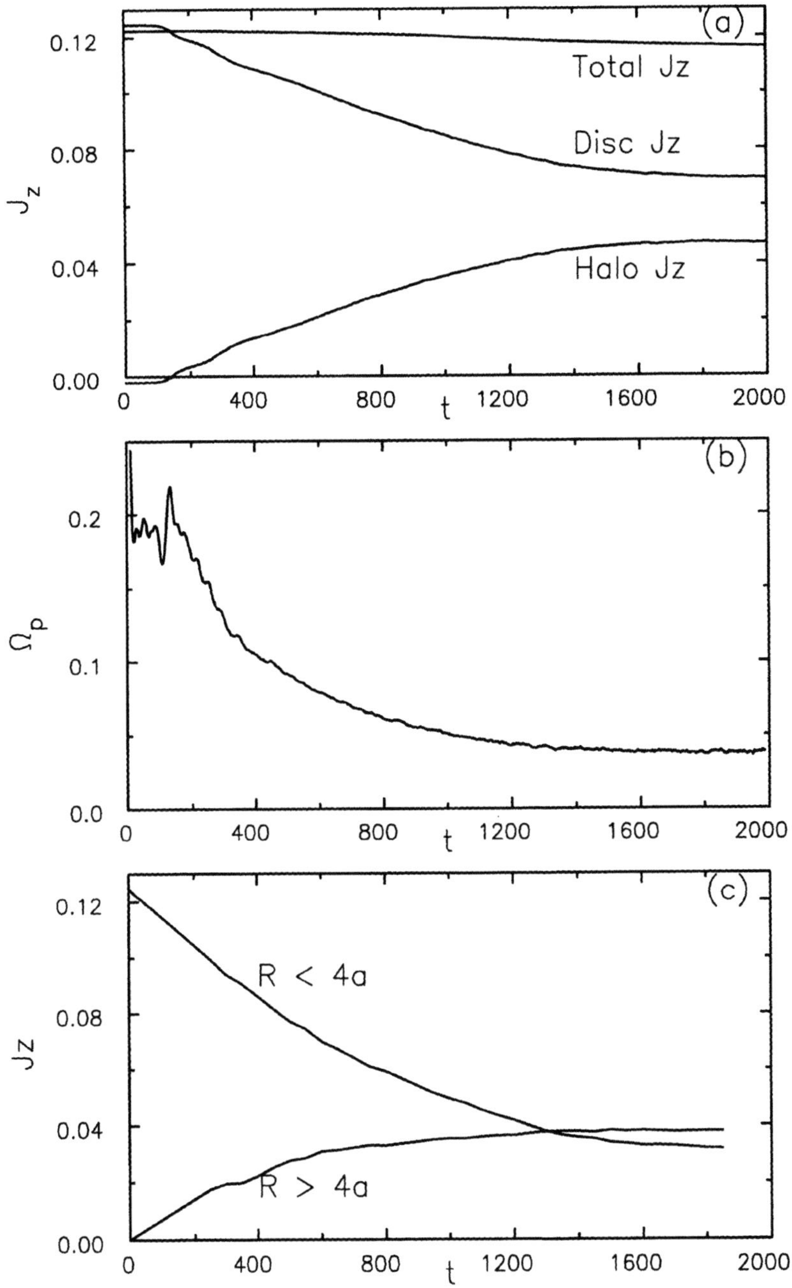

Fig. 2. **a** The time variations of the total angular momenta of the disc and halo, **b** of the bar pattern speed and **c** of the total angular momentum of particles in the inner and outer disc

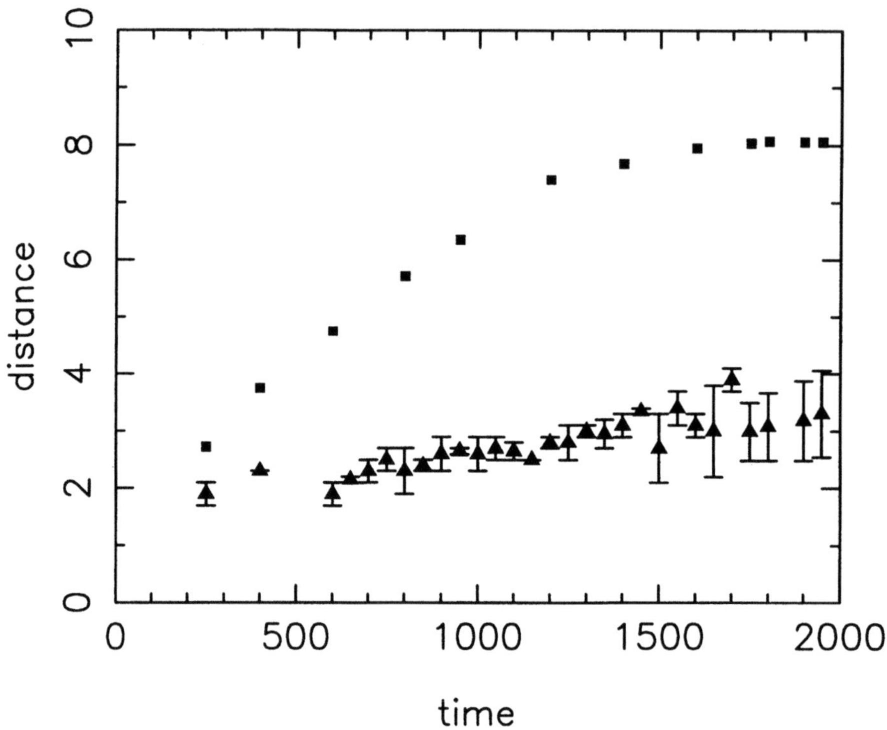

Fig. 3. The time variations of $D_{\rm L}$ (squares) and of $a_{\rm B}$ (triangles)

the visible bar which was estimated (Kent and Glaudell 1989) to end at about 50 arcsec from the centre. For this galaxy, therefore, $D_{\rm L}/a_{\rm B} = 1.4 \pm 0.3$.

All other techniques to estimate this ratio in galaxies are indirect. The best such evidence comes from the locations and shapes of dust lanes; the extensive survey of hydrodynamical bar flows by Athanassoula (1992) led her to conclude that $D_{\rm L}/a_{\rm B} = 1.1 \pm 0.1$ would best account for the dust lane morphology in galaxies. Her work was, however, restricted to rather artificial Ferrers bar models, which do not correspond well to the observed light distributions in bars. More recently Weiner (1996), P.A.B. Lindblad (this meeting) and others adopt mass models derived from the measured light distribution; preliminary results suggest that bars are indeed rotating rapidly, and that $D_{\rm L}/a_{\rm B} > 1.7$ seems to be firmly excluded.

There is evidence (Binney et al. 1991; Weiner and Sellwood 1996; Kalnajs this meeting) that the bar which is believed to reside in the Milky Way also has a high pattern speed, but the ratio $D_{\rm L}/a_{\rm B}$ is not yet firmly established.

Finally, it should be admitted that theoretical prejudice, which was originally the strongest "evidence" for fast bars (see Sellwood and Wilkinson 1993), has turned out to be almost worthless, since the bar in our simulation seems to survive quite happily with a low pattern speed!

While the above evidence could scarcely be described as overwhelming, it clearly favours values for D_L/a_B that are quite inconsistent with those we measure from our simulation, at least after the first few bar rotations.

4 What Could be Wrong?

One's first reaction to such a puzzling result is to question whether it needs to be taken seriously. Of course it is reassuring that the simulation behaved as theory had already predicted and that similar results are also being obtained by Athanassoula (1996) using a quite different N-body method. But perhaps the model differs from real galaxies in respects which cause it to severely overestimate the importance of dynamical friction, or that something has been omitted which would counteract the behaviour.

4.1 Friction Overestimated?

Chandrasekhar's formula (1), and Weinberg's analysis, indicates that the deceleration rate should be proportional to the bar mass and the halo density. Could either, or both, of these parameters be too large in our model?

Bar Strength. It is widely believed that bars are massive features, at least in some galaxies. They are observed to give rise to strong non-circular motions in many well studied cases: good examples of strongly non-axisymmetric gas motions are seen in NGC 5383 (Sancisi, Allen and Sullivan 1979; Duval and Athanassoula 1983), NGC 1365 (Jörsäter and van Moorsel 1995) and NGC 4123 (Weiner 1996). A similar streaming pattern in the stellar motions in NGC 936 was observed by Kormendy (1983). (It should be noted that such streaming patterns are easily masked in galaxies where the bar lies close to one of the principal axes of the projected disc.) Such large non-circular motions seem to indicate a strongly non-axisymmetric potential, which in turn implies that the bar has a significant mass compared with the axisymmetric components in the central parts of these galaxies.

In order to be more quantitative, we have made a comparison between Kormendy's (1983) slit observations of the barred galaxy NGC 936 with similar data taken from our model at $t = 250$ projected and inclined as NGC 936. (This time was chosen because D_L/a_B was close to the the value of 1.4 seen in NGC 936.) We estimate normalized $m = 2$ Fourier coefficients of both the radial and tangential velocities at fixed deprojected radii. Averaging over a range of radii near the end of the bar, we find we find the coefficients from our model are 2.8 and 1.7 times larger, for the azimuthal and radial components respectively, than the same values in NGC 936. We conclude that our bar is perhaps twice as strong as that in a typical barred galaxy, which may therefore have caused us to overestimate the spin down rate by a factor of two.

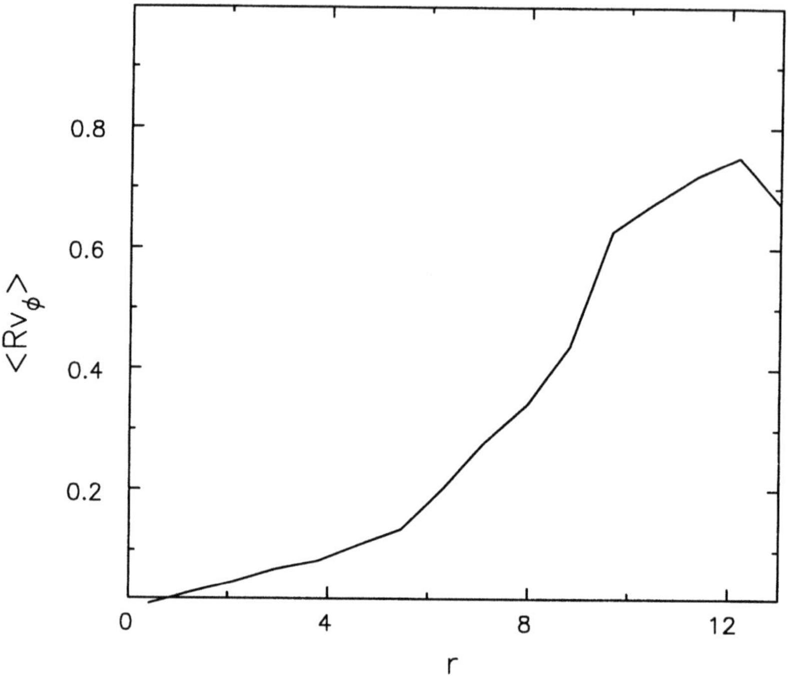

Fig. 4. The mean specific angular momenta of halo particles at $t = 2000$

Halo Mass. On the other hand, our halo is nowhere near massive or extensive enough to give rise to a flat rotation curve beyond the disc edge (Fig. 1). Halo particles that never come close to it would clearly be unaffected by the bar, but one would expect the halo out to radii several times the bar semi-major axis ($a_B \simeq 3$ for our bar) to be torqued up by the bar. Fig. 4 shows the mean specific angular momenta of halo particles at one instant late in the simulation, plotted as a function of radius, and indicates that particles far out in the halo have in fact gained disproportionately more angular momentum than those close in. Thus if we were to have run a simulation identical in most respects but having more mass in the outer halo to give a flat rotation curve, it seems likely that the bar would lose much more angular momentum. The under-massive halo of our model therefore gives too *little* dynamical friction.

Halo Core Radius. Although we do not expect the friction force on a strong bar to behave precisely as (1) would predict, that formula does suggest that the braking rate should scale with the density and depend on the halo velocity dispersion. For an isothermal halo with a core, the velocity dispersion (assumed isotropic) is largely set by the circular velocity at large radii; there is therefore little freedom to juggle this parameter for realistic halos. On the other hand, a halo having a larger core radius will have a weaker effect, but only to the extent that friction arises from the inner halo.

At the cost of eliminating any effective bulge component, we could decrease the central density of our halo (see Fig. 1). It cannot be decreased indefinitely, however, since the core radius of a realistic halo cannot be so large, relative to the disc scale, as to allow the rotation curve to decline significantly outside the disc. Thus observed asymptotically flat rotation curves require a minimum central halo density and a fixed velocity dispersion at large radii. Since we have already shown that halo mass at large radii takes up most of the angular momentum, we do not expect that a change to the central density, while keeping the halo mass fixed, will affect friction very much. Additional experiments to verify this expectation seem desirable.

Halo Rotation. Halos are not expected to have large angular momenta (e.g. Barnes and Efstathiou 1987). We have, nevertheless, tested the possibility that halo rotation could reduce the bar spin down rate by running two further simulations, identical in all respects except that in one case some fraction of the retrograde halo particles had their angular momenta flipped to give a total halo angular momentum about half the maximum possible. We found the bar pattern speed to drop by about the same amount in both and therefore conclude that dynamical friction is not significantly decreased by giving the halo even a large positive angular momentum.

4.2 Effects Omitted?

Secondary Bar Growth. Sellwood (1981) found that when a small bar formed within an extensive disc, it could grow in length due to trapping of additional stars into the bar as some angular momentum is removed by spirals in the outer disc. In his most extreme case, the bar's half-length approximately doubled from its initial value. Since the disc in our simulation was initially truncated at $R = 4$ and the bar which formed had a semi-major axis of fully half the distance to the initial edge, the scope for significant secondary bar growth is severely limited in our present model.

Could such substantial bar growth account for the small $D_{\rm L}/a_{\rm B}$ ratios of real galaxies? We do not think it likely for two main reasons: first, the bar would have to grow continually which requires incessant spiral activity in the outer disc. This is manifestly not happening now in the SB0 galaxy NGC 936; the bar in this galaxy is likely to have formed some time ago and with little sign of spiral activity in the outer disc, it cannot have grown much recently. Yet a low value of $D_{\rm L}/a_{\rm B}$ seems well established in this particular galaxy. The other reason is that secondary bar growth makes the bar longer and stronger, which would increase dynamical friction and therefore have a less than totally beneficial effect. Preliminary results from a further experiment seem to confirm that more extensive discs do not in fact lead to significantly smaller $D_{\rm L}/a_{\rm B}$ ratios.

Bar Spin-up. Our simulations are purely stellar and ignore the effects of gas. It is well known that the offset shocks on the leading side of the bar cause the gas to lose angular momentum. That angular momentum is, of course, given up to the bar. However, the amount of angular momentum is quite insignificant, since the gas mass is already small and the lever arm associated with it is short.

Radial inflows of gas are of slightly greater importance, however. Increases in the central mass concentration affect the potential in which the bar resides, and one consequence is an increase in the bar pattern speed (see also Kalnajs, this meeting). In Sect. 6.1 we give an example in which a substantial mass influx causes the bar pattern speed to rise by some 25%. This is helpful, but on its own, utterly inadequate to reconcile our simulation with observations.

5 Assessment

Thus far we have demonstrated that Weinberg's theoretical prediction of strong dynamical friction is at least qualitatively confirmed and that the bar is braked rapidly to an angular rate which is quite inconsistent with observed D_L/a_B ratios. We here list the possible solutions to this discrepancy between theory and observation that have occurred to us or been suggested by others.

1. Bars have low pattern speeds
2. Bars are weak
3. Bars grow in length as they slow down
4. Bars are spun up – e.g. by gas inflow
5. Bars have enormous effective moments of inertia
6. The halo co-rotates with the bar
7. Many halo particles are locked into resonance with the bar
8. Bars do not last long
9. Halos are not very massive

The entire problem hinges on alternative 1 being excluded. The evidence for fast bars (Sect. 3) is not as strong as we would wish, and we are uncomfortable that rather too many of our arguments rest on the assumption that the early-type SB0 galaxy NGC 936 is typical. The evidence for massive halos is strongest for unbarred, late-type spiral galaxies (see Sect. 7). More data confirming both a high pattern speed and a massive halo in several barred galaxies would be most welcome.

We have disposed of possibility 2 and argued that 3 and 4 are minor effects that could do little towards removing the discrepancy. The moment of inertia of the bar in our simulation, at least, is not large enough to prevent dynamical friction from slowing it; it seems unlikely that the structure of real bars is sufficiently different to change this conclusion. Alternative 6 also does not deserve lengthy consideration – the angular momentum of the halo would have to be inconceivably large.

Alternative 7 is somewhat more interesting. Dynamical friction in our simulation all but ceases while the bar rotated significantly, which seems to indicate that many halo particles have become trapped in resonances. This phenomenon deserves further investigation, but it is clear that it cannot provide a solution to our puzzle since friction ceases only after the bar pattern speed has dropped by a factor of five.

The remaining two alternatives are much more radical, but have to be contemplated since no other solutions seem tenable.

6 Transient Bars?

The possibility that bars could disappear before dynamical friction had sufficient time to slow them down was first suggested by Hernquist and Weinberg (1992). In order not to violate the bound of $D_L/a_B < 1.7$ suggested by observation, most bars would have to be destroyed quite quickly – within 10 rotations, judging from our simulation. Thus, to maintain the observed substantial fraction of galaxies containing strong bars (e.g. Sellwood and Wilkinson 1993), this idea requires bars to form and dissolve more than once over the lifetime of a galaxy. A second attraction of such a radical idea is that the fraction of galaxies containing strong bars, for which there is still no convincing explanation, represents a 30% duty cycle in the barred phase. Regenerating a bar in a disc where one has previously been destroyed presents a formidable problem, however.

6.1 Bar Destruction

Many authors have noted that bars are robust, long-lived systems that are not easily destroyed. Our simulation provides yet another example; despite having lost some 2/3 of its angular momentum and having reduced its pattern speed by a factor of 5, it remains a strong bar, as shown in Fig. 5. Thus our simulation excludes the possibility, left open by Weinberg's analysis, that bars simply would not survive such fierce braking.

There are just two known ways to destroy bars: one obvious way is to hit the bar with a companion, the other is to have a build-up of mass at the bar centre. As Athanassoula (this meeting) presents a major study of bar-satellite interactions, we do not discuss them here.

The effects of central mass concentrations have been explored extensively (Hasan and Norman 1990; Hasan, Pfenniger and Norman 1993; Wada and Habe 1992, Friedli and Benz 1993, 1995; Heller and Shlosman 1994; Norman, Sellwood and Hasan 1996). The idea here is that the gas driven towards the centre by the bar itself changes the gravitational potential within the bar to a sufficient extent that the main orbit family (Contopoulos' family x_1) becomes chaotic, and the regular part of phase space switches to the perpendicular x_2 family. A self-consistent bar can no longer survive once this happens, and the bar becomes a spheroidal bulge (Norman et al. 1996). The precise central mass and degree of

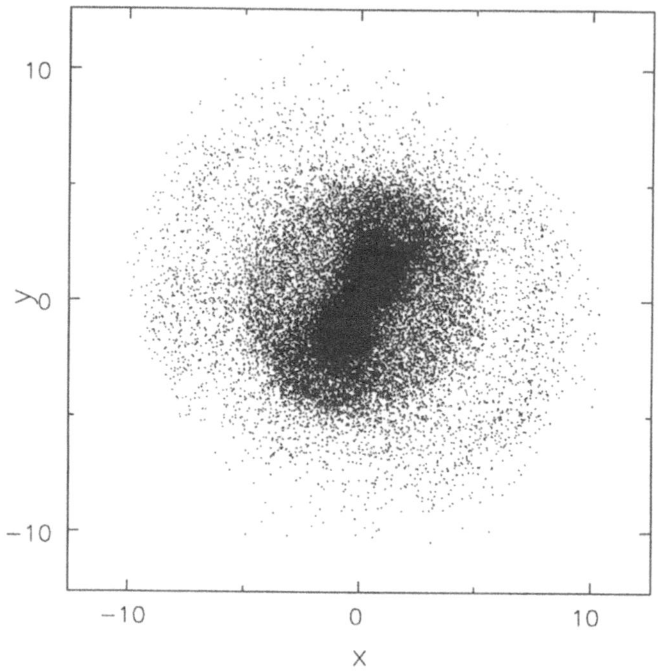

Fig. 5. The projected distribution of disc particles at $t = 2000$ showing that a strong, butterfly-shaped bar survives. We have not included any bulge/halo particles

concentration needed to achieve this has yet to be firmly tied down, but a few percent of the total mass of the disc seems ample.

Returning for a moment to the point made in Sect. 4.2, we present Fig. 6 to illustrate that the simple process of forming a central mass concentration increases the bar pattern speed. This result is taken from the 3-D simulation by Norman et al. in which the mass build-up was mimicked by simply contracting a rigid spherical mass component containing 5% of the disc and bulge mass. The increase in pattern speed (by some 25% in this case) therefore cannot have been caused by external torques and is simply a result of the changing internal structure of the bar.

6.2 Difficulties with Regenerating Bars

Destruction of a bar, either by the above mechanism or by interaction with a satellite, would leave the disc in a dynamically very hot state. The processes of both its formation and destruction would disturb the disc stars into quite markedly eccentric orbits, making the disc quite unresponsive to the kind of large-scale collective instability needed to reform a bar. Since only gas can cool,

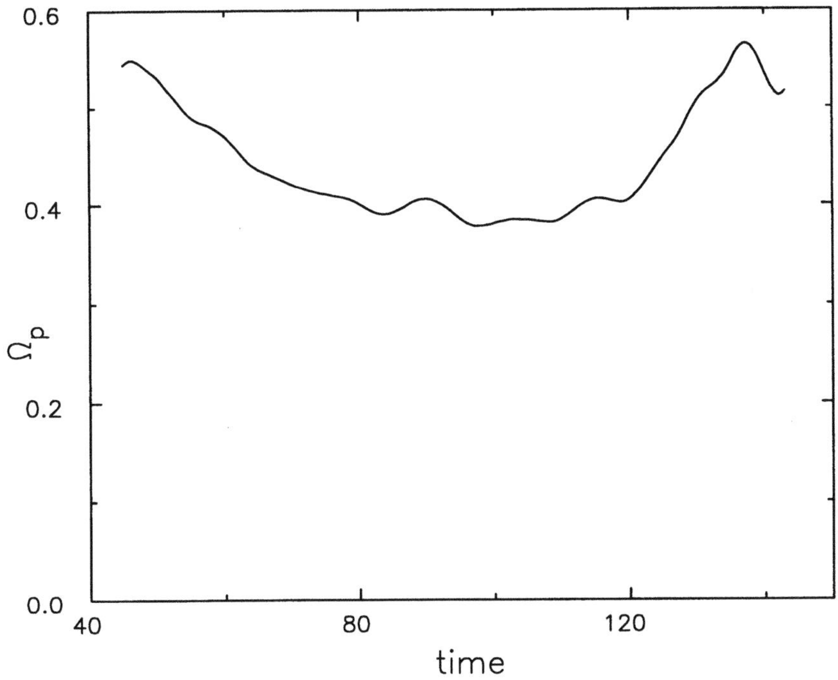

Fig. 6. The time dependence of the bar pattern speed in the 3-D model of Norman et al. (1996). The non-disc components of this model were rigid but the bar still slows through interactions with the outer disc until $t = 100$. Mass influx was mimicked by shrinking a rigid Plummer sphere component having 5% of the disc + bulge mass. The scale radius of this component decreased by a factor of 40 over the period $100 < t < 150$, which caused the pattern speed to rise before the bar dissolved at $t \simeq 130$

the galaxy would require a long recuperation period in which a large supply of fresh gas led to the formation of a substantial fraction of new stars on nearly circular orbits before the disc could become receptive to a new global instability. While the demands here seem excessive, this process could conceivably occur in the gas-rich late Hubble types; the theory would therefore appear to predict an increasing bar frequency along the Hubble sequence that is not observed (Sellwood and Wilkinson 1993).

Furthermore, if most bars are destroyed by central mass concentrations, the galaxy will be made more stable. A high central density is precisely what is required to stabilize a massive disc (Toomre 1981 and this meeting). This is not a watertight argument, since it is not clear that the galaxy would be absolutely stable no matter how cool the disc, and bars could also be triggered by interactions (e.g, Noguchi 1987), but it adds considerably to the difficulties faced by the recurrent bar idea.

Since both methods of bar dissolution make a bulge, bars in bulgeless galaxies must therefore be their first which, unless they are rotating slowly, would be required in this picture to be young. It is perhaps interesting that most bulgeless barred galaxies are low luminosity galaxies. Unfortunately, it is unclear what observational data on such galaxies would be required to test the prediction of dynamical youthfulness.

The idea of transient bars therefore faces extreme challenges. They would be reduced somewhat if one could argue that the first bar in a galaxy is braked by the halo, which is then sufficiently spun-up as to exert much weaker friction on a second bar. Some such wildly speculative idea is required if the regenerated bars alternative is to remain viable.

7 Low Mass Halos?

The final possible alternative is that galaxies with fast bars lack massive halos. The best evidence for massive halos comes from the extended, flat H I rotation curves in late-type, unbarred spiral galaxies (e.g. van Albada and Sancisi 1986). Occam's razor, together with current ideas of galaxy formation, suggest that all galaxies should have flat outer rotation curves, but the supporting observational data is still sketchy. Bosma (1992 and this meeting) concludes that barred galaxies generally do have extensive flat rotation curves. An exception for NGC 1365 is claimed by Jörsäter (this meeting, Jörsäter and van Moorsel 1995), but deprojection of the complex kinematic map of this strongly barred, asymmetric, and probably also warped galaxy is exceedingly difficult. The evidence for massive halos in early-type galaxies is also weak because they generally lack the gas disc which makes such a useful tracer of the potential in late-type systems. Van Driel and collaborators have attempted to address this issue by mapping the gas in those rare S0 galaxies that are relatively gas rich, finding some evidence for flat rotation curve at large radii in the case of NGC 4203 (van Driel et al. 1988).

We therefore think it likely that the circular velocity stays high at large radii in all galaxies, including those with fast bars. If this does not indicate a massive halo, then some alternative explanation for the phenomenon would need to be invoked (e.g. Milgrom and Bekenstein 1987).

8 Conclusions

The pattern speed problem presented by dynamical friction between a bar and bulge/halo is becoming rather insistent. Most possible solutions seem unattractive, some are excluded and others need to be stretched excessively. It is becoming increasingly difficult to find a tenable conventional explanation.

Acknowledgments. We would like to thank Scott Tremaine for a critical reading of the manuscript. This work was supported by NSF grant AST 93-18617 and NASA Theory grant NAG 5-2803.

References

Athanassoula, E. (1992): MNRAS **259**, 345
Athanassoula, E. (1996): in Proc. IAU Coll. 157 Barred Galaxies, eds R. Buta, D.A. Crocker, B.G. Elmegreen, PASP Conf. series **91**, p. 309
Barnes, J. E., Efstathiou, G. (1987): ApJ **319**, 575
Binney, J., Gerhard, O. E,, Stark, A. A., Bally, J., Uchida, K. I. (1991): MNRAS **252**, 210
Binney, J., Tremaine, S. (1987): Galactic Dynamics, Princeton University Press, Princeton
Bosma, A. (1992): in Morphology and Physical Classification of Galaxies, eds G. Longo, M. Capaccioli, G. Busarello, Kluwer, Dordrecht, p. 207
Chandrasekhar, S. (1943): ApJ **97**, 257
Combes, F., Debbasch, F., Friedli, D., Pfenniger, D. (1990): A&A **233**, 82
Duval, M. F., Athanassoula, E. (1983): A&A **121**, 297
Friedli, D., Benz, W. (1993): A&A **268**, 65
Friedli, D., Benz, W. (1995): A&A **301**, 649
Hasan, H., Norman, C. (1990): ApJ **361**, 69
Hasan, H., Pfenniger, D., Norman, C. (1993): ApJ **409**, 91
Heller, C., Shlosman, I. (1994): ApJ **424**, 84
Hernquist, L., Weinberg, M. D. (1992): ApJ **400**, 80
Jarvis, B. J., Freeman, K. C. (1985): ApJ **295**, 314
Jörsäter, S., van Moorsel, G. A. (1995): AJ **110**, 2037
Kent, S. M. (1987): AJ **93**, 1062
Kent, S. M., Glaudell, G. (1989): AJ **98**, 1588
Kormendy, J. (1983): ApJ **275**, 529
Little, B., Carlberg, R. G. (1991): MNRAS **250**, 161
Merrifield, M. R., Kuijken, K. (1995): MNRAS **274**, 933
Milgrom, M., Bekenstein, J. (1987): in IAU Symposium 117 Dark Matter in the Universe, eds J. Kormendy, G.R. Knapp, Reidel, Dordrecht, p. 319
Noguchi, M. (1987): MNRAS **228**, 635
Norman, C. A., Sellwood, J. A., Hasan, H. (1996): ApJ (to appear)
Prendergast, K. H., Tomer, E. (1970): AJ **75**, 674
Raha, N., Sellwood, J. A., James, R. A., Kahn, F. D. (1991): Nat **352**, 411
Sancisi, R., Allen, R. J., Sullivan, W. T. (1979): A&A **78**, 217
Sellwood, J. A. (1980): A&A **89**, 296
Sellwood, J. A. (1981): A&A **99**, 362
Sellwood, J. A., Merritt, D. (1994): ApJ **425**, 530
Sellwood, J. A., Wilkinson, A. (1993): Rep. Prog. Phys. **56**, 173
Tremaine, S., Weinberg, M. D. (1984a): MNRAS **209**, 729
Tremaine, S., Weinberg, M. D. (1984b): ApJ **282**, L5
Toomre, A. (1981): in Structure and Evolution of Normal Galaxies, eds S.M. Fall, D. Lynden-Bell, Cambridge University Press, Cambridge, p. 111
van Albada, T. S., Sancisi, R. (1986): Phil. Trans. R. Soc. London **320**, 447
van Driel, W., van Woerden, H., Gallagher, J. S., Schwarz, U. J. (1988): A&A **191**, 201
Wada, K., Habe, A. (1992): MNRAS **258**, 82
Weinberg, M. D. (1985): MNRAS **213**, 451

Weiner, B. J. (1996): in IAU Coll. 157 Barred Galaxies, eds R. Buta, D.A. Crocker, B.G. Elmegreen, PASP Conf. series **91**, p. 489

Weiner, B. J., Sellwood, J. A. (1996): in IAU Symp. 169 Unsolved Problems of the Milky Way, ed. L. Blitz, Kluwer, Dordrecht (to appear)

The Fate of Barred Galaxies in Interacting and Merging Systems

E. Athanassoula

Observatoire de Marseille, 2 Place Le Verrier, 13248 Marseille Cedex 04, France

Abstract. I use fully selfconsistent N body simulations to discuss two specific types of interactions between a barred target galaxy and a small companion.

In the first type of interactions an expanding ring is formed, in a way similar to the formation of ring galaxies from a nonbarred target. Depending on the impact angle and position, the center of the bar may be temporarily offset from that of the disc.

In the second type of interactions the companion is initially in a near-circular orbit in a plane either coinciding with the plane of the target disc, or at a small angle to it. The merging destroys the bar. The disc expands both in the vertical and the radial direction so that its shape remains that of a thin disc. If the orbit is at an angle with the target disc the latter undergoes an important tilt.

1 Introduction

It is by now well established that most galaxies are not isolated objects, but form parts of pairs, triplets, groups or clusters, and that this affects considerably their dynamical evolution (see e.g. papers in Wielen 1990, or Barnes and Hernquist 1992). Our own Galaxy, being part of a small group, is surrounded by a number of small satellites, of which the two most massive ones are the LMC and the SMC. Andromeda, the biggest galaxy in the Local Group, has also companions, of which M 32 is one of the most compact galaxies known. Studies of external galaxies also stress the existence of satellites around parent galaxies. Thus Holmberg (1969), using Palomar Sky Survey plates, found that spirals have between one and five satellites within a projected distance of ~ 50 kpc and brighter than roughly -10.5. Similarly Zaritsky et al. (1993) find that Sb/Sc galaxies have on average one galaxy brighter than -15 within a projected radius ~ 250 kpc.

The effects of interactions and mergings on the dynamics of disc galaxies has received a lot of attention in the last few years, particularly since computer advances now allow reasonable N-body simulations. Nevertheless the fate of bars in such interactions and mergings has been little studied so far. This paper will give preliminary results of a series of simulations designed to address this question.

2 Simulations

In all my simulations the target galaxy is barred. This was obtained by evolving a galaxy in isolation, starting with a bar unstable Kuzmin/Toomre disc embedded in a Plummer halo of somewhat less than twice its mass and twice its extent. The scale lengths of the disc and halo at the start of the simulation are equal to 1. and 5. computer units [1] respectively. More information on the initial conditions and set-up procedure have been given by Athanassoula, Puerari and Bosma (1996, hereafter APB). The evolution of this simulation was followed long after the bar formed so that comparisons between isolated and interacting discs at corresponding times would be possible.

Five different companions have been used, all of them initially Plummer spheres, covering a wide range of concentrations as well as masses, from $M_c/M_d = 0.1$, or $M_c/M_g = 0.035$ (where M_c, M_d and M_g respectively the mass of the companion, target disc and galaxy), to $M_c/M_d = 1.0$, or $M_c/M_g = 0.35$.

There is of course an infinity of possible companion trajectories, and I will here consider only two extreme types: For the first one the companion has initially a velocity perpendicular to the target disc, i.e. $v_x = v_y = 0$. Obviously this is not true anymore when the companion hits the disc, nevertheless I will, for brevity's sake, call such passages and simulations "perpendicular" or "initially perpendicular".

The second trajectory considered has a near-circular initial velocity in a plane which goes through the center of the target. I have so far considered only cases where this plane coincides with the plane of the disc or is at a small angle to it, not bigger than 15°. For brevity's sake I will call such passages "planar", although of course the orbit of the companion stays in a plane only if it was initially in the disc plane. As expected, the companion in such cases spirals inwards towards the center of the target, its distance from it decreasing initially slowly, and then, after it has reached higher density regions of the target, much faster.

The simulations were made using direct summation on our GRAPE 3AF (e.g. Ebisuzaki et al. 1993) system. Upon completion the survey will consist of 80 initially perpendicular orbits and an equal number of planar ones and uses 40 000 particles for the target galaxy. To date not all simulations have been carried out and a yet smaller fraction has been analysed, so the results presented here are necessarily preliminary. On the other hand some of the most interesting cases, like those discussed in the last section, were repeated using 120 000 particles for the target. The number of particles in the companion varies between 1 400 and 14 000 for the standard target disc and three times as many for the 120 000 particle target. The mass of all particles in the simulation is the same.

A wide variety of final configurations can be obtained from such encounters. Depending on the companion and orbit chosen, the encounter may result in a change of the pattern speed and amplitude of the bar, the formation of substructures (like a ring, lens, or bulge), the formation of asymmetries, the destruction of the bar, or other. Here I will only discuss two of the above possibilities.

[1] All numbers given in this paper are in computer units. They can be converted to kpc, $M_\odot$ and Gyr using the scaling given in APB.

3 Formation of Rings

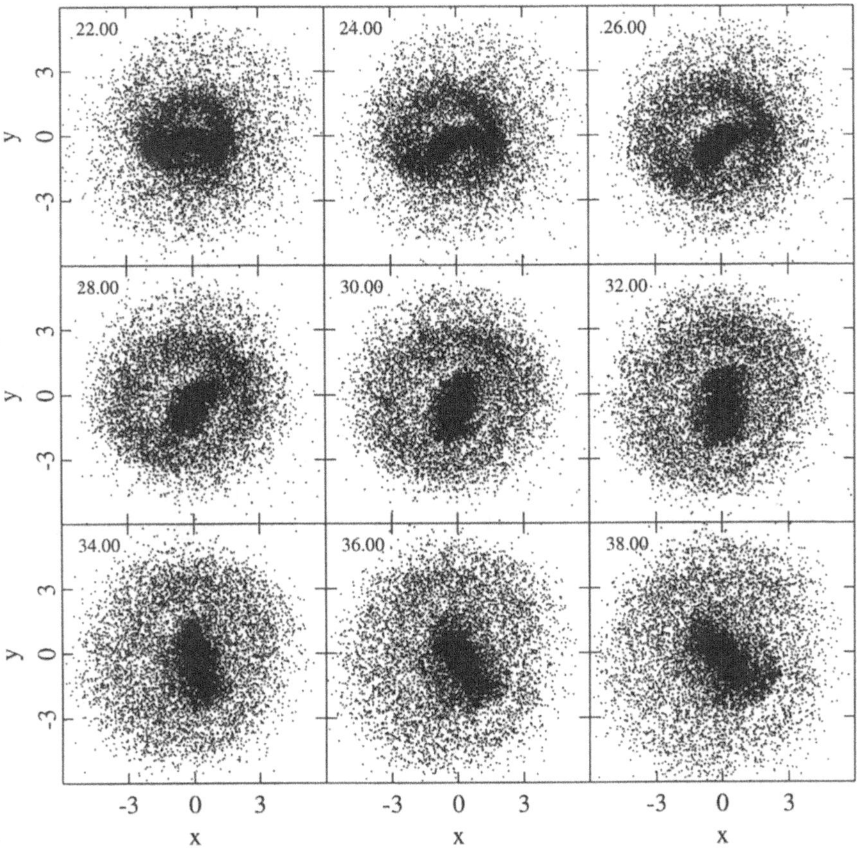

Fig. 1. Evolution of a ring formed by a central and vertical impact of a small companion. Time is given in the upper left corner of each panel

If a small galaxy hits the central regions of a disc, it will form a ring which is a density wave propagating outwards (Lynds and Toomre 1976; Theys and Spiegel 1976, 1976; Toomre 1978), provided of course that its relative mass and velocity are in a reasonable range of values. Selfconsistent simulations, sometimes including gas as well as stars, have permitted a thorough study of such interactions for the case of nonbarred target galaxies (Huang and Stewart 1988; Appleton and James 1990; Hernquist and Weil 1993; Struck-Marcell and Higdon 1993; Mihos and Hernquist 1994; APB), but not for barred galaxies, although the latter should also be victims of such impacts. The effect on barred targets was first addressed by APB, who presented three relevant simulations. The impacts they studied are either central, or near the bar major axis, and in all three cases the trajectory of the companion forms an angle of roughly 45° with the plane of the target. The impact drives the center of the bar away from the disc center, while the ring starts forming and expanding. During this stage the bar forms part of the ring. After some time the bar does not follow the expansion of the ring any more and falls slowly towards the disc center. Thus for a given time interval the result looks like a barred galaxy with a ring that does not touch the extremities of the bar.

Vertical and central impacts can lead to a different evolution, as is shown in Fig. 1. In this simulation a companion, of mass equal to one tenth the mass of the target galaxy, hits the disc at $t = 17$. At that time the bar puffs up into an oval, and then a ring detaches itself from its edge. In this case and for a short time the bar gets bent by the impact, but does not form at any time part of the ring as in the examples in APB. On the other hand at later times the result of the simulation is very similar to that of the examples in APB, with a ring whose diameter is larger than the bar major axis.

Several instances from the evolution of such rings, as e.g. the snapshot at $t = 30$ for the simulation shown in Fig. 1, resemble barred galaxies with stellar diffuse outer rings. Nevertheless it is easy to make the distinction between ring galaxies and ringed galaxies spectroscopically, since the rings of ring galaxies have a considerable expansion velocity, while the outer rings in barred galaxies do not.

4 Initially Near-Circular Trajectories at a Small Angle with the Plane of the Disc

Can bars be destroyed by interactions? Initially vertical passages, as those discussed in the previous section, can indeed achieve this (Athanassoula 1996, and in preparation), but amongst all simulations analysed so far those where the bar disappeared suffered also a very important thickening of the disc. Thus such interactions may not prove to be successful. In this section I will discuss a series of simulations with a different type of trajectory for the companion, namely such that the companion starts in or near the plane of the disc on a near-circular orbit. The aim will be to check the effect of such a trajectory on the bar and on the thickness of the disc. Such interactions, but for non-barred systems, have been recently discussed analytically by Tóth and Ostriker (1992), and with the help of fully selfconsistent simulations by Walker, Mihos and Hernquist (1996). One relevant simulation with a bar-unstable target was shown by Pfenniger (1991), unfortunately without much analysis. My simulations differ from it in that they have a live halo and in that the target is initially barred. I did not start from a non-barred disc galaxy, in which the bar would be triggered by the interaction itself, in order to avoid confusion between the effects of the overshoot and its aftermath on the one hand and the effects of the interaction on the other. The target galaxy is represented by 120 000 particles, out of which 42 000 rep-

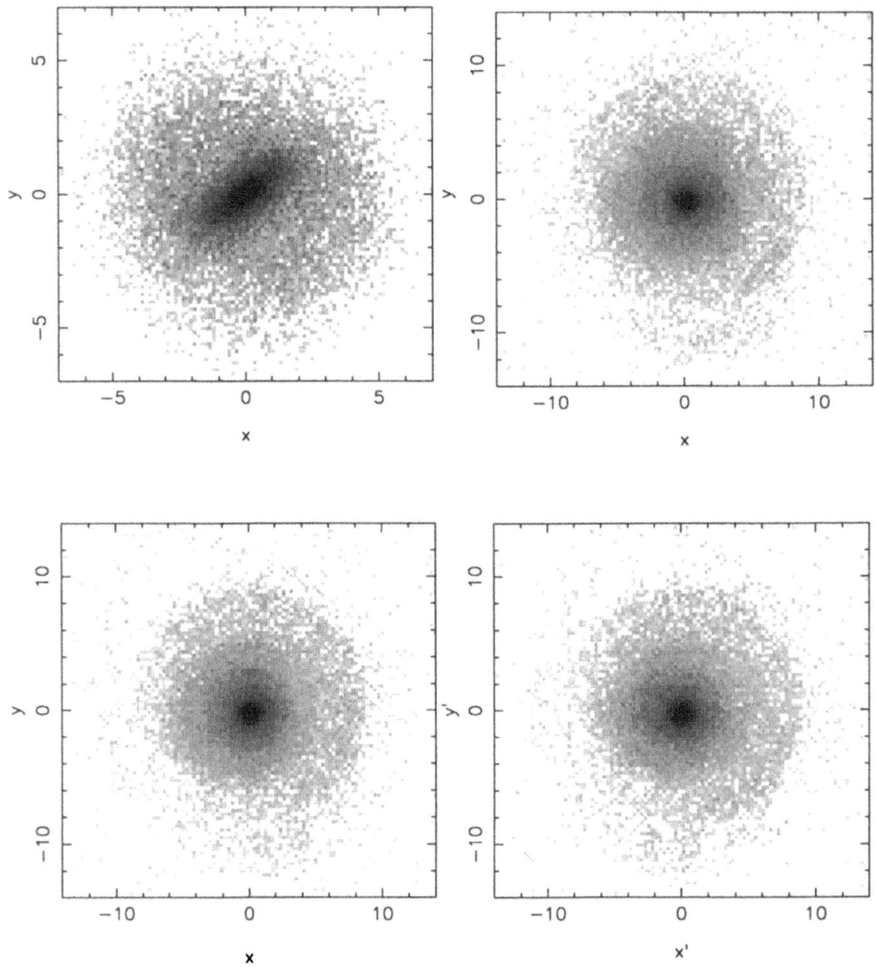

Fig. 2. Face-on view of the results of interactions discussed in Sect. 4. The scale in the greyscale plots is linear and the highest density parts are black and the lowest white. They were constructed using only the particles in the disc and companion, and not those in the halo. Note the difference in scale between the upper left panel and the other three

resent the disc. The companion has a mass equal to that of the disc, and is also represented by 42 000 particles.

Figs. 2 and 4 compare the results of three simulations. In the first one (upper left panels in both figures) the target galaxy is evolved in isolation. In the second one (upper right) it is perturbed by a companion placed initially in the plane of the disc at a distance 20% larger than the outer cutoff of the halo (i.e. 12 disc scalelengths) and with a near-circular velocity. In the third one (lower panels)

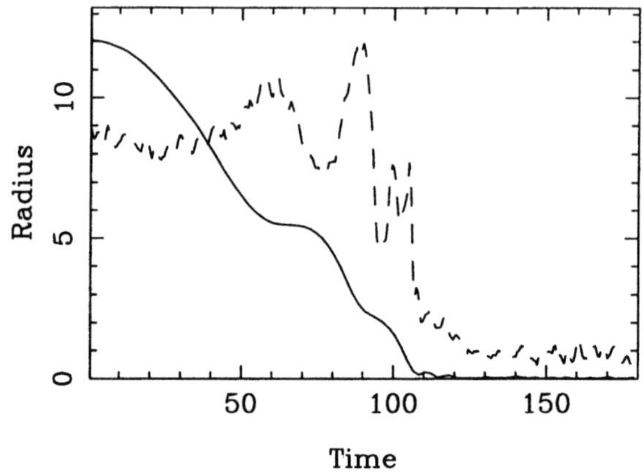

Fig. 3. Distance of the companion from the center of the target (full line) and $m = 2$ component of the density (dashed line) in arbitrary units. Both are given as a function of time

the companion is initially placed in a plane at an angle of roughly 15° with the plane of the disc. All three simulations have been carried out for the same time, sufficient for the companion to merge totally with the target and quasi-equilibrium to be reached. The face-on and edge-on views at this final time are compared in Figs. 2 and 4.

From the upper panels of Fig. 2 we see clearly that the in-plane interaction is able to destroy the bar. How that happens in time is given in Fig. 3 where I plot the distance of the companion from the center of the target as a function of time, as well as a measure of the $m = 2$ component of the density in the target. When the companion reaches the vicinity of the disc it starts inducing important, albeit temporary, increases and decreases of the bar amplitude. Note, however, that the bar is fully destroyed only after the companion reaches the center of the disc, presumably because it increases considerably the central concentration in the galaxy (Hasan and Norman 1990; Hasan, Pfenniger and Norman 1993; Friedli and Benz 1993; Friedli 1994; Sellwood 1996)

Fig. 4 shows that, as a result of the interaction, the disc has expanded considerably, as expected. This expansion concerns mainly the highest and the lowest density parts of the disc. An important point to note is that, due to the fact that the disc expands both radially and vertically, its shape remains that of a thin disc. Thus such interactions can destroy the bar without unduly fattening the disc. One could, however, argue that in-plane interactions should be relatively rare, the plane of the companion's orbit being in general at an angle with the plane of the target disc. It is thus essential to test whether the above result depends crucially on the fact that the plane of the companion's orbit coincides with that of the disc and whether orbits at an angle to it may cause more vertical heating, and larger relative thickening.

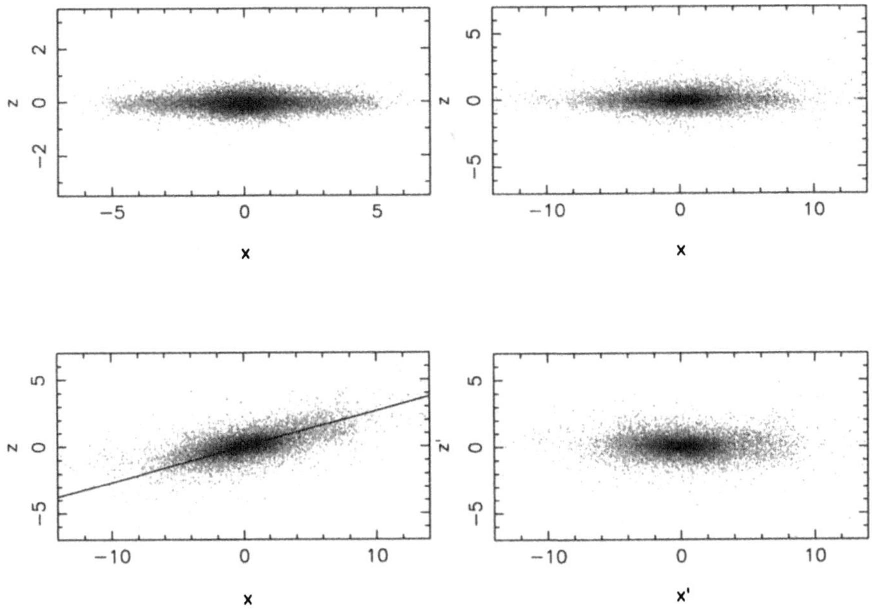

Fig. 4. Similar to Fig. 2, but the discs are now seen edge-on. The straight line in the lower left panel is at 15° to the x axis. The greyscales were constructed using only the particles in the disc, and not those in the halo or the companion, to show best its outline. Note the difference in scale between the upper left panel and the other three

The lower panels of Figs. 2 and 4 correspond to similar runs, but now the companion is initially placed in a plane at an angle of roughly 15° with the plane of the disc and intersecting it along the y axis. The left panel is in the standard frame of reference, i.e. where the disc was initially on the (xy) plane, and the right one is in a frame of reference $(x'y'z')$, determined by the three principal axes of the moment-of-inertia tensor of the disc. This frame gives a better estimate of the intrinsic thickness of the disc. As in the previous case the bar was destroyed by the merging. The plane of the disc undergoes an important tilt, so that an important percentage of the vertical energy of the satellite is converted to coherent motions in the disc, and the vertical heating is not as high as a straightforward application of the Tóth and Ostriker analysis would have given. As in the previous example the disc has again expanded both radially and vertically, in a way that its shape remains that of a thin disc.

The companion looses only a small fraction of its mass. Most of it settles at the center of the target, contributing to a bulge population. Thus such mergings can drive an evolution of disc galaxies along the Hubble sequence.

These simulations, albeit preliminary, show that it is possible for a companion to destroy the bar without destroying or unduly thickening the disc. A more quantitative analysis, together with the results of other similar simulations will be given elsewhere.

Acknowledgements. I would like to thank Albert Bosma for interesting discussions and Jean-Charles Lambert for his help with the GRAPE software and with the administration of the runs. I would also like to thank the INSU/CNRS and the University of Aix Marseille I for funds to develop our GRAPE facility.

References

Appleton, P.N., James, R.A. (1990): Dynamics and Interaction of Galaxies, ed. R. Wielen, Springer, Berlin, p. 200
Athanassoula, E. (1995): in Proc. IAU Coll. 157 Barred Galaxies, eds R. Buta, D.A. Crocker, B.G. Elmegreen, PASPC **91**, p. 309
Athanassoula, E., Puerari, I., Bosma, A. (1996): preprint (APB)
Barnes, J.E., Hernquist, L. (1985): ARA&A **30**, 705
Ebisuzaki, T., Makino, J., Fukushige, T., Taiji, M., Sugimoto, D., Ito, T., Okumura, S.K. (1993): PASJ **45**, 269
Friedli, D. (1994): in Mass-Transfer Induced Activity in Galaxies, ed. I. Shlosman, Cambridge University press
Friedli, D., Benz, W. (1993): A&A **268**, 65
Hasan, H., Norman, C. (1990): ApJ **361**, 69
Hasan, H., Pfenniger, D., Norman, C. (1993): ApJ **409**, 91
Hernquist, L., Weyl, M. (1993): MNRAS **261**, 804
Holmberg, E. (1969): Ark. Astron. **5**, 305
Huang, S., Stewart, P. (1988): A&A **197**, 14
Lynds, R., Toomre, A. (1976): ApJ **209**, 382
Mihos, J.C., Hernquist, L. (1994): ApJ **437**, 611
Pfenniger, D. (1991): in Dynamics of Disc Galaxies, ed. B. Sundelius, Göteborg Univ., p. 191
Struck-Marcell, C., Higdon, J.L. (1993): ApJ **411**, 108
Sellwood, J.A. (1995): in Proc. IAU Coll. 157 Barred Galaxies, eds R. Buta, D.A. Crocker, B.G. Elmegreen, PASPC **91**, p. 259
Theys, J.C., Spiegel, E.A. (1976): ApJ **208**, 650
Theys, J.C., Spiegel, E.A. (1977): ApJ **212**, 616
Toomre, A. (1978): in Proc. IAU Symp. 79 The Large Scale Structure of the Universe, eds. M.S. Longair, J. Einasto, Reidel, Dordrecht, p. 109
Tóth, G., Ostriker, J.P. (1992): ApJ **118**, 106
Walker, I., Mihos, C., Hernquist, L. (1996): preprint
Wielen, R. (ed.) (1990): Dynamics and Interactions of Galaxies, Springer, Berlin, Heidelberg
Zaritsky, D., Smith, R., Frenk, C., White, S. (1993): ApJ **405**, 464

H I in Barred Spirals

Albert Bosma

Observatoire de Marseille, 2 Place Le Verrier, F-13248 Marseille Cedex 4, France

Abstract. The main results of H I studies of barred spiral galaxies are summarized. The question of dark matter in barred spirals is raised, and possible differences with respect to ordinary spirals are discussed.

1 Introduction

The inhomogeneous distribution of the main kinematic tracers, ionized and neutral hydrogen gas makes the determination of the global gas kinematics of barred spirals much more difficult than that of ordinary spirals. For a lot of galaxies the best one can hope for is that the H I data provide a global rotation curve, while the Hα data provide an estimate of the amount of peculiar motions directly attributable to the action of the bar. Only for a few galaxies complete kinematical information is available, e.g. NGC 1365, for which the Stockholm group has patiently collected a huge dataset over the years.

Most of the older studies emphasize the presence of non-circular motions, and indeed it took a while to sort out the differences between non-circular motions due to in-plane oval or bar distortions from those due to warping of the H I disk. The distinction can only be made on the basis of collective effects: the effect of in-plane oval orbits leads to a velocity field where the kinematical major and minor axes are not perpendicular, while the effect of tilted rings leads to a velocity field where the kinematical axes stay perpendicular (cf. Bosma 1978, 1981). Of course the distinction is not clearcut, since one can imagine a warped barred spiral. If the distortions are in the inner parts, I tend to favour the oval distortion interpretation, since strong optical warps are absent in most unperturbed edge-on galaxies. As is clear from the literature, even today there are suggestions based on velocity fields for strong warps within the optical image (as e.g. advocated for NGC 1365 in this meeting), where a simpler interpretation can be had if the optically visible material is thought to be planar.

We will briefly review the results for the bright barred spirals, and ignore the small asymmetric bars in late type magellanic irregulars since their dynamics do not pertain much to the topic of this meeting.

2 H I Distributions and Velocity Fields

2.1 Barred Galaxies with Dust Lanes

The prediction that the dust lanes constitute the loci of shocks (cf. Prendergast 1962, unpublished) has been difficult to corroborate. Only for two barred galaxies a direct long slit emission line spectrum has shown the strong discontinuity in the velocities expected across a dust lane (Pence and Blackman 1982 for NGC 6221 and Lindblad and Jörsäter 1988 for NGC 1365). Indirect evidence comes from the spatial coincidence of enhanced radio continuum emission and the dust lanes in M 83 and NGC 1097 (Ondrechen and van der Hulst 1983; Ondrechen 1985). Recent CO data show evidence for a strong concentration of molecular material in some dust lanes (e.g. Handa et al. 1990).

In the late seventies the northern hemisphere barred spiral NGC 5383 served as a prototype for detailed studies attempting to model non-circular motions due to a bar in a two-dimensional velocity field. Hα data by Peterson et al. (1978) showed strong deviations from circular motions and additional H I data by Sancisi, Allen and Sullivan (1979), which have much lower angular resolution, nevertheless permitted to determine a rotation curve. Several detailed models were constructed (e.g. Huntley 1978; Sanders and Tubbs 1980; Duval and Athanassoula 1983), which reproduce the characteristics of the non-circular motions. A skewing of the isovelocity contours around the systemic velocity towards the bar, and a corresponding S-shape distortion in the kinematical major axis are the main features. Contrary to warps, the kinematical minor axis is not perpendicular to the kinematical major axis. The amplitude of the non-circular motions does not only depend on the strength of the bar, but also on angular resolution and the viewing angle (cf. Fig. 41 of Athanassoula 1984).

Improvements in the gas flow codes (e.g. van Albada 1985), and extensive calculations of the orbital families in the corresponding potentials, have lead to the formulation of strict conditions for which shocks resembling the observed dust lanes can occur (Athanassoula 1992a,b). These include the realization that the presence of an x_2 family of orbits is not only necessary, but that a sizeable extent is needed. However, when the shock velocities are very high, the amount of shear is so large that molecular clouds would not have the time to assemble (cf. Tubbs 1982; Athanassoula 1992b). This might explain the large variety in appearance of the dust lanes and OB-associations : the absence of the latter could just indicate the regions of high shear so that the dust lanes are there, but not the molecular gas and their associated regions of star formation.

Several southern hemisphere galaxies have now been studied in detail. The best data is for NGC 1365. The Hα data by Teuben et al. (1986) clearly indicate that an inner Lindblad resonance is present in this galaxy, in agreement with the gas flow studies. H I studies are available from Ondrechen and van der Hulst (1989) and Jörsäter and van Moorsel (1995). The latter have been complemented by optical spectra, and modelled in detail with a potential derived from near infrared imaging by Lindblad et al. (1996).

NGC 1097 shows strong H I emission in the outer arms, and weak emission in the bar region (cf. Ondrechen et al. 1989). The velocity field indicates deviations from circular motions in the bar region, as in NGC 5383, as well as strong local deviations associated with the outer arms. Unfortunately, the presence of a companion perturbs the outer disk of this galaxy, hence no detailed model has yet been attempted. The continuum emission shown by Ondrechen and van der

Hulst (1983) indicates not only shock regions in the straight dust lanes leading the bar, but, at the SE side, also at the edge of the bar making a sharp angle with the main dust lane. This behaviour of the shocks can also be seen in some of the simulations by Athanassoula (1992b).

A study concerning NGC 1300 (England 1989, 1990), who finds that the H I here also is very strongly concentrated in the arms, has only obtained a qualitative agreement between the modelling and the data, partly due to an unfavourable viewing angle of the bar.

2.2 Barred Galaxies with Inner Rings

These galaxies are less well studied, in part because the bar itself is devoid of Hα and H I emission. A H I study of NGC 1398 (Moore and Gottesman 1995) only allows an attempt to derive a pattern speed from the rotation curve data, but no unique answer has been obtained. Likewise, NGC 3992, studied by Gottesman et al. (1984) and modelled by Hunter et al. (1988), does not have H I emission in the central parts including the bar, although there is normal H I emission in the disk. The velocity field does show the expected deviations from circular motion. On the other hand, Broeils (1992) presents a H I study of NGC 6674, amidst several studies of ordinary spiral galaxies. The bar in this galaxy seems relatively weak, although there is some sign of non-circular motion due to the bar.

Ryder et al. (1996) studied the galaxies NGC 1433 and NGC 6300, and again try to derive pattern speeds, on the assumption of rings being associated with resonances. For NGC 1433, which has several rings, the solution seems to be clearcut, but the determination of the inclination makes the result uncertain.

2.3 Ordinary Galaxies with Outer Rings and Oval Distortions

A few ordinary spirals (i.e. not classified as barred spirals) are identified as having oval distortions, based on distinct signatures in their velocity fields (cf. Bosma 1978, 1981).

NGC 4736, already a favourite object for study during P.O. Lindblad's early numerical work (Lindblad 1960, 1974), is one of the best examples. The early H I data (Bosma et al. 1977b) showed H I in an inner ring, an outer ring, and in the main body in between. Since the kinematical axes were not perpendicular and since the major axis changes its position angle with radius, it was concluded that the main body has an oval shape. More recent data have by and large confirmed this picture (Mulder and van Driel 1993). Infrared data reveal an additional small bar inside the inner ring (cf. Möllenhof, Matthias and Gerhard 1995). Thus this galaxy has two pattern speeds, one associated with the inner bar, and the other one with the oval. A similar situation exists for NGC 1068 (cf. Bosma 1992).

New VLA B-array 21-cm H I data of NGC 4151 by Mundell et al. (priv. comm.) fully confirm the main result from previous work by Davies (1973), Bosma et al. (1977a) and Pedlar et al. (1992), that the main body of this galaxy is a fat bar. Boksenberg et al. (1995) find from HST data evidence for a central mass concentration of about 10^9 $M_\odot$ in the inner few arcsec. Compared to the total mass of $2 \cdot 10^{11}$ $M_\odot$ ($H_0 = 50$ km s^{-1} Mpc^{-1}) enclosed within the outer ring radius, this means that a central mass concentration of 0.5 % of the total mass out to the outer radius, or equivalently, of order 2% of the disk mass within the bar radius, has not managed to scatter the x_1-orbits of the bar sufficiently to decay it.

The data for NGC 4258 (van Albada 1980) show strong indications for the presence of a bar, and the outer arms could then be thought of as a pseudoring. This is most clearly visible in the image of the H I distribution. The kinematics show all the signs for the presence of a bar. The interpretation for this galaxy has been obscured somewhat by the issue of the anomalous arms. Evidence for a central black hole in this galaxy has been presented by Miyoshi et al. (1995).

Also in this category is NGC 1808, which appears to have a fat bar containing spiral filaments, and two outer arms forming a pseudo ring, which makes this a good case for a Θ-galaxy (in other words, an oval distortion). Koribalski et al. (1993) construct a tilted ring warp model to their H I data. However, the higher angular resolution (but poorer sensitivity) H I data by Saikia et al. (1990) quite clearly show that there are significant non-circular motions entirely consistent with the idea of the main body of this galaxy being a fat bar.

Finally, NGC 7217, discussed in detail by Verdes-Montenegro et al. (1995) and Buta et al. (1995), presents the simultaneous presence of three rings (a nuclear ring, an inner ring and an outer ring). The H I is predominantly in the outer ring, but there is little deviation from axial symmetry. A single pattern speed of 86 km s^{-1} kpc^{-1} fits the positions of the rings as resonances, yet no clear-cut bar has been found in optical images. It is possible to develop a scenario in which the bar in this galaxy could have decayed, while the rings still persist (cf. Athanassoula 1996).

2.4 Barred Galaxies with Outer Rings

Van Driel (1987, cf. van Driel and van Woerden 1991 for a summary) made an extensive study of early type disk galaxies, some of them being barred galaxies with outer rings. In most cases the H I is almost absent from the bar, and mainly concentrated in the outer ring, although sometimes extending beyond it.

For true SB0 galaxies there seems little relation between the H I emission and the main optical image : the H I is often mostly in an annulus outside the optical disk, and the kinematics indicate that its spatial orientation differs from the main body. This holds for NGC 2787 (Shostak 1987), NGC 4262 (Krumm et al. 1985) as well as for several ordinary S0 galaxies (cf. van Driel and van Woerden 1991). For NGC 3941 (van Driel and van Woerden 1989), however, the outer H I ring is more or less in the same plane as the optical disk.

For slightly later types most of the H I does coincide with an outer optical ring. In NGC 1291 and NGC 5101 little or no H I emission is found in the bar region itself (cf. van Driel, Rots and van Woerden 1988). NGC 1291 is seen very face-on, and the H I is mainly concentrated in the outer ring, with some extended emission beyond it. The velocity field suggests that this outer H I is in a warped disk. The inner edge of the H I ring coincides with the inner edge of the outer ring seen in optical images. In the central parts very strong X-ray emission has been detected (Bregman, Hogg and Roberts 1995). For NGC 5101 the results are rather similar, with extended H I emission beyond the optical ring. Here the velocity field is more regular, and a rotation curve could be derived (see below).

For the rather obscured SA(r)0/a galaxy NGC 7013 Knapp et al. (1984) find the H I emission mainly in two rings, and a velocity field which shows the characteristics of an oval distortion, even though the authors postulate an inner ring out of the main disk plane. Work on a few later type ringed SB galaxies has been reported by van Driel and Buta (1991). In the (R')SB(rs)ab galaxy NGC 2273 the H I is mainly in the outer ring and the rotation curve is flat. In the (R)SB(rs)bc galaxy NGC 6217 there is H I in the inner parts, mainly in the inner ring, and in the outer ring and beyond. The kinematical major axis does not coincide with the major axis of the outer ring, indicating a possible intrinsic elongation of the latter.

2.5 SBc Galaxies

For these galaxies, the bar does not extend to the radius beyond which the rotation curve flattens off. Such is the case for NGC 925 (Wevers et al. 1986), NGC 1073 (England et al. 1990), NGC 3359 (Ball 1986, 1992) and NGC 4731 (Gottesman et al. 1984). In these galaxies the bar is not very strong, not even in H I , and the bulk of the H I is found in the disk. These galaxies prove difficult to model, and the only attempt made is for NGC 3359 (Ball 1992). However, a model with only the observed bar taken into account does not describe correctly the deviations in the velocity field and the presence of the stronger spiral arms.

3 Dark Matter in Barred Spiral Galaxies

The absence of a Keplerian drop-off in the rotation curve at large galactic radii constitute the most solid evidence for the presence of dark matter in spiral galaxies. For recent reviews on this subject, see e.g. Bosma (1995) and Sackett (1995). Due to the inhomogeneous nature of the H I and Hα emission in barred spirals, and the presence of strong non-circular motions, their rotation curves are not determined with the same accuracy as those for ordinary spirals. Yet it remains interesting to see whether the behaviour of the rotation curves of barred spirals is different from those of ordinary spirals. The notion is sometimes held that barred spirals are the result of the bar instability in disk galaxies (which could well be true), and thus might not have a lot of dark matter (which is not necessarily true). This arises from misunderstanding one aspect of the historical development of the dark matter problem: initially, a dark halo was proposed to help cure the bar instability in disks. What matters here is the dark matter in the halo acting on the disk. Any dynamically hot component inside the disk radius will help, but any halo mass outside the disk radius will not contribute. Thus the stabilizing influence of a bulge should not be underestimated. Likewise, the presence of a hot disk (perhaps dark) may help to cure the bar instability (cf. Ostriker and Peebles 1973, Athanassoula and Sellwood 1986).

In an early compilation (Bosma 1978, 1981), the oval galaxies NGC 4151, NGC 4258 and NGC 4736, and the barred spirals NGC 5383 and M 83 do not stand out in the behaviour of their rotation curves : they are about as flat as the others, and if all the mass is in the disk, local mass-to-light ratios become very high in the outer parts (cf. Bosma and van der Kruit 1979). The shapes of the rotation curves of barred spirals observed more recently do not differ much from those observed for ordinary spirals : the predominant characteristic is that most rotation curves are rising, flat or slightly declining, just like the rotation curves of ordinary galaxies. In most of the modelling work a dark halo component has been included. For a few objects, in particular NGC 3992 (Gottesman et al. 1984) and NGC 1073 (England et al. 1990), the rotation curve might be abruptly declining in the outer parts, but the uncertainties are rather large due to a poor signal-to-noise ratio. On the other hand, the extreme warp around M 83 outlined by Rogstad et al. (1974) suggests the occasional presence of an extended H I envelope also around barred spirals.

The RSB(rs)0/a galaxy galaxy NGC 5101 (van Driel, Rots and van Woerden 1988) is a good candidate for not needing a dark halo. The H I is distributed mainly in the outer ring and beyond it. The velocity field is rather regular, and permits the derivation of a rotation curve, though the inclination correction is rather uncertain. The rotation curve declines, and can be fitted entirely with an exponential disk with a scalelength corresponding to the optical scalelength derived from the B-band photometry given in Lauberts and Valentijn (1989). However, since outer rings tend to be blue, the B-band scalelength may overestimate the scalelength of the bulk of the mass in the disk. Thus a K-band image of NGC 5101 seems needed to assess whether the rotation curve for NGC 5101 really can be fitted without the need for a dark halo. Van Driel et al. (1988) remark that a warp of only 8° in the outer H I disk is needed to keep the rotation curve flat.

Jörsäter and van Moorsel (1995) argue for a Keplerian decline of the rotation curve in NGC 1365. To achieve this, they propose that the inclination of the galaxy increases drastically from 40° to 55° over the outer third of the optical disk. Such a large warp within the optical radius is seldom seen in other galaxies (cf. Sanchez-Saavedra et al. 1990), and one can only imagine how odd this galaxy will look if the inner parts are seen edge-on. In view of the presence of non-circular motions due to the bar, their treatment of the two-dimensional velocity field may not be correct. If no warp is assumed, the rotation curve declines much less, and the conclusions about the dark matter in NGC 1365 would be different.

4 Concluding Remarks

As function of Hubble type, the H I distribution in barred disk galaxies thus shows a progression, in the sense that for early Hubble types, SB0/a - SBa, the H I is predominantly distributed in the outer ring and sometimes beyond, for intermediate types, SBb, the H I is frequently absent from the central parts, but otherwise distributed throughout the disk, while for SBc and later there is also H I in the central parts. The rotation data are in general consistent with the need for dark matter in and around barred spirals, just as for ordinary spirals.

Acknowledgments. Thanks are due to Lia Athanassoula for frequent discussions about barred spirals.

References

Athanassoula, E. (1984): Phys. Reports **114**, 320
Athanassoula, E. (1992a): MNRAS **259**, 328
Athanassoula, E. (1992b): MNRAS **259**, 345
Athanassoula, E. (1996): in Proc. IAU Coll. 157 Barred Galaxies, eds R. Buta, D.A. Crocker, B.G. Elmegreen, PASPC **91**, p. 309
Athanassoula, E., Sellwood, J.A. (1986): MNRAS **221**, 213
Ball, R. (1986): ApJ **307**, 453
Ball, R. (1992): ApJ **395**, 418
Boksenberg, A. et al. (1995): ApJ **440**, 151
Bosma, A. (1978): Ph.D. thesis, University of Groningen
Bosma, A. (1981): AJ **86**, 1825
Bosma, A. (1992): in Morphological and Physical Classification of Galaxies, eds G. Busarello et al., Dordrecht, Reidel, p. 207
Bosma, A. (1995): in Dark Matter, eds S.S. Holt, C.L. Bennett, New York, AIP, p. 111
Bosma, A., Ekers, R.D., Lequeux, J. (1977a): A&A **57**, 97
Bosma, A., van der Hulst, J.M., Sullivan III, W.T. (1997b): A&A **57**, 373
Bosma, A., van der Kruit, P.C. (1979): A&A **79**, 281
Bregman, J.N., Hogg, D.E., Roberts, M.S. (1995): ApJ **441**, 561
Broeils, A.H. (1992): Ph.D. thesis, University of Groningen
Buta, R. et al. (1995): ApJ **450**, 593
Davies, R.D. (1973): MNRAS **161**, 25P
Duval, M.F., Athanassoula, E. (1983): A&A **121**, 297
England, M. (1989): ApJ **337**, 191
England, M. (1990): ApJ **344**, 669
England, M., Gottesman, S.T., Hunter, J.H. (1990): ApJ **348**, 456
Gottesman, S.T., Ball, R., Hunter, J.H., Huntley, J.M. (1984): ApJ **286**, 471
Handa T. et al. (1990): PASJ **42**, 1
Hunter, J.H. et al. (1988): ApJ **324**, 721
Huntley, J.M. (1978): ApJ **225**, L101
Jörsäter, S., van Moorsel, G.A. (1995): AJ **110**, 2037
Koribalski, B., Dahlem, M., Mebold, U., Brinks, E. (1993): A&A **268**, 14

Knapp, G.R., van Driel, W., Schwarz, U.J., van Woerden, H., Gallagher, J.S. (1984): A&A **133**, 127
Krumm, N., van Driel, W., van Woerden, H. (1985): A&A **144**, 202
Lauberts, A., Valentijn, E.A. (1989): The Surface Photometry Catalog of the ESO-Uppsala Galaxies, Garching-bei-München : ESO
Lindblad, P.A.B., Lindblad, P.O., Athanassoula, E. (1996): A&A in press
Lindblad, P.O. (1960): Stockholm Obs. Ann. **21**, no. 4
Lindblad, P.O. (1974): in IAU Symp. 58 The Formation and Dynamics of Galaxies, ed. J.R. Shakeshaft, Reidel, Dordrecht, p. 399
Lindblad, P.O., Jörsäter, S. (1988): in Proc. 10th European Regional IAU Meeting Evolution of Galaxies, ed. J. Palous, p. 289
Miyoshi et al. (1995): Nat **373**, 127
Möllenhof, C., Matthias, M., Gerhard, O.E. (1995): A&A **301**, 359
Moore, E.M., Gottesman, S.T. (1995): ApJ **447**, 159
Mulder, P.S., van Driel, W. (1993): A&A **272**, 63
Ondrechen, M.P. (1985): AJ **90**, 1474.
Ondrechen, M.P., van der Hulst, J.M. (1983): ApJ **269**, L47
Ondrechen, M.P., van der Hulst, J.M. (1989): ApJ **342**, 29
Ondrechen, M.P., van der Hulst, J.M., Hummel, E. (1989): ApJ **342**, 39
Ostriker, J.P., Peebles, P.J.E. (1973): ApJ **186**, 467
Pedlar, A., Howley, P., Axon, D.J., Unger, S.W. (1992): MNRAS **259**, 369
Pence, W.D., Blackman, C.P. (1984): MNRAS **207**, 9
Peterson, C.J., Rubin, V.C., Ford, W.K., Thonnard, N. (1978): ApJ **219**, 31
Rogstad, D.H., Lockhart, I.A., Wright, M.C.H. (1974): ApJ **193**, 309
Ryder, S. et al. (1996): ApJ in press
Sackett, P.D. (1995): in IAU Symp. 173 Gravitational Lensing, in press
Saikia, D.J. et al. (1990): MNRAS **245**, 397
Sanchez-Saavedra, M.L., Battaner, E., Florido, E. (1990): MNRAS **246**, 458
Sancisi, R., Allen, R.J., Sullivan III, W.T. (1979): A&A **78**, 217
Sanders, R.H., Tubbs, A.D. (1980): ApJ **235**, 803
Shostak, G.S. (1987): A&A, **175**, 4
Teuben, P.J., Sanders, R.H., Atherton, P.D., van Albada, G.D. (1986): MNRAS **221**, 1
Tubbs, A.D. (1982): ApJ **255**, 458
van Albada, G.D. (1980): A&A **90**, 123
van Albada, G.D. (1985): A&A **142**, 491
van Driel, W. (1987): Ph.D. thesis, University of Groningen
van Driel, W., Buta, R.J. (1991): A&A **245**, 7
van Driel, W., Rots, A.H., van Woerden, H. (1988): A&A **204**, 39
van Driel, W., van Woerden, H. (1989): A&A **225**, 317
van Driel, W., van Woerden, H. (1991): A&A **243**, 71
Verdes-Montenegro, L., Bosma, A., Athanassoula, E. (1995): A&A **300**, 65
Wevers, B.H.M.R., van der Kruit, P.C., Allen, R.J. (1986): A&AS **66**, 505

H I Observations of a Sample of Barred Spirals

Helmuth Kristen

Stockholm Observatory, S-13336 Saltsjöbaden, Sweden

Abstract. We select a sample of barred spiral galaxies in order to test previous findings concerning NGC 1365. We discuss results regarding NGC 1300 and NGC 1365, as well as new VLA[1] observations of NGC 613, NGC 1350, and NGC 2263. The H I halos of the investigated objects are found to be compact. The effect of companions on the H I extent is illustrated.

1 Introduction

The Stockholm Galaxy Group has through a number of years conducted a case study of the barred spiral galaxy NGC 1365. The study consists of multiple wavelength observations in combination with numerical modeling, a prime tool for the investigation of galactic dynamics (see these proceedings). This galaxy has several characteristics: Strong non-circular motions indicate a strong bar. There is a large decline in rotational velocity at large radii. Furthermore, it lacks an extended H I halo, i.e. the extent of the neutral atomic hydrogen (H I) corresponds to the size of the optical disc.

By extending the acquired methods to a carefully selected sample of barred spirals, our aim is to test the universality of these findings. In Sect. 4 we discuss H I observations of NGC 1365 as well as NGC 613, NGC 1300, NGC 1350, and NGC 2263 (Table 1), all part of the mentioned sample. The extent of the H I is particularly interesting since it has in some cases been found to trace the dark matter (cf. Broeils 1992, hereafter B92).

2 Selection

We selected a sample with the following criteria:

- Type: Barred systems with numerical Hubble type 2-5 (SBab to SBc), as defined in the Third Reference Catalogue of Bright Galaxies (de Vaucouleurs et al. 1991, hereafter RC3)
- Inclination $i \geq 20°$, in order to select systems suitable for velocity mapping.
- $D(0) \geq 2'$, i.e. a diameter/beam ratio $\gtrsim 6$ in VLA C-array, to provide sufficient spatial resolution.

[1] The VLA is operated by Associated Universities, Inc., under contract with the National Science Foundation

These criteria were applied to the electronic version of the RC3, extracting 261 out of the over 23 000 entries. Moreover, in order to avoid non-selfconsistent systems, we checked against optical irregularities, indicative of recent interaction. In addition we checked against companions, since there might be considerable H I interaction without any obvious perturbations in the optical (e.g. Yun et al. 1994). We inspected $1° \times 1°$ fields extracted from the Digitized Sky Survey (Association of Universities for Research in Astronomy, Inc. 1993, 1994), in combination with the multi-wavelength information available on the NASA Extragalactic Database.

It should be noted that the aim of this study is to investigate the similarities and differences of a number of comparable objects, not to select a sample that is 'complete' in a statistical sense.

Table 1. Global parameters

	NGC 613	NGC 1300	NGC 1350	NGC 1365	NGC 2263
R.A. (2000)	$1^h34^m17^s.5$	$3^h19^m40^s.8$	$3^h31^m08^s.4$	$3^h33^m36^s.6$	$6^h38^m28^s.3$
Dec. (2000)	-29°24′58″	-19°24′41″	-33°37′44″	-36°08′17″	-24°50′49″
Revised type (1)	.SBT4..	.SBT4..	PSBR2..	.SBS3..	PSBR2..
T (Hubble type)(1)	4.0	4.0	1.8	3.0	2.1
V_{21} (km s^{-1})	1475±5 (1)	1573±10 (2)	1890±9 (1)	1632±5 (4)	2743±2 (5)
Position Angle (°)	120 (1)	267±2 (2)	0 (1)	220 (4)	143 (1)
Inclination (°)	41 (1)	35±5 (2)	58 (1)	55 (3)	54 (1)
$\int S\,dv$ (Jy km s^{-1}) (6)	45.8 ± 8.6	29.3 ± 7.5	22.6 ± 8.3	141.0 ± 10.8	-

(1) de Vaucouleurs et al. (1991)
(2) Lindblad, P.A.B et al. (1996a)
(3) Lindblad, P.O. (1978)
(4) Jörsäter and v. Moorsel (1995)
(5) Bottinelli et al. (1992)
(6) Huchtmeier and Richter (1989)

3 H I Observations and Reductions

Observational parameters are listed in Table 2. All data were edited and calibrated within AIPS (Astronomical Image Processing System - developed by the NRAO). Data were Hanning-smoothed, either online or following observations.

Preliminary maps were used to determine the line-free channels. With the AIPS task UVLIN we performed a linear fit to the visibilities of the line-free channels with a subsequent continuum subtraction of the line channels in the UV plane.

Table 2. VLA observations

	NGC 613	NGC 1350	NGC 2263
Date	10/94	10/94	10/94
Array configuration	CnB	CnB	CnB
Length (minutes)	330	248	325
R.A. pointing center (2000)	$1^h34^m15^s.0$	$3^h31^m10^s.0$	$6^h38^m30^s.0$
Dec. pointing center (2000)	-29°24′30″	-33°38′00″	-24°50′30″
Velocity band center (km s^{-1})	1475	1890	2740
Velocity range (km s^{-1})	1145 - 1805	1560 - 2220	2410 - 3070
Velocity resolution (km s^{-1})	10.42	10.42	10.42
Number of channels	63	63	63
Beam FWHM (″)	22.13×14.33	22.29×13.79	23.33×14.19

Normal weighting was used for mapping in order to improve the signal-to-noise ratio by suppressing longer baselines; a trade-off against the higher resolution provided by uniform weighting. The AIPS task MX was used for mapping.

De-convolution was accomplished by cleaning (Högbom 1974) using the APCLN task in AIPS. An initial cleaning of the complete maps was performed in order to locate the areas containing flux in each of the velocity channels. These areas were then used to define 'clean-boxes' in a second deconvolution run, cleaning the channel maps down to 2σ. Since the initial clean-run already delineates the areas of detected flux, the concept of clean-boxes only marginally influences the extent of H I. Total fluxes were estimated combining the cleaned and residual fluxes, not with equal weight, but with a weight factor of the ratio clean beam size/dirty beam size, since beam sizes differ both between dirty and clean maps, and between different channel maps (Jörsäter and v. Moorsel 1995, hereafter JvM). The results are data cubes with true units Jy/clean beam. In the case of existing single dish values, fluxes were found to converge well.

A cube spatially convolved to a much lower resolution was used for defining an upper threshold for blanking areas, preventing noise from influencing the subsequent moment calculations. The total flux in the blanked cubes only differed from the non-blanked by the order of one percent, so virtually all flux was conserved in the process. Finally moment maps were created by the XMOM task in AIPS.

4 Results and Discussion

We here discuss previous observations of NGC 1300 (England 1989) and NGC 1365 (JvM) as well as the new observations of NGC 613, NGC 1350 and NGC 2263 (see previous section and Fig. 1).

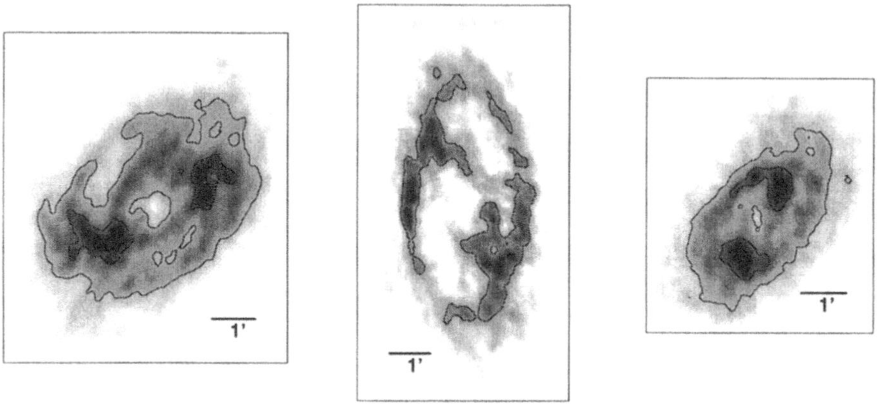

Fig. 1. VLA H I observations. From left to right: NGC 613, NGC 1350, and NGC 2263. H I contours correspond to 1 and 3 times the 100 K km s^{-1} inclination-corrected isophotes

4.1 Bar Strength

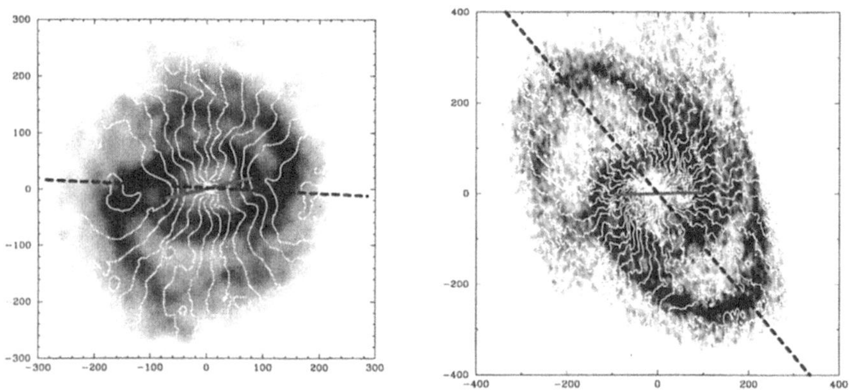

Fig. 2. H I maps of NGC 1300 (left) and NGC 1365 (right). The combined H I and optical velocity fields are superposed. The dashed black line indicates the *P.A.*, the solid grey line shows the orientation and length of the bar. Note the 'Z'-shaped distortion of the isovelocity curves across the bar region in NGC 1365

In two cases so far, NGC 1300 (Lindblad and Kristen 1996) and NGC 1365 (Lindblad et al. 1996b), we have performed gas dynamical modeling in order to investigate such elusive entities as bar strength, pattern speed, and the location of resonances associated therewith. NGC 1300 and NGC 1365 are similar in a number of aspects. Both are barred spirals of intermediate type. They feature a prominent bar with offset dust lanes, a compact H I halo, and a depletion of H I in the bar region. The latter has made us combine the H I velocity field with optical measurements (Fig. 2).

Deviation from circular motion in the bar region is a measure of bar strength. However, alignment of the bar with the position angle (*P.A.*) may hide asymmetries in the isovelocity curves. The absence of major asymmetries in the velocity field of NGC 1300 differs significantly from the strong 'Z'-shaped distortion of the isovelocity curves in NGC 1365. In this case, our simulations indicate that in NGC 1300 the bar potential is indeed weaker than in NGC 1365. Furthermore, in NGC 1300, the close alignment of the bar with the major axis, combined with the lower H I resolution, in effect hides the non-circular velocities. This illustrates the importance of conducting dynamical simulations when interpreting observations.

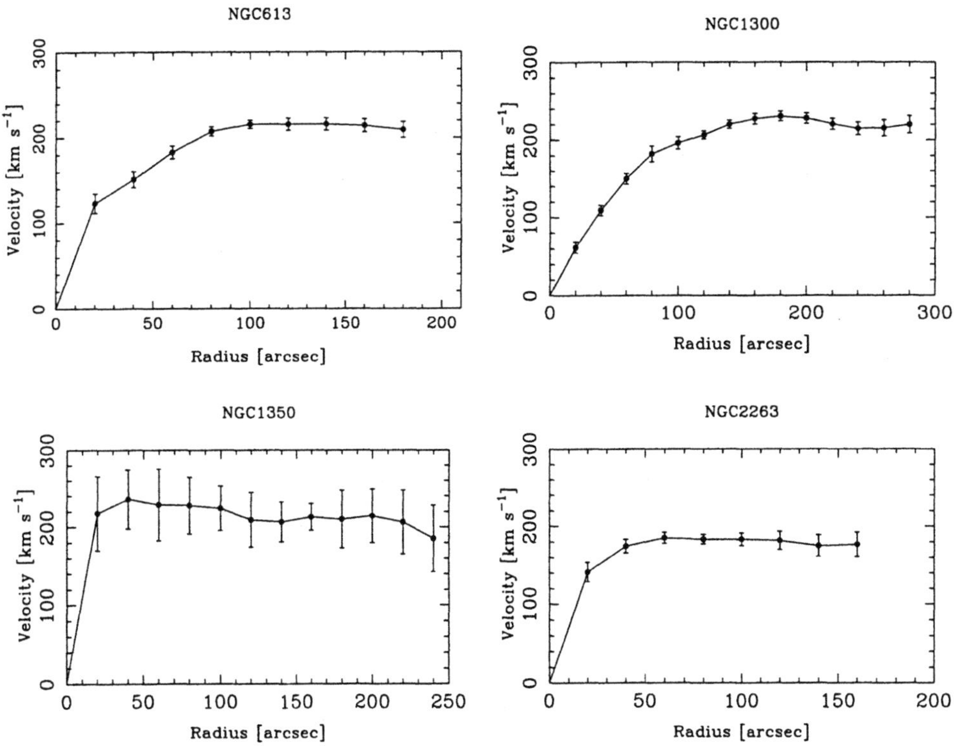

Fig. 3. Rotational velocities of NGC 613, NGC 1300, NGC 1350, and NGC 2263. The error bars indicate the r.m.s. of the velocity fit

4.2 Rotation Curves

NGC 1365 features a 40% decline in rotational velocity at large radii. JvM fit a $3.9 \cdot 10^{11}$ $M_\odot$ Keplerian to the outer part of the rotation curve. In contrast, NGC 613, NGC 1300, NGC 1350, and NGC 2263 only show weak declines (Fig. 3). Rotation curves were extracted by fitting the velocity fields with tilted ring models. This was done with the AIPS task GAL.

4.3 The H I Diameter

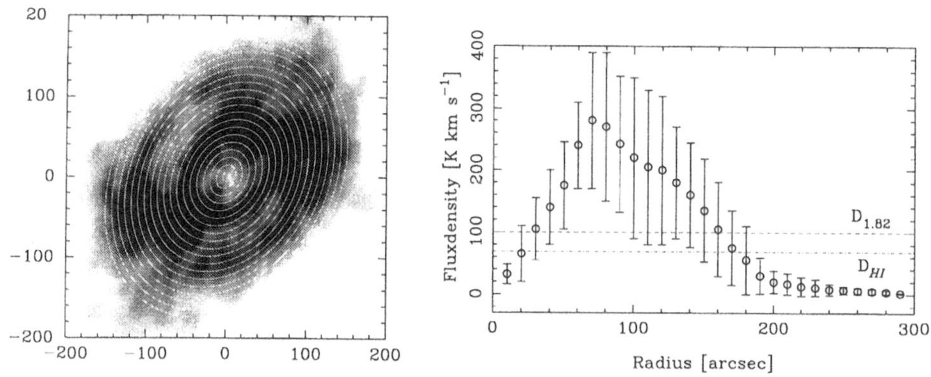

Fig. 4. Illustration of the H I diameter measurement in the case of NGC 613. **Left:** H I map of NGC 613. The inclination corrected H I flux is averaged along 10″-thick concentric rings indicated by the overlay. The units on the axes are arcseconds offset from the optical centre. **Right:** The resulting flux as a function of radius. We determine the H I diameter at $D_{1.82}$ and $D_{\rm HI}$ (see text). The bars indicate the r.m.s. of the flux in each ring; a measure of the patchiness of the H I

The flux as a function of radius is determined by averaging the total flux in inclination corrected 10″ thick concentric rings. In order to facilitate comparison with literature, we determine the diameter of the H I face-on flux density isophotes at two levels: $D_{1.82}$ (corresponding to 1.82×10^{20} atoms cm^{-2}, 100 K km/s or 1.44 $M_\odot$ pc^{-2}), and $D_{\rm HI}$ (1 $M_\odot$ pc^{-2}). The results are given in Table 3. The error estimate of the listed H I isophote diameter is an assessment of the uncertainty associated with the patchiness of the H I distribution (Fig. 4). $D(0)$, as listed in RC3, is the optical diameter of the face-on-corrected 25 $m_{\rm B}$ arcsec^{-2} isophote along the major axis, compensated for extinction. The average diameter ratios $D_{1.82}/D(0)$ and $D_{\rm HI}/D(0)$ for the five galaxies investigated here, are 1.1 ± 0.1 and 1.2 ± 0.1 respectively (1σ r.m.s.). Bosma (1981, hereafter Bos81) and B92 find $D_{1.82}/D(0) = 2.2 \pm 1.1$ and $D_{\rm HI}/D(0) = 1.8 \pm 0.4$ respectively, for samples of various types. This indicates compactness of the H I halo in the five galaxies discussed here, admittedly a small sample.

Table 3. Results

	$D_{1.82}$	D_{HI}	Optical Diameter $D(0)$	$D_{1.82}/D(0)$	$D_{HI}/D(0)$	$\int S\,dv$ (Jy km s^{-1})
NGC 613	5.4′± 0.3′	5.8′± 0.3′	5.5′± 0.1′ (1)	1.0± 0.1	1.1± 0.1	37.6 (3)
NGC 1300	7.0′± 0.3′	7.5′± 0.3′	6.2′± 0.3′ (1)	1.1± 0.1	1.2± 0.1	43.2 (2)
NGC 1350	5.7′± 0.3′	6.0′± 0.3′	5.3′± 0.1′ (1)	1.1± 0.1	1.1± 0.1	14.4 (3)
NGC 1365	12.3′± 0.6′	12.7′± 0.6′	11.2′± 0.1′ (1)	1.1± 0.1	1.1± 0.1	161.5 (2)
NGC 2263	4.2′± 0.2′	4.6′± 0.2′	3.2′± 0.1′ (1)	1.3± 0.1	1.4± 0.1	18.8 (3)

(1) de Vaucouleurs et al. (1991)
(2) Jörsäter and v. Moorsel (1995)
(3) This study

For comparison, Bos81 provides a $D_{1.82}/D(0)$ of 1.9 ± 0.9 for SA + SAB galaxies and 2.7 ± 2.0 (1.9 ± 0.7 disregarding NGC 3109) for SBs. B92 gives a $D_{HI}/D(0)$ of 1.9 ± 0.7 for SA + SABs and 1.6 ± 0.3 for SBs. Thus, Bos81 and B92 show no clear connection between diameter ratio and the presence of bars.

Broeils sample. 'Sample1' in B92 consists of short aperture synthesis H I observations of about 50 spiral galaxies of various types. Discerning which galaxies in the Broeils sample are isolated by applying the same criteria as in Sect. 2, indicates a spread in H I diameters in the presence of a companion (Fig. 5).

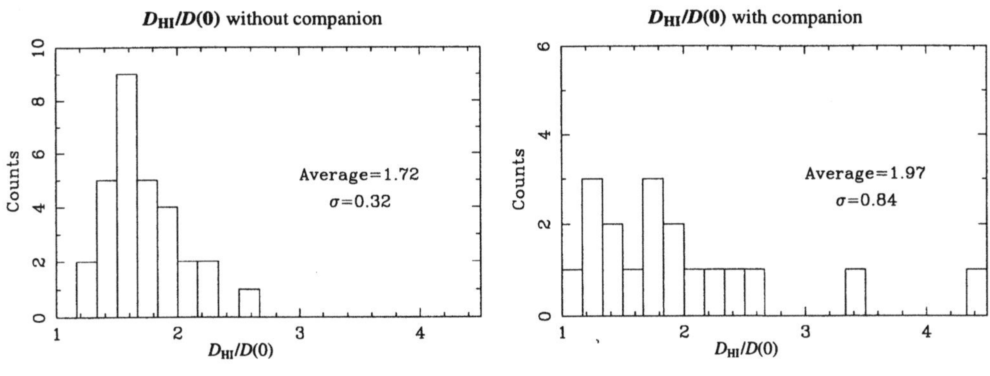

Fig. 5. The Broeils sample divided into galaxies without and with detected companions, respectively

5 Conclusions

- Among the five barred spiral galaxies discussed here, only NGC 1365 features a strong decrease in rotational velocity at large radii.
- The five investigated galaxies all have compact H I halos. A natural next step is to further investigate a possible connection between the H I halo in isolated spirals and the development of bars.
- The extent of H I seems to be influenced by companions. We thus conclude that companions are an important aspect of understanding the H I distribution in galaxies.

Acknowledgements. We thank Martin England for generously providing us with his H I data of NGC 1300, and Gustaaf v. Moorsel for performing the corresponding re-reductions.

The Digitized Sky Survey was produced at the Space Telescope Science Institute (STScI) under U.S. Government grant NAG W-2166. The images on these disks are based on photographic data obtained using the UK Schmidt Telescope.

The Nasa Extragalactic Database (NED) service (ned@ned.ipac.caltech.edu) is made possible by Science Operations Branch, Astrophysics Division, Office of Space and Science Applications, National Aeronautics and Space Administration.

References

Bosma, A. (1981): AJ **86**, 1825, (Bos81)
Bottinelli, L., Durand, N., Fouqué, P. et al. (1992): A&AS **93**, 173B
Broeils, A. (1992): PhD Thesis, University of Groningen, (B92)
England, M., N. (1989): ApJ **337**, 191
Huchtmeier, W., K., Richter , O.,-G. (1989): in H I Observations of Galaxies, Springer Verlag, New York
Högbom, J., A., Baschek, B. (1974): A&AS **15**, 417
Jörsäter, S., van Moorsel, G. (1995): AJ **110**, 2037, (JvM)
Lindblad, P. A. B., Kristen, H. (1996): A&A, in press
Lindblad, P. A. B., Kristen, H., Jörsäter, S., Högbom, J., A. (1996a): A&A, in press
Lindblad, P. A. B., Lindblad, P. O., Athanassoula, E. (1996b): A&A, in press
Lindblad, P., O. (1978): eds. A. Reiz and T. Andersen, in Astronomical Papers dedicated to Bengt Strömgren, Copenhagen University Observatory, p. 402
de Vaucouleurs, G., de Vaucouleurs, A., Corwin, H., G. Jr., Buta, R., J., Paturel, G., Fouqué, P. (1991): Third Reference Catalogue of Bright Galaxies, Springer Verlag, New York
Yun, M., S., Ho, P. T. P., Lo, K., Y. (1994): Nat **372**, 530

Hydrodynamical Simulations of the Barred Spiral Galaxy NGC 1365

P.A.B. Lindblad[1], P.O. Lindblad[1] and E. Athanassoula[2]

[1] Stockholm Observatory, S-133 36 Saltsjöbaden, Sweden
[2] Observatoire de Marseille, 2 Place Le Verrier, F-13248 Marseille Cedex 4, France

Abstract. We perform two-dimensional, time dependent, hydrodynamical simulations of the gas flow in a potential representing the barred spiral galaxy NGC 1365 using the FS2 code originally written by G.D. van Albada. We find good agreement between observations and models, both concerning the morphology and the velocity field. Contrary to observations, the pure bar perturbed models cannot drive an inner arm across corotation. Models having both a bar and spiral perturbing potential reduce this problem, thus suggesting the existence of massive spiral arms in NGC 1365.

1 Introduction

1.1 Hydrodynamical Simulations of Barred Spiral Galaxies

Several authors have explored the field of gas dynamics in barred systems using different approaches. One of the aims of these investigations was to compare the model gaseous response, due to some assumed underlying stellar gravitational field, with observed gas density distribution and kinematics of barred galaxies. The gas is known to respond in a highly non-linear way, and therefore should give clues to dynamical parameters like the mass distribution, existence and positions of principal resonances and thereby the pattern speed.

Sanders and Huntley (1976) came to the conclusion that the natural response of any differentially rotating gaseous disc to an oval-like perturbation in an axisymmetric force field is to form a two-armed trailing spiral wave. The streamlines of the gas in their calculations are consistent with particle orbit theory in the epicyclic approximation far from resonances.

Besides the gaseous spiral arms, one often observes in barred galaxies narrow offset dust lanes, situated at the leading edges of the bar (assuming trailing spiral arms). Prendergast (1962, unpublished) suggested that these features are due to shocks in the gas. Similar shocks show up quite naturally in gas dynamical calculations (e.g. Athanassoula 1992).

Direct comparisons with the observed H I morphology and velocity fields have been done for e.g. NGC 1300 (England 1989), NGC 3359 (Ball 1992). In these simulations the observed bar was used to constrain the mass distribution of the perturbation. Their major drawback is that they are not able to reproduce the observed spiral pattern without adding an *ad hoc* oval component to the model mass distribution in order to maintain the spiral response out to large radii.

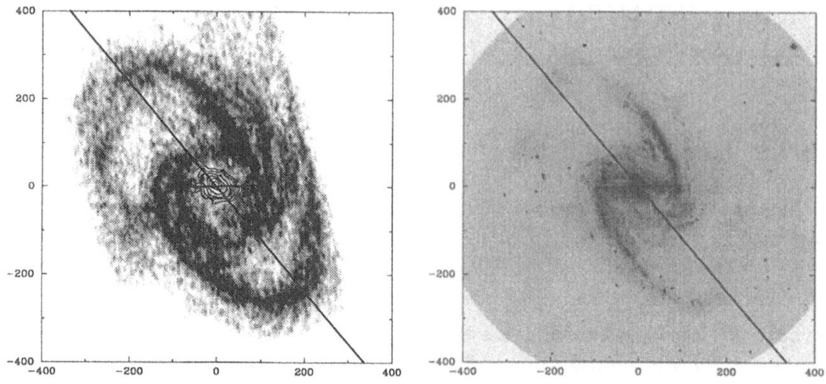

Fig. 1. The total H I column density map of NGC 1365 from JvM, with the CO (2-1) contours from Sandqvist et al. (1995) overlaid (**left**), together with the optical IIIa-J ESO 3.6 m prime focus plate of NGC 1365 (**right**). The line of nodes and the bar major axis are marked as the straight lines running through the centre

1.2 The Galaxy NGC 1365

The high resolution VLA H I total column density map of NGC 1365 from Jörsäter and van Moorsel (1995, hereafter JvM), together with the CO (2-1) contours from Sandqvist et al. (1995), and an optical image of NGC 1365 are reproduced in Fig. 1. The bar in NGC 1365 is strong, with prominent dust lanes running along its front edges. Both the optical and H I spiral arms are well developed and have a tendency to turn inwards at the outer edge of the galaxy. Their structure appears at first look very symmetric. A closer look reveals multiple spiral arms and significant deviations from strict bisymmetry. The CO molecular gas is strongly concentrated to the nucleus and, further out, tends to be aligned with the dust lanes along the bar.

The velocity field in the intermediate and outer regions of NGC 1365, is obtained from the H I observations, and in the bar region from optical long slit measurements (Jörsäter et al. 1984; Lindblad et al. 1996, hereafter L96). The combined optical and H I velocity field is presented in Fig. 2. A mosaic image of NGC 1365 in the J-band is used to derive the gravitational potential of the bar and of the main spiral structure.

The aim of the present paper is to compute the gas response to a model potential, obtained from a combination of a J-band image and the observed rotation curve of NGC 1365. Our goal is to let the input parameters for the calculations be determined as far as possible by observations. The observations will also serve as the data set with which the results of the computations will be compared.

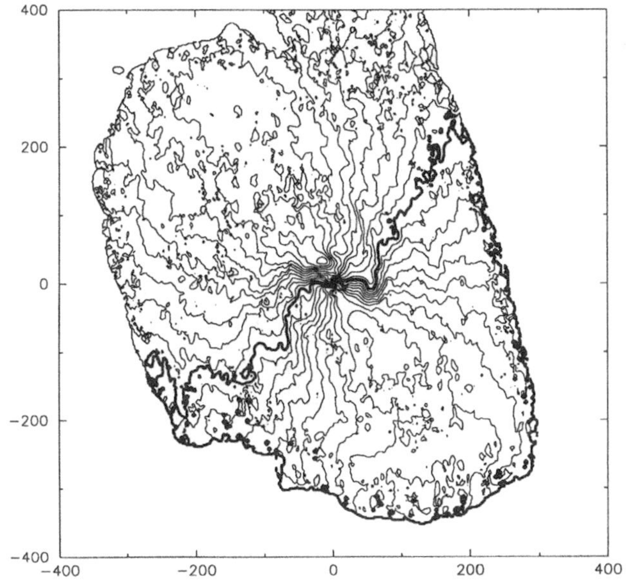

Fig. 2. The observed H I and optical hybrid radial velocity field, after subtraction of the systemic velocity, $V_{sys} = +1630$ km s^{-1}. The contour interval is 20 km s^{-1} and the zero radial velocity is marked by the thick contour

2 Input Potential

2.1 Rotation Curve and Axisymmetric Potential

The H I rotation curve, derived by JvM, is the basis for our input rotation curve for the region $120'' < R < 400''$. Thus the warp advocated by JvM has been taken into account and our models have been projected onto the sky using the variable warp orientation parameters, $PA_{\rm lon}(R)$ and $i(R)$, as derived by JvM.

In strongly non-axisymmetric galaxies like NGC 1365 the azimuthally averaged observed rotation curve is not a good description of the underlying axisymmetric potential in the bar region since the streaming is highly non-circular. By altering the numerical values for the model input rotation curve until the observed velocity and morphological constraints were met, we obtained the final rotation curve.

2.2 Perturbing Potential

We will perform two different types of simulations: one where the perturbing potential arises from the bar only, and one where we include a spiral component. We base the observed rotation curve on the axisymmetric potential as described

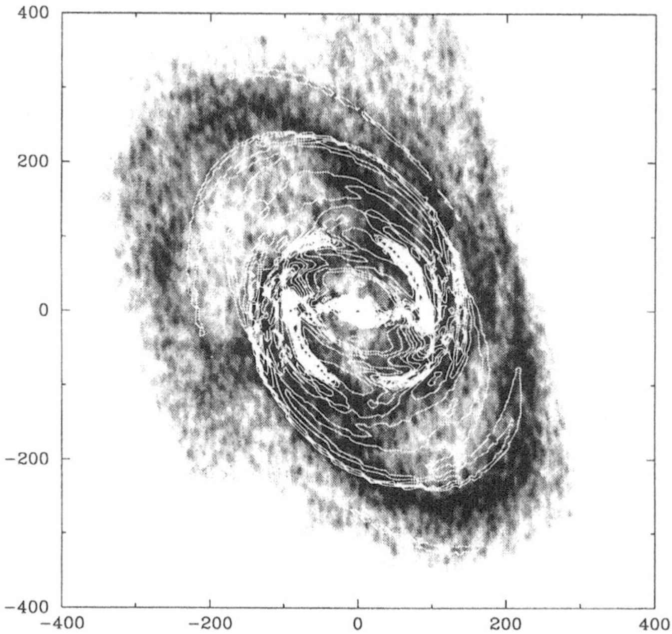

Fig. 3. The BM density contour map overlaid on the total column H I density map

above, while using the observed J-band surface brightness to extract the perturbing potential. Since there are two different origins for the potentials, we need to couple them in some way.

To obtain this coupling we will assume that the observed axisymmetric Fourier component of the J-band surface brightness represents the axisymmetric mass distribution giving the rotation curve in the radial interval $120'' < R < 200''$. The mass to luminosity ratio, M/L_J, obtained this way defines the relative bar potential $A_{bar} = 1.0$, which is kept as a variable parameter thus allowing for different M/L_J in the disc and the bar.

3 Results

3.1 A Bar + Disc Model; BM

We will here present what we consider to be a rather successful bar + disc model, hereafter BM.

The model BM has a pattern speed of $\Omega_p = 20$ km s^{-1}kpc^{-1} giving the main resonance radii $R_{ILR} = 27''$, $R_{CR} = 145'' = 1.21 R_{bar}$, and $R_{OLR} = 216''$. We have chosen a relative bar amplitude of $A_{bar} = 1.2$, and the axisymmetric forces in the model are represented by the adopted rotation curve.

Outer Arms and CR. In Fig. 3 we see the gas response of BM, as a contour plot on the gray scale H I total column density map. The outer SW H I arm is

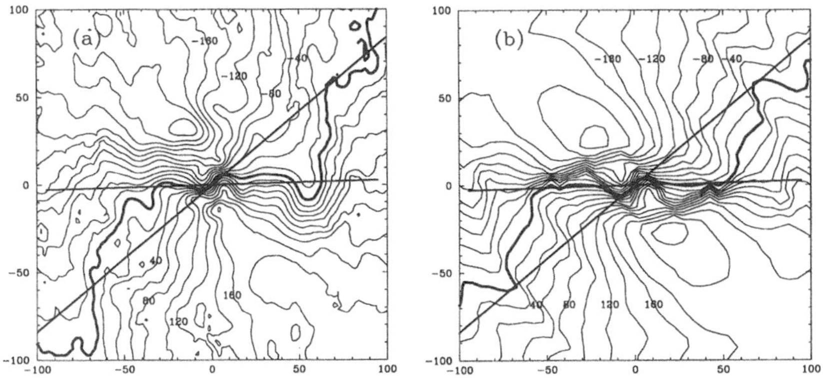

Fig. 4. The optical radial velocity field in the inner $R < 100''$ of NGC 1365 (**a**), and the corresponding BM velocity field (**b**). The bar major axis (PA=92°) and the galaxy minor axis (PA=130°) are marked by the straight lines. The thick contour corresponds to the zero velocity, and the contour interval is 20 km s^{-1}

well fitted by the model out to a radius of $\sim 240''$. The fit to the NE H I arm is of lesser quality, but still acceptable. Outside this radius the model arms bend inwards earlier than the H I arms. The discrepancy can be explained in terms of the warped disc, which makes the determination of the observed apparent rotation curve somewhat uncertain here, and also complicates the projection of a model to the same orientation as NGC 1365 (but see Sect. 3.3).

The outer model arms form just outside CR, where the orbit orientation has become perpendicular to the bar major axis, and start to twist towards being parallel to the bar major axis when outside the OLR. This twisting is responsible for the orbit crowding outlining the outer arms in BM.

Inner Arms at the End of the Bar. BM has inner trailing arms emanating from the ends of the bar at $R \sim 110''$. The radial position, along the bar major axis, for this arm feature does not depend noticeably on the pattern speed and bar potential amplitude, but is sensitive to the adopted physical length of the bar. Increasing the bar length pushes the position of the arm outwards in the galaxy, i.e. the inner arms are directly associated with the ending of the bar.

The inner H I arms basically show a ring or square structure approximately at the radius of model CR. This structure is reproduced to a large extent by BM. However, there is H I gas tracing the inner bright optical arms across CR in contradiction to the inner arms in model BM, where the arms tend to become parallel to the CR circle as it approaches CR. Thus, in our bar + disc models the observed behaviour cannot be reproduced. In order to drive a gaseous arm across CR we adopt a perturbing potential having both a bar and a spiral component. The results of this approach are discussed in Sect. 3.2.

The Bar Region and Offset Gas Lanes. The morphology in the bar region of NGC 1365 is dominated by the offset gas lanes situated on the leading side of the bar major axis and by the CO concentration in the central region. The observed dust lanes in NGC 1365 are fairly well matched by the corresponding gas lanes of model BM. The model gas lanes require an ILR to be present, in good agreement with the results of Athanassoula (1992).

The observed and BM radial velocity fields for $R < 100''$ are shown in Fig. 4. Outside $R \sim 10''$, the observed contours are consistent with motion along twisted elliptical orbits, inclined with respect to the bar major axis, as found in BM. The observed zero contour follows the bar major axis out to $\sim 50''$, again consistent with the streaming in BM. Outside $R \sim 50''$ the BM model orbits become increasingly more circular when approaching the projected bar end at $\sim 90''$, which is reflected by the zero velocity contour approaching the minor axis of the galaxy. This trend is clearly seen in the observed velocity field.

3.2 A Bar + Disc + Spiral Model; BSM

The J-band image shows that the bright star forming optical arms are superposed on stellar density enhancements. This is in agreement with the conclusions of Elmegreen and Elmegreen (1985) from their sample of 16 barred galaxies. In order to investigate the effects of an additional spiral perturbing potential we extract a representative spiral potential from the J-band image, and compute a new set of models containing both a bar and a spiral potential.

The main parameter values for the best fit bar + disc + spiral model (hereafter BSM) are: $\Omega_{\mathrm{p}} = 18$ km s^{-1}kpc^{-1}, $A_{\mathrm{bar}} = 1.2$ and $A_{\mathrm{spiral}} = 0.3$. The rotation curve is the same as for model BM.

A value of $A_{\mathrm{spiral}} = 0.3$ means that the M/L_{J} ratio for the spiral luminosity is 30% of the M/L_{J} ratio for the bar , and the pattern speed $\Omega_{\mathrm{p}} = 18$ km s^{-1}kpc^{-1} gives the main resonance radii at $R_{\mathrm{ILR}} = 30''$, $R_{\mathrm{CR}} = 157'' = 1.31 R_{\mathrm{bar}}$ and $R_{\mathrm{OLR}} = 232''$.

The gas response of BSM is presented in Fig. 5 as a contour plot, overlaid on the gray scale H I total column density map.

The problem of driving the inner spiral arm across CR is reduced when including the spiral component. In BSM the inner arm follows the observed H I and optical arms out to approximately $R = 190''$, thus across model CR at $R = 157''$, and then turns towards being parallel to the outer main arm. This turning of the model arm is not consistent with observations, where the optical bright arm joins the main H I arm at $R \sim 250''$. The reason for this discrepancy could be that our surface density Fourier components are truncated at $R \sim 200''$, which is the limit of our J-band image.

3.3 Model Using a Modified Rotation Curve; BSM2

There is one region where the fit of models BM and BSM to the H I density is not satisfactory and that is the outermost part, $R > 240''$. This could have been

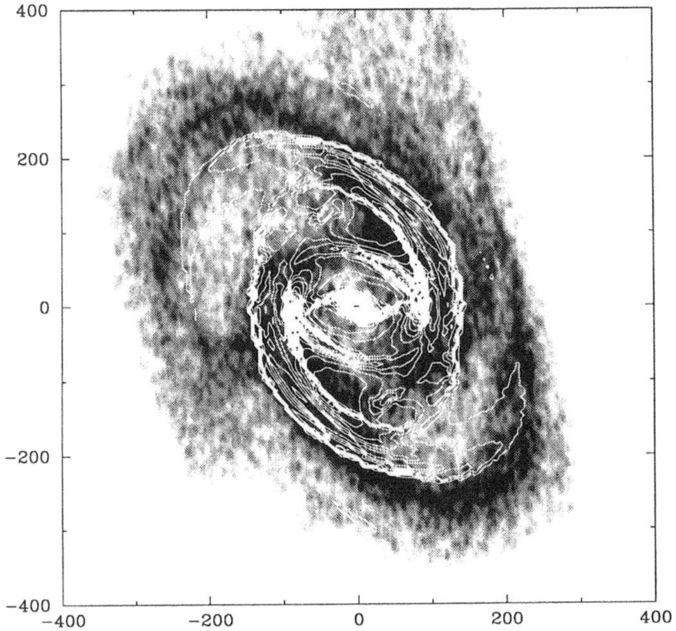

Fig. 5. The BSM density contour map overlaid on the total column H I density map

expected since both part of our data and our numerical code are not adequate for that region. Our code is planar and two-dimensional, while NGC 1365 seems to have a sizeable warp in its outer parts, which also affects the derivation of the rotation curve. To check what effect a small change in the rotation curve could have on the model, we ran some models with a somewhat less abrupt fall of the rotation curve thereby lowering the effect of the warp. Figure 6 shows the density response for a model fairly similar to BSM, hereafter BSM2, overlaid on the H I density gray scale plot. The rotation curve has been raised by less than 10 km s^{-1} (i.e. within the observational uncertainties) in the region between $220'' < R < 340''$. One can see a very substantial improvement of the fit of the outer southern arm in the outer parts, proving that such a fit is indeed possible. It is thus possible to obtain a satisfactory fit of the outer arm using only the bar and spiral forcing rotating at the same pattern speed.

4 Conclusions

We show that the density and velocity structure of NGC 1365 may be reproduced on most scales with a rather simple model. The bar potential shape, and the rotation curve outside $R > 120''$, have values directly from observations. In the inner region $R < 120''$ we use a rotation curve that is consistent with the observed optical slit velocities to the first order, when the perturbations of the bar have been taken into account. The pattern speed has a value in BM and

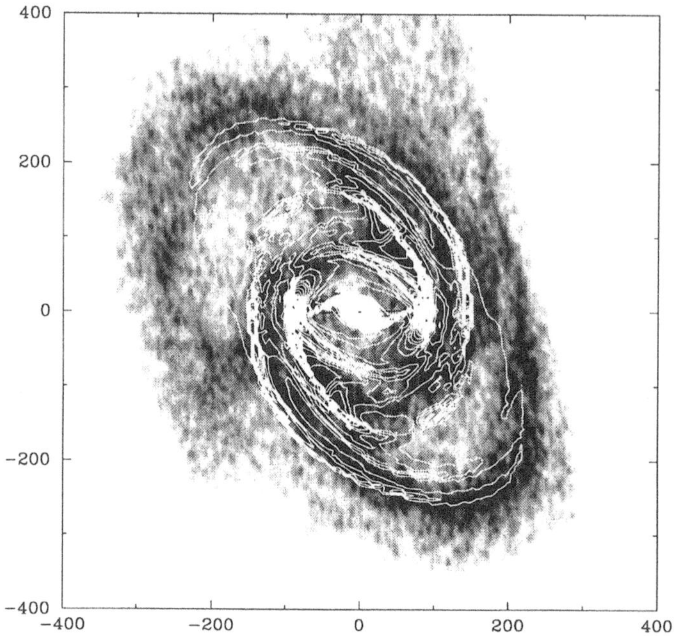

Fig. 6. The BSM2 density contour map overlaid on the total column H I density map

BSM that puts CR at a radius $1.21R_{\rm bar} < R_{\rm CR} < 1.31R_{\rm bar}$, where $R_{\rm bar} = 120''$ is the optical semi major axis of the bar. This is in agreement with the value $R_{\rm CR} = (1.2 \pm 0.2)R_{\rm bar}$ found by Athanassoula (1992). The spiral potential in the models has the same spatial location and spatial shape as that of the observed bright optical arms in the *J*-band image. The better fit achieved with BSM compared to BM suggests the presence of massive spiral arms in NGC 1365.

References

Athanassoula, E. (1992): MNRAS **259**, 328 and 345
Ball, R. (1992): ApJ **395**, 418
Elmegreen, B.G., Elmegreen, D.M. (1985): ApJ **288**, 438
England, M.N. (1989): ApJ **344**, 669
Jörsäter, S., Peterson, C.J., Lindblad, P.O., Boksenberg, A. (1984): A&AS **58**, 507
Jörsäter, S., van Moorsel, G. (1995): AJ **110**, 2037 (JvM)
Lindblad, P.O., Hjelm, M., Högbom, J., Jörsäter, S., Lindblad, P.A.B., Santos-Lleó, M. (1996): A&AS, in press (L96)
Sanders, R.H., Huntley, J.M. (1976): ApJ **209**, 53
Sandqvist, Aa., Jörsäter, S., Lindblad, P.O. (1982): A&A **110**, 336
Sandqvist, Aa., Jörsäter, S., Lindblad, P.O. (1995): A&A **295**, 585

Evolution of Galaxies Along the Hubble Sequence

Daniel Pfenniger

Geneva Observatory, University of Geneva, CH-1290 Sauverny, Switzerland

Abstract. Today we have numerous pieces of evidence suggesting that galaxies evolve dynamically along the Hubble sequence through various processes, sometimes over much shorter periods than the standard galaxy age of $10 - 15$ Gyr. Linking this to the known mass components provides new indications on the nature of the galaxy dark matter. Bounds on the amount of dark stars can be given along the spiral sequence, and the existence of large quantities of yet undetected dark gas appears as the most natural option. The recent recognition of the fractal structure of the cold interstellar gas turns out to provide not only a possible solution why much gas can be very cold and clumpy down to very small scales, so hard to detect, but also points toward the generally ignored but fundamental problem of applying statistical mechanics concepts to systems with long range interactions.

1 Introduction

Until recent years the concepts prevailing in understanding galaxy evolution have been largely dominated by the ELS scenario (Eggen, Lynden-Bell and Sandage 1962). This scenario favors a nearly *synchronous* and rapid (≈ 100 Myr) formation of the galaxies at a particular early time in the universal expansion. As a consequence, the galaxy properties of today, such as their typical scale and mass, must depend directly on the initial conditions fixed by the physical state of the Universe at the "galaxy formation epoch".

In this context the only subsequent significant evolution in galaxies to be discussed was then the slow changes in the stellar populations. The 2-body relaxation was the only dynamical evolution process to consider, but was known as much too slow to be relevant in galactic systems. Hence dynamical processes could be ignored.

Since then major progresses of direct relevance, in observations as well as in theory, occurred, which resulted in a gradual shift in the meaning of galaxy evolution. It suffices to recall that basically all the advances in non-linear dynamics, including the recognition of the fundamental rôle of chaos, have been made since, and computer simulations of galaxies had then just begun, in particular the pioneer simulations of P.O. Lindblad (1960).

Today, the ELS scenario is no longer tenable as such for several reasons explained below, even though this is still not perceived so in many fields loosely connected with stellar and galactic dynamics. This comes partly from the ELS scenario itself which led people believing that after the "galaxy formation epoch" dynamics could be safely ignored. Galactic dynamics remained perhaps underdeveloped, and instead much more efforts were invested in understanding the stages preceding the hypothetic brief "galaxy formation epoch".

A disagreement between the ELS scenario and more recent works appears in simulations of the Universe at several 100 Mpc scale (see e.g. White 1994). In such simulations nothing like homogeneous collapses do occur, instead hierarchical clustering proceeds at all the computable scales with different speeds. This implies that galaxies, exactly as stars, form in different regions of the sky *asynchronously*. The formation process covers several dex of time-scales, so not only the free-fall time, but merging and later infall epochs belong to it. For many galaxies the formation/evolution should be considered as not terminated even now. The galaxy age looses its original meaning because the aging, traced by various observables, may occur at widely different speeds.

A central aspect not considered in the ELS picture is the likely coupling of dynamics with star formation and stellar activity. In recent years the large far-infrared emission of spirals, which is even largely dominating the light in starburst galaxies, was still found substantial particularly over the "late" part of the spiral sequence. The far-infrared emission, which comes mostly from UV and visible stellar light absorbed and recycled in the FIR by dust, is consistent with the also recent recognition of the partial opacity of the optical region of spirals (e.g. Davis et al. 1993). It turns out that the total stellar power is comparable to the power that dynamics can exchange (Pfenniger 1991b).

The coupling of star formation and dynamics via a feed-back mechanism in the disk has been discussed several times (Quirk 1972; Kennicutt 1989). The interesting aspect of this coupling is that the systematic global properties of galaxies are then no longer necessarily linked to the initial conditions of formation. As for stars, the galaxy properties are then more directly dependent on their internal small scale physics, i.e. star formation and ISM physics. This may solve the old problem of the absence of galactic scale structure at the radiation-matter decoupling epoch. The galactic scale would result mainly from the proper balancing of star formation effects and dynamics during the active star formation phases.

The Hubble sequence represents most probably an incomplete sample. In particular many low surface brightness galaxies may well be missed (e.g. Bothun et al. 1990). However, within several tens of Mpc much of the galaxy mass should be detected once stars shine, since the stellar giant population is directly detectable for several Gyr. Thus, whenever fast morphological changes do occur to normal galaxies, they imply mostly changes of type *within the Hubble sequence*.

The general properties of the Hubble sequence have been progressively accumulated (e.g. de Vaucouleurs 1959; Broeils 1992; Roberts and Haynes 1994; Zaritsky et al. 1994), and are well known by now. In order to extract useful information from this "zoo", we must consider only the most general properties, keeping in mind that galaxies form a variety of objects with different ages and aging speeds. For example mergers certainly cause morphological changes at speeds which depend on the fortuitous strength of the interactions.

In isolated gravitating systems the scale should not matter. Yet, the total galaxy mass is in average increasing by a factor 100 from Sm to Sa, with large fluctuations. On the other hand, during hierarchical clustering (see e.g. White 1994) the initial large fluctuations condense faster than the small ones. So the mass trend is more indicative about the history of clustering than about the specific evolution of isolated galaxies. Because of the large mass fluctuations for a given type, to understand the evolution of galaxy *types* irrespective of the total mass, one should therefore consider quantities relative to the mass.

The following list summarizes the main trends and scale-independent properties along the spiral sequence from Sm to Sa, that are useful for the discussion.

Specific kinetic energy	$\nearrow$	$V_{\max} \approx 70 \rightarrow 300$ km s^{-1}
Bulge-disk ratio	$\nearrow$	$L_{\rm sph}/L_{\rm tot} \approx 0 \rightarrow 0.6$
Symmetry	$\nearrow$	
Detected gas	$\searrow$	$M_{\rm HI+H_2}/M_{\rm tot} \approx 0.10 \rightarrow 0.07$
		$M_{\rm HI+H_2}/M_{\rm stars} \approx 1.4 \rightarrow 0.1$
Dark matter	$\searrow$	$M_{\rm dark}/M_{\rm lum} \approx 10 \rightarrow 1$
Metallicity	$\nearrow$	$12 + \log({\rm O/H}) \approx 8.3 \rightarrow 9$

2 Sense of Evolution from Irreversible Processes

In fact, already a systematic sense of the evolution is clear by making a list of the major irreversible processes known in spirals. Each of them gives a possible criterion and a sense of aging.

1) The energy dissipation in gravitating systems is measured by the present amount of kinetic energy, which equals the minimum energy the system had to release in order to reach the present bounded state. In spirals the rotation speed is an excellent indicator of the dissipated energy since the rotation curves vary slowly with radius. Because disks are systems having lost a maximum of energy while conserving angular momentum, further energy dissipation necessarily implies dissipation of angular momentum, which is best achieved by some mass transport and breaking of axisymmetry. Bars and spiral arms are just a manifestation of this necessity. The energy factor already indicates clearly that the Sa side of the spiral sequence is energetically more evolved than the Sm side.

2) Building a central bulge or spheroid from the heating of the disk, by whatever process (as reviewed in next Sect.), is also an irreversible process, because once stars initially in a disk are heated up, there is no way to cool them back toward circular orbits. Thus from the stellar dynamical point of view Sa galaxies with big bulges are more evolved than bulgeless Sm galaxies.

3) Overall if galaxian shapes are some form of attractor, the degree of organization and symmetry toward these shapes is a sign of evolution. Clearly the spiral sequence looks increasingly organized and regular in the sense Sm to Sa, which is also the sense in which the spiral pitch angle decreases.

4) The transformation of gas into stars in cold clouds is mainly an irreversible process, since star formation locks most of the mass in stars for time-scales longer

than the galaxy ages. So the ratio of star to gas is a tracer of evolution, and Sa's are thus more evolved in this respect than Sm's. In this context, it is remarkable that the dark matter fraction varies systematically along the spiral sequence, decreasing to non-problematic values on the Sa side.

5) Obviously related to the previous criterion, the nucleosynthesis within stars is also irreversible, the more metal rich and dusty galaxies have a longer history through the internal activity of their stars. Sa's are more enriched than Sm's, which again indicates a sense of evolution.

In summary, the only consistent sense of aging along the spiral sequence is from Sm to S0. This is consistent with the hierarchical clustering scenario in which massive lumps collapse faster than light ones, explaining why Sa's are simultaneously more massive and more evolved than Sd's. The important consequence is that proto-galaxies would then be mostly bulgeless gas rich disks like Sm-Sd's, or even pure gas disks, instead of the galaxies with a bright initial bulge as envisioned in the ELS scenario.

3 Evolution from Dynamics and Observations

Contemporary to the ELS scenario, Safronov (1960) and Toomre (1964) realized the unexpected fact that gaseous and stellar disks with too close to circular motion are gravitationally unstable with respect to radial perturbations. So energy dissipation with angular momentum conservation brings first collapsing gravitating systems toward disk shapes with an increasing fraction of kinetic energy in rotational motion. But subsequent dissipation brings disks ineluctably toward a global instability. In brief, a reason had been found that disk galaxies may be dynamically unstable, and so may evolve with dynamical time-scales.

Shortly hereafter computer simulations of stellar disks (e.g. Miller and Prendergast 1968; Hohl and Hockney 1969) made it possible to simulate the non-linear phase of disk instability. They showed a systematic tendency to produce a robust bar. These results illustrated an example of major and fast morphology change (within a couple of rotational periods) of galaxy type from non-barred to barred[1].

The other significant proposition in the 70's came from Toomre and Toomre (1972) in which ellipticals may result from the merging of spirals or other ellipticals. Another case of a major and fast change of galaxian morphology was put forward. Despite much resistance this scenario appears today as the most natural way of forming ellipticals, although it is not necessarily the only one. In fact, the more a galaxy is violently shaken, for whatever reason, the more it ends up like an elliptical.

In the 80's the bar phenomenon was investigated in more depth. From observational material, Kormendy (1982) pushed forward the idea of secular evolution

[1] Incidentally, the emergence of the still alive idea of hot and round dark matter halos to prevent the formation of bars appears today to have started on a pseudo-problem: already then, but even more later, it was known that most spirals are barred.

in barred galaxies. The reason why bars do exist in the first place was understood by studies of their periodic orbits in the plane (Contopoulos 1980; Athanassoula et al. 1983) and in 3D (Pfenniger 1984). It was discovered that bars may evolve into boxy bulges via vertical resonances boosting bending instabilities transverse to the plane (Combes and Sanders 1981; Combes et al. 1990; Raha et al. 1990). It became also clear that chaotic orbits are playing an important role in bars (Pfenniger 1984, 1985). Later it was understood how the accretion of only a few percents of mass within the Inner Lindblad Resonance (ILR), either by dissipation of gas (Hasan and Norman 1990; Pfenniger and Norman 1990) or by dynamical friction of galaxy satellites (Pfenniger 1991a), may rapidly destroy a bar into a spheroidal component of similar size. This led to an increased confidence that many of the complex non-linear events and morphologies observed in galaxies, as well as in N-body simulations, can be interpreted, and even predicted, via the knowledge of the underlying periodic orbits (Pfenniger and Friedli 1991). These studies of bars showed that even isolated galaxies must also be seen as dynamical evolving structures, with possible short evolution phases.

In recent years more work has completed the above picture. Secondary bars (Friedli and Martinet 1993), gaseous and star formation effects (e.g. Friedli and Benz 1993, 1995), interactions with external galaxies and mergers (e.g. Barnes 1992) continue to be investigated. In any case, these additional complications make it even harder to freeze galaxy morphologies beyond a few Gyr. The obvious requirement is then to understand the time-sequence of galaxy morphologies.

Independently of dynamics, several observational results strongly indicate rapid galaxy evolution, either secularly or by bursts:

1) An often quoted alternative scenario to ELS is the one of Searle and Zinn (1978). From halo stellar cluster abundances these authors arrived to the conclusion that the Milky Way stars did form inside-out over several Gyr.

2) In galaxy clusters the Butcher-Oemler (1978) effect (Rakos and Shombert 1995) shows that galaxies are increasingly bluer at higher redshifts. Furthermore, the morphology-radius relationship (Dressler 1980; Whitmore 1993) shows that the majority of galaxies at the cluster periphery are spirals, as in the field, but these are replaced by lenticular and then ellipticals at smaller radii. To a dynamicist this relationship means that spirals do not survive one or a few center crossings, because galaxies must move within a cluster, and on rather elongated orbits. If correct, this relationship reveals directly the spiral morphological evolution caused by environmental disturbances. Either spirals end in part as ellipticals, and/or they are largely dissolved and contribute to the hot cluster gas and its metallicity. Studies of compact groups and mergers also show that interactions boost internal evolution (e.g. Mendes de Olievera and Hickson 1994).

3) Finally, the high resolution images of the Space Telescope show directly the galaxy morphologies at high redshifts (Dressler et al. 1994). Several Gyr in the past galaxies were more frequently irregular and disk-like than bulge-like, contrary to expectations from the ELS scenario.

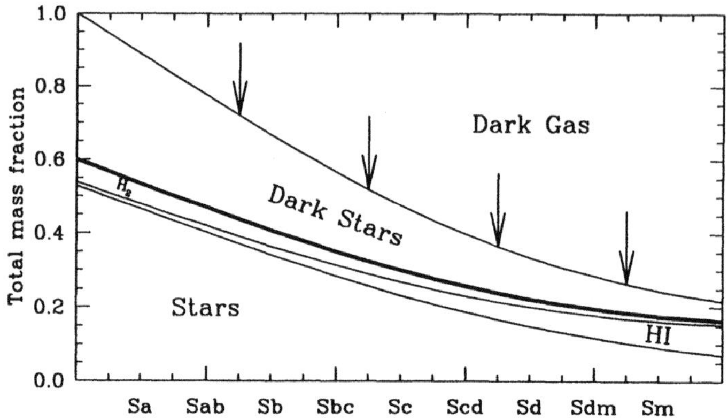

Fig. 1. Composition diagram along the spiral sequence. The thick line separates the detected matter from the dark matter. Here the dark stellar-like objects (DS) are assumed to be proportional to the stars. The arrows indicate that the DS mass is an upper limit and the dark gas a lower limit

4 Constraints from an Evolutive Spiral Sequence

If we take the published data about the ratios of the different known matter forms along the spiral sequence (e.g. Broeils 1992 for the stars and H I; Young and Knezek 1989 for H_2 derived from CO emission), we can solve for the mass fraction of each component: stars, H I, H_2 from CO, and the rest – dark matter (for more detail, see Pfenniger 1996). We obtain the composition diagram of the spiral sequence shown in Fig. 1.

Now, if spirals do evolve along the sequence from Sm to S0, then several constraints follow (Pfenniger et al. 1994). We must consider that stars or jupiters are made from the gas and lock most of it for $\gg 5-10$ Gyr, and that for dynamical reasons little accretion (less than a few % in mass) can occur transversally to a stellar disk without heating it to too high values (Tóth and Ostriker 1992)[2].

Most of the usual dark matter candidates such as CDM particles, neutrinos, jupiters, brown dwarfs, and black-holes, are inconsistent with the spiral sequence properties including dynamical evolution, mainly because the transformation of galaxies by dynamical evolution from one Hubble type to the next (including the changes of stars and dark matter fraction) can be fortuitous and sporadic. Particularly during mergers, sensible transformation processes of the above mentioned dark matter candidates should remain unrelated to large-scale dynamical changes. For example it would appear ad-hoc that decaying neutrinos

[2] But gas accretion is more acceptable at the disk periphery since angular momentum can be absorbed there, fitting with the generally warped H I disks. However, if any later massive accretion would occur in the outer disks during evolution, then the mass fraction in the outer disks would increase from Sm to S0, contrary to observations: the rotation curves change from raising to declining when proceeding from Sm to S0 (e.g. Broeils 1992).

(Sciama 1990) would adapt their decay rate to the fast transformation implied by a merger, or that mergers would be able to evaporate selectively *collisionless* particles, preferentially the dark ones.

But since the star fraction increases from Sm to Sa in a proportion exceeding the detected gas content, everything happens as if dark matter is transformed into stars, i.e. a substantial fraction of dark matter should be dark gas. We have explained elsewhere (Pfenniger and Combes 1994) the many reasons why today we can consider this conservative candidate, for long totally neglected, as worth to be more investigated. As summarized in next Sect., we extrapolate towards small scale ($\sim$ 100 AU) and colder ($\sim$ 3 K) temperature the observed fractal structure of molecular clouds in the optical disks, and apply this to outer H I disks.

Others did also arrive to the conclusion that much more gas should exist in spirals. Toomre (1981) argued that the cold dynamics and chaotic structure of Sc's require much more gas than observed to maintain their structure. Also, just to sustain the present star formation rates over several Gyr, Sc's require much more gas (Larson et al. 1980). This is the "gas consumption problem".

However, not all the dark matter in spirals is necessarily dark gas. Just the fact that S0's and ellipticals are dynamically evolved systems, apparently at the end of the gas-to-star transformation process, with about 40% of dark matter rather hints toward the presence of non-gaseous components such as white dwarfs, brown dwarfs or jupiters, called dark stars (DS). Another perfectly plausible "candidate" can be simply an increasing fraction of dust toward earlier types, increasing the mass-to-light ratio of the optical disks. Anyway, if we assume that a special population of DS pre-exists a galaxy (the same rôle would be played by CDM particles), its maximal fraction is determined by the final S0 stage. Before the final stage the difference must be gas to make future stars. If instead we assume that DS's or dust form proportional to the stars as galaxies evolve, the fraction of dark gas needed to form these DS's or normal stars hidden by dust is correspondingly increased. For the Milky Way (an SBbc) the fraction of DS's or obscured stars would be at most 40% in the first case, and at most 20% in the second case (see Fig. 1). The second case is more compatible with the present loose constraints from the micro-lensing experiments toward the LMC.

5 Nature of the Interstellar Gas

So, do we understand the interstellar gas at all? Actually the theory of the interstellar gas is still in a very primitive shape; it was known long ago that the typical state of the interstellar gas is fundamentally *inhomogeneous* in density and temperature. Supersonic turbulence is the rule, a poorly understood state.

In the recent years observations have shown that the interstellar gas is better described with fractal geometry. Molecular clouds display a fractal structure over at least 4 dex, from 0.01 pc up to $\sim$ 100 pc (Scalo 1985; Falgarone et al. 1991). This is very different from the classical description of the gas with a mass spectrum of fiducial homogeneous clouds ($dN/dM \sim M^{-1.7}$ as for a stellar population). In molecular clouds clumps are made of N sub-clumps, and so on, with N observed to be $\sim 5 - 10$. Unlike the classical view, the *spatial positions* of the clumps at a given scale are then correlated, not random; clumps are fragmented in smaller, denser, and *colder* sub-clumps.

Gravity is estimated to be important throughout the scales of the observed fractal. Anyway, when stars form gravity is important also at much smaller scale. In some cases, such as in the nearest planetary nebula (Helix), cold and dense "cometary globules" as small as the solar system are directly resolved (Meaburn 1992); this shows that $\lesssim 5$ K clumps that small can exist. Then the Larson (1981) size-line-width relation ($\sigma \propto r^{0.3-0.5}$) in molecular clouds allows us to derive the fractal dimension D of the mass distribution ($M(r) \propto r^D$): $D \approx 1.6 - 2.0$.

A fractal dimension below 2 has major implications for interpreting observations with devices having a restricted dynamical range of measurable column density (typically $< 2 - 3$ dex, Scalo 1990). Such instruments are likely to be blind to the higher density and colder material covering only a tiny fraction of the sky. In a physical fractal with $D < 2$ the projection properties depend directly on the smallest scale at which the fractal behavior stops.

By extrapolating the observed properties of the fractal cold gas down to the coldest admissible temperature around 3 K, the smallest scale at which gas can gravitationally fragment is of the order of the solar system size. Since there is no reason why the observed clumpiness in nearby molecular clouds should not be universal, we consider it reasonable to assume that outer H I disks may be also just the warm "atmosphere" of a colder and massive molecular component at the bottom scale of a fractal distribution. This solution fits well the known proportionality of H I and dark matter in spirals.

Indeed, the reason we are able to observe molecular clouds in CO is that they lie in the optical disks where stars have enriched the ISM. In contrast, outside the optical disks, the quasi-absence of CO emission in H I disks does not imply the absence of H_2, while the widespread presence of rapidly cooling H I in low excitation regions ask for a general explanation for their mere survival over many Gyr.

Broadly, the fractal interstellar gas looks like a system in a phase transition with long range correlations. Phase transitions are states for which thermodynamical principles fail. Thermodynamicists (e.g. Jaynes 1957) are well aware that systems dominated by gravity (i.e. systems that are Jeans unstable) are incompatible with the axioms of thermodynamics. Long-range interactions prevent energy to be extensive (proportional to the volume), a basic hypothesis in statistical physics. Oddly, this fundamental restriction is mostly ignored by astrophysicists. When thermodynamic concepts are applied to gravitating systems, paradoxes, such as negative specific heats, immediately occur (Lynden-Bell and Lynden-Bell 1977). Strictly, such a concept as fundamental as the temperature can not be used, since it is undefined. Further, hydrodynamics in galaxies is also on shaky foundations because hydrodynamical equations require not only differentiable flows, so no fractal distributions, but also that the higher moments of Boltzmann's equation must be closed with an equation of state, which remains undefined if the proper gas statistical physics is yet unknown. The purpose of these remarks is to point out the present huge ignorance about cosmic gas in every situation where gravity dominates the interactions and no local thermal equilibrium (LTE) is established (gravitating optically thin media).

6 Conclusions

Galactic dynamics, and in particular the study of bars, forces us to see disk galaxies as evolving structures, with possible short time-scales. Taking into account today's observational and theoretical constraints, the only possible sense of evolution is from Sm to S0. During galaxy evolution everything happens as if dark matter in galaxies is transformed into stars, that is, dark gas is required.

A possible solution to the dark matter problem in galaxies is that gas in outer disks clumps along a fractal structure down to solar system sizes. The smallest clumps are then very dense and cold which makes them presently hard to detect. A sizable amount of dark or obscured stars can also be argued just from the fact that evolved galaxies (Sa-S0's) still contain about 40% of dark matter.

Further understanding about galaxies, such as the conditions of star formation, requires a better characterization of the fractality of the cold gas. Obviously, such advances would also have deep consequences for cosmology. Unfortunately the extent of our ignorance about the fundamental statistical processes active in gravitating gases has been and still is insufficiently appreciated. I hope to complete soon a work bringing some insight into this problem.

References

Athanassoula, E., Bienaymé, O., Martinet, L., Pfenniger, D. (1983): A&A **127**, 349
Barnes, J.E. (1992): ApJ **393**, 484
Bothun, G.D., Shombert, J.M., Impey, C.D., Schneider, S.E. (1990): ApJ **360**, 427
Broeils, A. (1992): Dark and Visible Matter in Spiral Galaxies, PhD Thesis, Univ. Groningen
Butcher, H., Oemler, A. (1978): ApJ **219**, 18
Combes, F., Debbasch, F., Friedli, D., Pfenniger, D. (1990): A&A **233**, 82
Combes, F., Sanders, R.H. (1981): A&A **96**, 164
Contopoulos, G. (1980): A&A **81**, 198
Davies, J.I., Phillips, S., Boyce, P.J., Disney, M.J. (1993): MNRAS **260**, 491
de Vaucouleurs, G. (1959): in Handbuch der Physik LIII, Astrophysik IV: Sternsysteme, ed. S. Flügge, Springer-Verlag, Berlin, p. 275
Dressler, A. (1980): ApJ **236**, 351
Dressler, A., Oemler, A., Butcher, H., Gunn, J.E. (1994): ApJ **430**, 107
Eggen, O.J., Lynden-Bell, D., Sandage, A.R. (1962): ApJ **136**, 748 (ELS)
Falgarone, E., Phillips, T.G., Walker, C.K. (1991): ApJ **378**, 186
Friedli, D., Benz, W. (1993): A&A **268**, 65

Friedli, D., Benz, W. (1995): A&A, **301**, 649
Friedli, D., Martinet, L. (1993): A&A **277**, 27
Hasan, H., Norman, C. (1990): ApJ **361**, 69
Hohl, F., Hockney, R.W. (1969): J. Comput. Phys. **4**, 306
Jaynes, E.T. (1957): Phys. Rev. **106**, 620
Kennicutt, R.C. (1989): ApJ **344**, 685
Kormendy, J. (1982): in 12th Advanced Course Swiss Soc. Astr. Astrophys., Morphology and Dynamics of Galaxies, eds. L. Martinet, M. Mayor, Geneva Obs., 115
Larson, R.B. (1981): MNRAS **194**, 809
Larson, R.B., Tinsley, B.M., Caldwell, C.N. (1980): ApJ **237**, 692
Lindblad, P.O. (1960): Stockholm Obs. Ann. **21**, No. 3-4
Lynden-Bell, D., Lynden-Bell, R.M. (1977): MNRAS **181**, 405
Meaburn, J., Walsh, J.R., Clegg, R.E.S., Walton, N.A., Taylor, D., Berry, D.S. (1992): MNRAS **255**, 177
Mendes De Oliveira, C., Hickson, P. (1994): ApJ **427**, 684
Miller, R.H., Prendergast, K.H. (1968): ApJ **151**, 699
Pfenniger, D. (1984): A&A **134**, 373
Pfenniger, D. (1985): A&A **150**, 112
Pfenniger, D. (1991ab): in Dynamics of Disc Galaxies, ed. B. Sundelius, Göteborg U., (a) p. 191, and (b) p. 389
Pfenniger, D. (1996): in Third Paris Cosmology Colloquium, eds. H.J. de Vega, N. Sánchez, World Scientific, Singapore, in press
Pfenniger, D., Combes, F. (1994): A&A **285**, 94
Pfenniger, D., Combes, F., Martinet, L. (1994): A&A **285**, 79
Pfenniger, D., Friedli, D. (1991): A&A **252**, 75
Pfenniger, D., Norman, C.A. (1990): ApJ **363**, 391
Quirk, W.J. (1972): ApJ **176**, L9
Raha, N., Sellwood, J.A., James, R.A., Kahn, F.D. (1991): Nat. **352**, 411
Rakos, K.D., Schombert, J.M. (1995): ApJ **439**, 47
Roberts, M.S., Haynes, M.P. (1994): ARA&A **32**, 115
Safronov, V.S. (1960): Annales d'Astroph. **23**, 979
Scalo, J.M. (1985): Protostars and Planets II, eds. D.C. Black, M.S. Matthews, Univ. Arizona Press, Tucson, p. 201
Scalo, J. (1990): in Physical Processes in Fragmentation and Star Formation, eds. R. Capuzzo-Dolcetta et al., Kluwer, Dordrecht, p. 151
Sciama, D.W. (1990): ApJ **364**, 549
Searle, L., Zinn, R. (1978): ApJ **225**, 357
Toomre A. (1964): ApJ **139**, 1217
Toomre A. (1981): in The Structure and Evolution of Normal Galaxies, eds. S.M. Fall, D. Lynden-Bell, Cambridge Univ. Press, p. 111
Toomre, A., Toomre, J. (1972): ApJ **178**, 623
Tóth, G., Ostriker, J.P. (1992): ApJ **389**, 5
White, S.D.M. (1994): in Formation and Evolution of Galaxies, Les Houches Lectures, astro-ph/9410043
Whitmore, B.C. (1993): in Physics of Nearby Galaxies, Nature or Nurture?, eds. T.X. Thuan, C. Balkowski, J.T.T. Van, Editions Frontières, Gif-sur-Yvette, p. 425
Young, J.S., Knezek, P.M. (1989): ApJ **347**, L55
Zaritsky, D., Kennicutt, R.C., Huchra, J.P. (1994): ApJ **420**, 87

Rings, Lenses, Nuclear Bars: the Fundamental Role of Gas

Francoise Combes

DEMIRM, Observatoire de Paris, 61 Av. de l'Observatoire, F–75014 Paris, France

Abstract. Spiral galaxies evolve through angular momentum transfer from the center to outer parts; stars are participating to this evolution but quickly heat up through gravitational instability to reach a hot quasi steady-state in the absence of gas. The latter is then the motor of evolution, and is at the origin of the formation of secondary structures, as rings, lenses or bars within bars. The evolution can be self-regulated, since too much gas concentration weakens the bar, and star-formation locks the gas in the disk, slowing the gas inflow to the nucleus. We show how fast-rotating nuclear features can halt the evolution during a short time.

1 Introduction

That the gas is fundamental for the evolution and morphology of spiral galaxies is quite obvious in observations. As soon as a spiral galaxy becomes deficient in gas, it also becomes smooth and featureless, like a lenticular. Even in a very small amount, a few percent of the total mass, the gas component is necessary to maintain the spiral structure (e.g. Sellwood and Carlberg 1984; Toomre 1990). For the latter to be constantly renewed and to explain the frequency of observed spirals, a disk has to be considered still growing at the present epoch through gas accretion. Secondary structures, as rings and nuclear bars are observed in gaseous spiral galaxies (e.g. Buta 1986, 1992); rings are the sites of enhanced star-formation, they are conspicuous by their blue colours (Buta and Crocker 1991, 1993).

While the gravitational instabilities of a purely stellar disk have been widely studied, mainly through simulations, and begin to be well understood, the situation is much more open for a self-gravitating disk composed of stars and gas. Even when a stellar disk is stabilized against axisymmetric instabilities (Toomre criterion, 1964), it can be the site of spiral and bar instability (e.g. Sellwood and Wilkinson 1993), together with z-instabilities (e.g. Combes et al. 1990a). These heat considerably the stellar disk, that can then no longer sustain spiral structure. Instabilities can be suppressed by reducing the effective self-gravity of the stellar disk, either through the addition of a hot bulge or halo (Ostriker and Peebles 1973), or through disk heating, i.e. increasing the velocity dispersion (Athanassoula and Sellwood 1986).

The behaviour of a self-gravitating disk of both stars and gas, even if the latter represents only a small fraction of the total mass, presents much more variety and complexity. Due to the dissipation, the gaseous component remains cool, and evacuates the heating due to the gravitational instabilities: spiral structure can then be continuously renewed (e.g. Sellwood and Carlberg 1984), even in the stellar disk. Criteria for stability are more complex, the gravitational coupling between gas and stars making the ensemble unstable, even when each component would have been separately stable (e.g. Jog and Solomon 1984; Bertin and Romeo 1988; Elmegreen 1994). Gravity torques exerted on the gas produce strong radial flows, and gas can accumulate at Lindblad resonances and form several rings (e.g. Schwarz 1981; Combes 1988). The strong gas concentration in the nucleus can destroy stellar bars (Friedli and Benz 1993), after having sometimes triggered a nuclear bar within the main one (Friedli and Martinet 1993). This bar destruction can be explained by an increase of chaotic orbits in the stellar bar (Hasan and Norman 1990). The copious inflow of gas into the nuclear region may be one of the main sources of nuclear activity in galaxies (Phinney 1994).

When the gas possesses too much self-gravity, however, it forms lumps through Jeans instability, and the lumpiness of the gas can scatter the stars, randomize their motions, and prevent any bar formation (Shlosman and Noguchi 1993). While the gas is triggering bar instability when it represents only a few percent of the total mass, it can play the inverse role when its mass is above $\approx 10\%$.

Galaxy evolution is therefore a complex problem, involving many interacting physical components: collisionless stars, dense clouds, diffuse gas. The total system is governed by self-gravity, dissipation, phase transitions between the various components: dense clouds heated in diffuse gas, or star-formation locking up the dissipative component, and re-injecting energy in the medium. Along the evolution, matter tends to concentrate towards the center, after the angular momentum has been transported outwards. This is done essentially through spiral waves developing in the disk, since viscous torques are inefficient. We show here how the gas component has a fundamental role in this transport, essentially because it is a cold component, very sensitive to spiral waves. Unfortunately, the detailed physics of the insterstellar medium is not yet well known, and there are many reasons to suspect that the galaxy evolution depends significantly on the gas behaviour.

Observations show that the gas component is a multi-phase medium, spanning a wide range of densities and temperatures. To drastically simplify, we can consider two distinct forms - dense, cold and clumpy molecular clouds and warm diffuse gas - with different spatial distribution but having comparable total masses. The two aspects of the gas component have been studied in the literature: ensembles of interstellar clouds that undergo collisions, but do not behave as a fluid (Schwarz 1981, 1984; Roberts and Hausman 1984; Combes and Gerin 1985). The collisions take place mainly along spiral arms. This method describes essentially the molecular gas component. The response of the diffuse gas in barred potentials was also explored (Huntley et al 1978; Roberts et al. 1979; Sanders and Tubbs 1980; van Albada and Sanders 1982; Prendergast 1983; van Albada 1983; Contopoulos et al. 1989; Athanassoula 1992). In these calculations the gas was considered as a fluid, submitted to pressure forces, and undergoing shocks. In most cases, the gas self-gravity was not considered. Already in these two kinds of work, results appear to be very different: resonant rings are easily formed in the first point of view (dense gas clouds), while thin shocks along the bar can form only in the diffuse gas.

In this review, we first describe how the angular momentum can be transfered in a disk galaxy, then compute quantitatively the efficiency of the phenomenon, through an efficient "gravitational viscosity". It is shown that viscous torques are negligible at a galactic scale with respect to gravity torques. Gas inflow towards the center helps the formation of lenses through bar weakening, and we show that this is a self-regulating process. Also the formation of nuclear bars or nuclear waves ($m = 1$ or $m = 2$) can halt for some dynamical times the rapid gas inflow.

2 Angular Momentum Transfer

This is the basis of the secular evolution, we will describe the relevant phenomena successively in the stellar and gas components.

2.1 Stellar Component

In a pioneering paper, Lynden-Bell and Kalnajs (1972) have shown how spiral waves can carry angular momentum (A.M.), and that only trailing spirals can carry it outwards, which explains the predominance of trailing waves in observations. For a steady wave, stars can exchange angular momentum at resonances only: they emit A.M. at inner Lindblad resonance, while they absorb at corotation and outer resonance. They also show that, while stars do not gain or lose angular momentum on average away from resonances, they are able however to transport angular momentum as lorries in their orbiting around the galactic center. When they are at large radii, they gain A.M., while they lose some at small radii: even if the net balance is zero, they carry A.M. radially, and the sense of this lorry transport is opposite to that of the spiral wave. This phenomenon can then damp the wave, if the amplitude is strong enough. Lynden-Bell and Kalnajs (1972) noticed that this damping phenomenon became negligible for small wavelengths, i.e. when $kr \gg 1$.

This A.M. transport has been investigated recently by Zhang (1996), who puts forward another point of view: due to the phase-shift between the stellar density and the potential of the spiral wave, gravity torques are exerted by the wave on the basic state stars, and stars gain or lose angular momentum, even away from resonances. This is the consequence of a kind of dissipational process, corresponding to small-angle scatterings of neighboring stars in the spiral arm. This process transforms ordered motions into disordered ones (resulting in increased velocity dispersion, or large epicycle amplitudes), and when collective effects are taken into account, secular modifications of the stellar orbits can result: they can lose energy and angular momentum inside corotation.

This might appear incompatible with a quasi-steady wave, since in the corotating frame, the Jacobi integral is conserved. However in the presence of a collective dissipation process induced by the spiral potential/density phase-shift, this conservation law, which is derived for the non-self-consistent orbital response in an applied spiral potential, is found to be violated in the N-body simulations. Zhang (1996) has shown that the stellar density azimuthal profile steepens with time during the initial wave growth process, indicating the presence of large-scale gravitational shocks, which is likely to be the underlying mechanism responsible for the spiral-induced collective dissipation. The result is a secular and global redistribution of mass: stars inside corotation lose A.M. and move to smaller radii, which steepens the radial density profile in the disk. This has been observed during many N-body simulations of a stellar disk, that has obtained a long-lived spiral pattern (e.g. Donner and Thomasson 1994).

The A.M. transport is not only due to propagating wave packets, of negative A.M. density, that transport angular momentum outwards, with their group velocity directed inward (Toomre 1969). The spiral patterns formed in the N-body calculations on unperturbed disks appear to be unstable spiral modes, which remain for much longer than a wave travel time, and do not disappear through winding at ILR, but through heating of the disk (Donner and Thomasson 1994).

Expressions for the Torque and Phase-Shift. It is obvious that the torque exerted by the wave on particles is a second order term, since the tangential force $\frac{\partial V_1}{\partial \theta}$ is first order, and has a net effect only on the non-axisymmetric term of the surface density $\Sigma_1(r, \theta)$. The torque $T(r)$ applied by the wave on the disk matter in an annular ring of width dr, can be expressed by:

$$T(r) = r\mathrm{d}r \int -\Sigma_1(r, \theta) \frac{\partial V_1}{\partial \theta} \mathrm{d}\theta$$

(Zhang 1996).

On that formula, it is easy to see that the torque vanishes if the density and potential spiral perturbations are in phase. But there must be a phase-shift in the general case, according to the Poisson equation. The potential is non-local, and is influenced by the distant spiral arms. So the sign and amplitude of the phase-shift depends on the radial density law of the perturbation. It has been shown by Kalnajs (1971) that the peculiar radial law of $r^{-3/2}$ for an infinite spiral perturbation provides exactly no phase shift. The phase-shift is such that the spiral density leads the potential if the radial falloff is slower than $r^{-3/2}$, and the reverse if it is steeper.

Now in a self-consistent disk, the phase-shift given by the Poisson equation must agree with that given by the equations of motion. Through the computation of linear periodic orbits in the rotating frame, Zhang (1996) has found that the forcing consists of two terms in quadrature, and that the phase shift δ of the orbit orientation with respect to forcing potential has the expression:

$$\tan(m\delta) \approx \frac{-2\Omega}{(\Omega - \Omega_{\mathrm{p}})kr}$$

This shows that in the WKBJ approximation, the phase shift is negligible ($kr \gg 1$), and that it changes sign at corotation. The fact that the torque is a non-linear effect, and vanishing in the WKBJ approximation, may explain why it was neglected before, but it has a quite important effect in galactic conditions. Inside corotation, we expect that the density leads the potential, and the contrary outside corotation. These predictions have been confirmed through N-body simulations by Zhang (1996), as for the sign of the phase-shift on each side of the corotation. The amplitude of the phase-shift appears quite high, up to 20°. The existence of the phase-shift is no surprise of course, but the amplitude of δ was not suspected to be so high.

I have reproduced a comparable N-body simulation (2D polar grid PM with 10^5 particles), and reported the Fourier analysis results in Fig. 1. In this simulation, initial parameters were chosen such as to stabilize the stellar disk with respect to bar formation, through the presence of an analytical bulge component of mass equal to the mass of the self-gravitating stellar disk. A bar developed yet, but was delayed until $2 \cdot 10^9$ yr, i.e. $\approx$ 20 dynamical times. In the mean time, a spiral wave developed, and remained for several dynamical times. Its power spectrum revealed a well-defined pattern speed, at least up to corotation ($\approx$ 7 kpc). We can see in Fig. 1 that the phase-shift between stellar density and potential is indeed quite high, up to 28° inside corotation, where it is the most meaningful.

Exchange of A.M. Between Particles and Wave. Previous studies had neglected the exchange of energy and angular momentum between particles and the wave outside resonances (Lynden-Bell and Kalnajs 1972; Goldreich and Tremaine 1979). This interaction is mediated by the graininess of the particles, which under the local gravitational instabilities in the spiral arms, can scatter particles and produce dissipation (Zhang 1996). When this was not taken into account, it was believed that the torque induced by the phase-shift was exactly compensated by the advective term of torque coupling between the wave and particles, which implied no net angular momentum exchange between the wave and the basic state stars.

It was found in Zhang (1996, II) that the total torque coupling integral of an unstable spiral mode has a characteristic bell shape, with the peak of the bell near the corotation radius. Due to this particular shape of the torque-coupling integral, an unstable spiral mode is able to deposit negative angular momentum in every annulus inside corotation, en route of the outward angular momentum transport, and positive angular momentum outside corotation. This radial profile of the total torque coupling is in fact what allowed the spontaneous growth of a trailing spiral mode. Moreover, as the wave reaches nonlinear regime, an increasing fraction of the deposited angular momentum by the spiral wave is

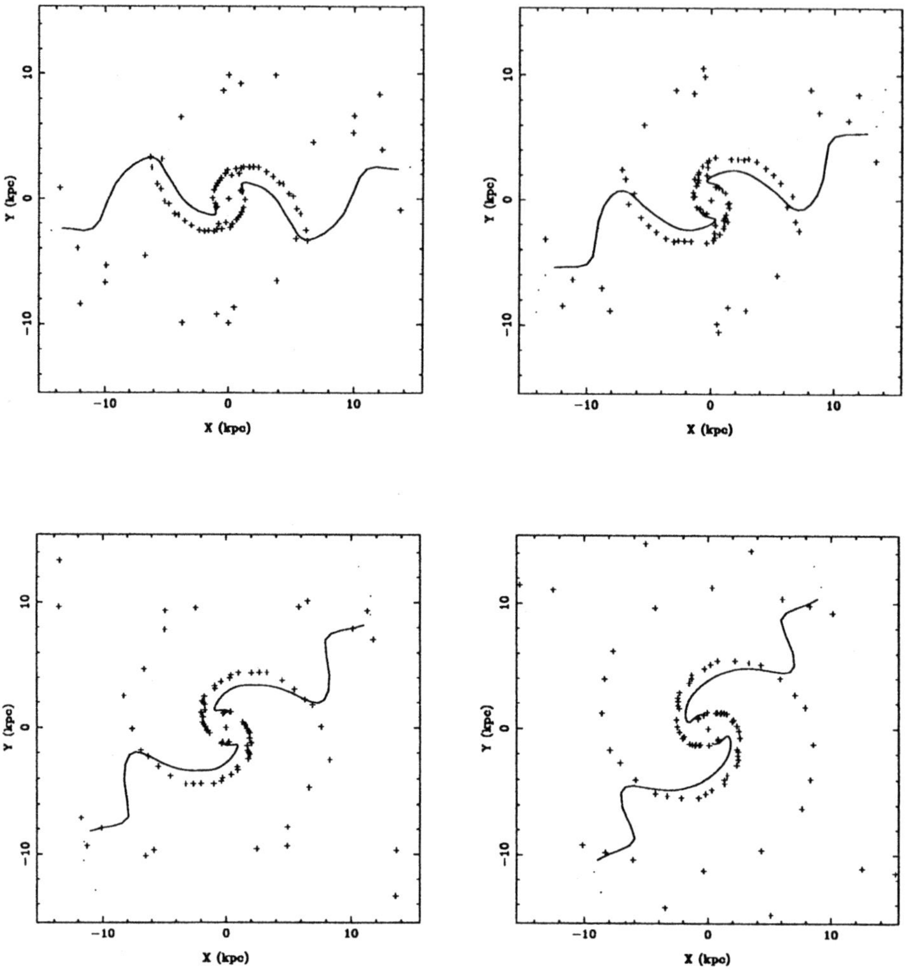

Fig. 1. Results of the Fourier analysis of the stellar density (crosses) and potential (full lines) in a purely stellar N-body simulation, while a spiral structure rotates in the disk. Plotted here are the phases of the pattern at each radius. The density leads the potential almost everywhere inside corotation. The perturbation was weaker outside corotation

being channeled to the basic state by the spiral gravitational shock; and finally, at the quasi-steady state of the wave, all of the deposited angular momentum goes to the basic state, since by the very definition of the quasi-steady state the wave should not grow any further. We can note that this collective dissipation process and angular momentum transport, induced by the phase-shift between density and potential, is inexistent for a bar. The latter can therefore be robust and long-lived in a collisionless ensemble of particles (galaxy disk without gas).

It is interesting to note with Zhang (1996, II) that the wave damping through

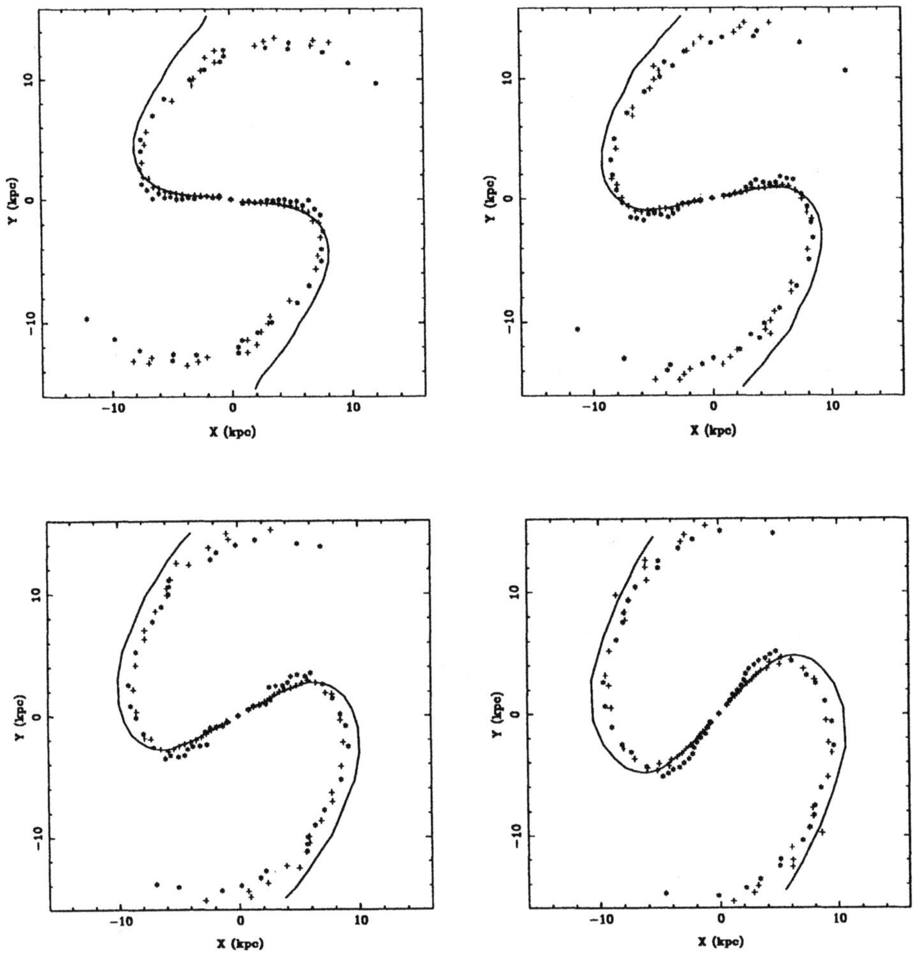

Fig. 2. Same as Fig. 1 for an N-body simulation taking into account gas and stars, and where a bar develops. The phases of the gas pattern are indicated by asterisks, while crosses indicate the stars, and the full lines the potential. Now the potential is in phase with the stellar density in the bar, and leads the density in the spiral outside corotation

collective dissipation processes makes spiral structure join the class of problems of non-equilibrium phase transitions, where an ordered structure can form spontaneously, and increase the speed of entropy evolution in the underlying systems.

2.2 Gas Component

The mechanism of the A.M. transfer for the gas is the same that was previously described: the gas settles in a spiral structure which is not in phase with the potential. Gravity torques are exerted by the wave on the gas. The dissipation here is of course different, since the gas radiates away its energy. The gas component is then maintained cool and responsive to new gravitational instabilities. This is the source of more drastic secular evolution, with the possibility of the whole gas component inflowing towards the center.

The gas response in a barred potential has been tackled through many simulations by several authors (e.g. Sanders and Huntley 1976; Schwarz 1981) and can be understood in terms of periodic orbit families in a bar potential (e.g. Combes 1988). The sign of gravity torques also change at corotation. They can change also at ILR, according to the shape of the $\Omega - \kappa/2$ precession-rate curve.

Figure 2 shows the relative phases of the gas and stellar component with respect to the total potential, in an N-body simulation, while a bar was developing, with its external spiral structure at the same pattern speed. There is now a negligible phase-shift of the stellar density with respect to the potential in the bar, while there is a slight gas phase-shift. In the arms, outside corotation, the potential leads both components, but the phase-shift is always larger for the gas.

What is the actual role of dissipation? In an axi-symmetric galaxy, viscous torques would also transfer the A.M. of the gas towards the outer parts, but the time-scale would be longer than a Hubble time at large radii (Lin and Pringle 1987a). Gravity torques are then directly responsible for the gas inflow, and dissipation prevents the gas heating, and maintains the non-axisymmetric features required for gravity torques (Combes et al. 1990b). That is why the global gas behaviour should not depend much on the gas hydrodynamics, and the point of view adopted: gas as a fluid or as ballistic colliding clouds. In details it is not quite true, the two approaches lead to different results: the characteristic shocks along the dust lanes in a barred galaxy are reproduced only with the fluid approach, and all the variety of resonant rings are better obtained with the sticky particles approach. The reality must be in between, when the multi-phase property of the gas is taken into account. We want now to quantify the effects of gravity torques by estimating an equivalent "viscosity" parameter.

3 Gravitational Viscosity

Already Lin and Pringle (1987a) proposed to define a "viscosity" parameter to account for the redistribution of A.M. due to gravitational instabilities. The length scales L of the disturbances that can grow through self-gravity are in the range $L_J < L < L_{\rm crit}$, where L_J is the Jeans length $\approx \frac{v_s^2}{G\Sigma}$ (v_s is the velocity dispersion and Σ the surface density), and $L_{\rm crit}$ is the size beyond which the shear stabilizes: $L_{\rm crit} \approx \frac{G\Sigma}{\Omega^2}$ (Ω is the rotational frequency). The size over which the angular momentum can be transferred is then of the order of $L_{\rm crit}$, and the time-scale is $\approx \Omega^{-1}$, so that they prescribe an equivalent "gravitational viscosity" of

$$\nu_{\rm eff} \approx \frac{L_{\rm crit}^2}{\Omega^{-1}} = Q^{-2} v_s^2 \Omega^{-1}$$

This formulation, analogous to the α parameterization of Shakura and Sunyaev (1973), with $\alpha \approx Q^{-2}$ allows them to find similarity solutions, where Σ is a power law in radius.

In real galactic disks, star formation must be taken into account, and can play a regulating role. There is no doubt that the star-formation rate is related to the gravitational instabilities, as noticed by Kennicutt (1989), and therefore that the time-scale for star formation t_* is related to the viscous time t_ν. If both time-scales are assumed of the same order $t_* \approx t_\nu$, Lin and Pringle (1987b) demonstrate that the final stellar surface density is exponential, meeting remarkably the observations.

In a recent work (von Linden et al. 1996) we follow this approach through N-body simulations with stars and gas. By monitoring $\Sigma(r,t)$ we can derive the equivalent viscosity quantitatively. We use the formalism of viscous torques (e.g. Frank et al. 1985), but since the torques do not correspond to a physical shear, their sign has not to align to the usual one, and the equivalent "viscosity" can sometimes be negative. Figure 3 shows the estimation of the radial velocity v_r and the equivalent viscosity ν as a function of radius, at a given epoch.

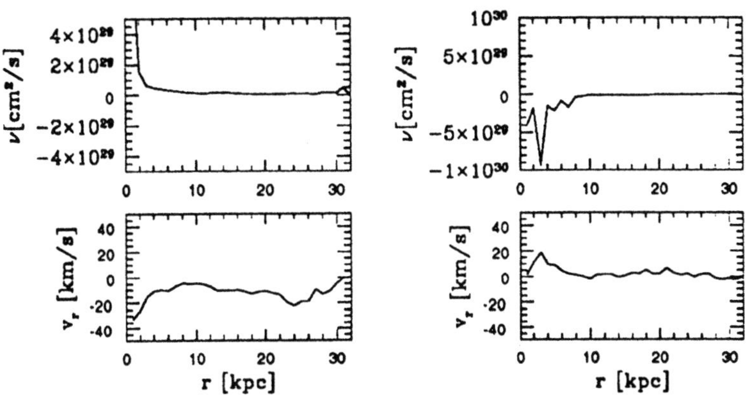

Fig. 3. Example of radial behaviours of gravitational viscosity and radial velocity at a given epoch (from von Linden et al. 1996)

The results of the monitoring $\Sigma(r,t)$ of the gas and/or the stars showed at first an oscillating behaviour (von Linden et al. 1996), on time-scales of the dynamical time ($\sim 10^8$ yr). This behaviour was interpreted as due to the crossing of the disk by successive waves, and was averaged out on longer time-scales, to measure the net gain of mass within a given radius (see Fig. 4). The average viscosity is then positive inside corotation. From the evolution of viscosity as a function of time, it is easy to notice the bar formation and subsequent destruction through gas mass concentration in the center (large positive maximum). The time-scale for gravitational viscosity can then be estimated as a function of radius:

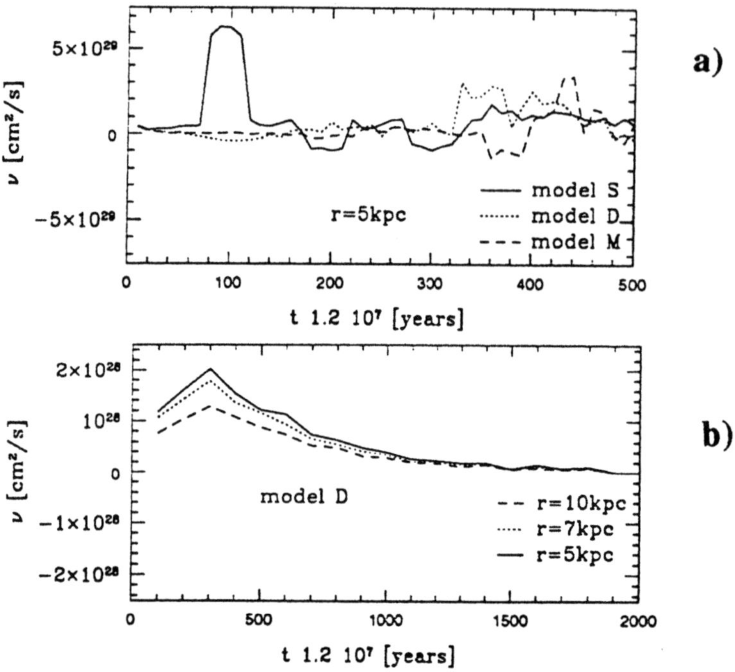

Fig. 4. **a** Viscosity measured from the gas as a function of time at 5 kpc, for three different models (a strong bar developed in model S). **b** Viscosity measured from the stars for three different radii (from von Linden et al. 1996)

$$t_{\rm vis} \sim \frac{r^2}{\nu} \sim \frac{r}{v_r} \sim 3 \cdot 10^9 (\frac{r}{10 \text{ kpc}})^2 (\frac{\nu}{10^{28} \text{ cm}^2\text{s}^{-1}})^{-1} \text{ years}$$

at a radius of 10 kpc we get, with an average $\nu = 1.0 \cdot 10^{27}$ cm^2 s^{-1} or 3 kpc km s^{-1} for the standard disk-dominated model, in which a bar developed, $t_{\rm vis}(10 \text{ kpc}) = 3. \cdot 10^{10}$ years, and at a radius of 5 kpc, since the efficient ν is 100 times larger, $t_{\rm vis}(5 \text{ kpc}) = 3. \cdot 10^8$ years, while the dynamical timescale $t_{\rm dyn} = \frac{r}{v_\phi}$ ranges from 10^7 years to 10^8 years, typically over the disk.

The turbulent viscous timescale is

$$t_{\rm turb-vis} = 8.25 \cdot 10^7 \frac{1}{\alpha} (\frac{r}{\text{pc}}) \text{ years}$$

(Larson 1988) assuming that the α–parameter is equal to 1 at a radius of 10 kpc $t_{\rm turb-vis}(10 \text{ kpc}) = 8.25 \cdot 10^{11}$ years and $t_{\rm turb-vis}(5 \text{ kpc}) = 4.12 \cdot 10^{11}$ years.

Viscous torques are therefore more than 2 orders of magnitude less efficient than gravity torques at large radii. The efficiency of gravitational viscosity increases considerably towards the galaxy center; it becomes several orders of magnitude higher than the turbulent viscosity efficiency. The gravitational viscosity time-scale becomes of the same order as the dynamical time at $r = 3$ kpc.

4 Lenses

With such a large gravitational viscosity in a barred galaxy, the gas is driven inwards very quickly and accumulates in the center. When the central mass reaches a few percents of the total mass, the bar weakens and evolves into a lens.

4.1 Observations

Lenses are slightly elliptical components, with a sharp outer edge. They form a conspicuous wiggle in the brightness radial profile, as demonstrated in NGC 1553 by Freeman (1975). Their dynamics has been studied by Kormendy (1977, 1979, 1984): they are quite hot in the central parts, even hotter than the bulge; but they are cold at the edge, which explains their sharp cut-off. Lenses are associated to bars; Kormendy (1979) found that $\sim 54\%$ of SB0-a galaxies have lenses. When the bar and the lens are found to co-exist in a galaxy, the bar fills the lens in one dimension. From the statistics on their sizes L, we know that on average $L/r = 1.3$ and $R/L = 1.8$, where r and R are the radii of inner and outer rings (Athanassoula et al. 1982). This suggests that the lens edge is near corotation.

4.2 Lens Formation

A few mechanisms have been proposed to explain the lens component, but none has been developed enough to meet all observational constraints. Kormendy (1982) has described a global and fruitful evolutionary scheme, developing several possibilities of secular evolution processes driven by the bar. One of these processes was the evolution of bars into a more axisymmetric component such as a lens (Kormendy 1979, 1982). But this transformation should occur in a time-scale much shorter than a Hubble time, since half of all SB0-a galaxies contain both lenses and bars, and must stop before complete dissolution of the bar. The bar was assumed to interact with another component, letting some stars escape from the bar. Since the lens phenomenon is associated with early-type, it was natural to propose an interaction with the bulge. If the bulge is rotating fast enough, some angular momentum can flow from the bulge to the bar, which destroys it (the bar is a wave with negative angular momentum). The process reduces bulge rotation, and is therefore self-regulated. The interaction with a non-rotating halo would have produced the opposite effect.

A completely different idea is to suppose that the lens is a primary component, and may form early during the star-formation history (Bosma 1983): the sharp cut-off would then be a star-formation threshold. Athanassoula (1983) suggests that lenses form just like bars, through gravitational instabilities; the difference in eccentricities would come from the initial conditions, lenses forming out of a much hotter population. There remains to be found why two distinct populations of stars exist initially in the disk, why the edge of lenses are so sharp and cold, and why the lens components are still almost as flat as disks.

Along the lines of Kormendy (1982), I propose here another hypothesis, that the lenses are secondary components coming from the regulation of the bar strength due to a central mass concentration. Many studies have been published on the influence of a central mass concentration (or black hole, BH) on the bar secular evolution (Norman et al. 1985; Hasan and Norman 1990; Pfenniger and Norman 1990; Nishida and Wakamatsu 1996; Hasan et al. 1993). When the BH scatters particles at a velocity

$$v \approx (GM_{\rm bh}/a_{\rm bh})^{1/2} \approx V_{\rm rot}$$

then particles can escape the center and be re-distributed, their orbits become stochastic for high enough BH masses. The bar favours elongated orbits in the center, with less angular momentum, which are more sensitive to the central BH than circular orbits. When the central stars, with the less angular momentum have escaped, the efficiency of the BH decreases, and the bar destruction progressively switches off. Through orbit computations, Hasan and Norman (1990) have determined the percentage of the remaining regular x_1 orbits, as a function of the BH mass. This percentage falls significantly when the central mass reaches 5% of the total mass: the percentage of stochastic orbits is so large that the bar is weakened, then destroyed for a mass fraction of 10%. Through self-consistent stellar simulations, Nishida and Wakamatsu (1996) have shown that a bar still forms with the same amplitude in a galaxy with an initial mass concentration as high as 5%, but its strength weakens in the long run, while a bar with no central mass concentration keeps its amplitude in a steady state through a Hubble time.

What can play the role of the central mass concentration, as high as 5% of the total mass, i.e. $5 \cdot 10^9$ M$_\odot$? There cannot exist so massive black holes in most normal spiral galaxies. In fact, a very concentrated bulge could be the scattering mass, and therefore the intuition of Kormendy that the lens originates from bar-bulge interaction turns out to be true (although not for the right mechanism). Indeed, the order of magnitude given above concerning the scattered velocities leads to a required M/r ratio comparable for the bulge and the disk, i.e. $(M/r)_{\rm b} \approx (M/r)_{\rm d}$, or in terms of the surface densities $\mu_{\rm b} \approx 10\mu_{\rm d}$, if we assume a scale-length an order of magnitude lower for the bulge (cf. Courteau et al. 1996). In terms of magnitude, if the central surface brightness of disks are around 21.6 mag arcsec^{-2}, that of the bulge should be at least 19.1 mag arcsec^{-2}, which is often observed in early-type galaxies. More specifically, barred galaxies with bulges reaching such a high concentration should be accompanied by a lens. Another possibility is that the strong gas infall driven by the bar produces itself a sufficient mass concentration, as in the simulations; in practice new stars are formed in those condensations (cf. hot spots in nuclear rings), which would also join the bulge later on.

This lens formation mechanism is part of a self-regulated process, based on gas accretion: the torques due to the strong initial bar drive the gas inwards in a short time-scale. This increases the central mass concentration, up to the point where the bar weakens; there are then two ILRs and the aligned x_1 regular orbits are less preponderant. The torques weaken and the gas infall is slowed down or halted; the gas coming from the outer parts of galaxies has now time to form stars in the disk, which re-establishes the mass balance between the nucleus and the disk. The central mass concentration loses its dynamical efficiency, and a new bar can form through gravitational instability, which closes the cycle. To have several such cycles in a Hubble time requires however a substantial mass accretion, i.e. important gas reservoirs in the outer parts of galaxies (cf. Pfenniger, Combes and Martinet 1994). Another possible loop of the cycle is provided by the decoupling of a second bar (Friedli and Martinet 1993), since it can delay for a while the gas infall (Shaw et al. 1993; Combes 1994).

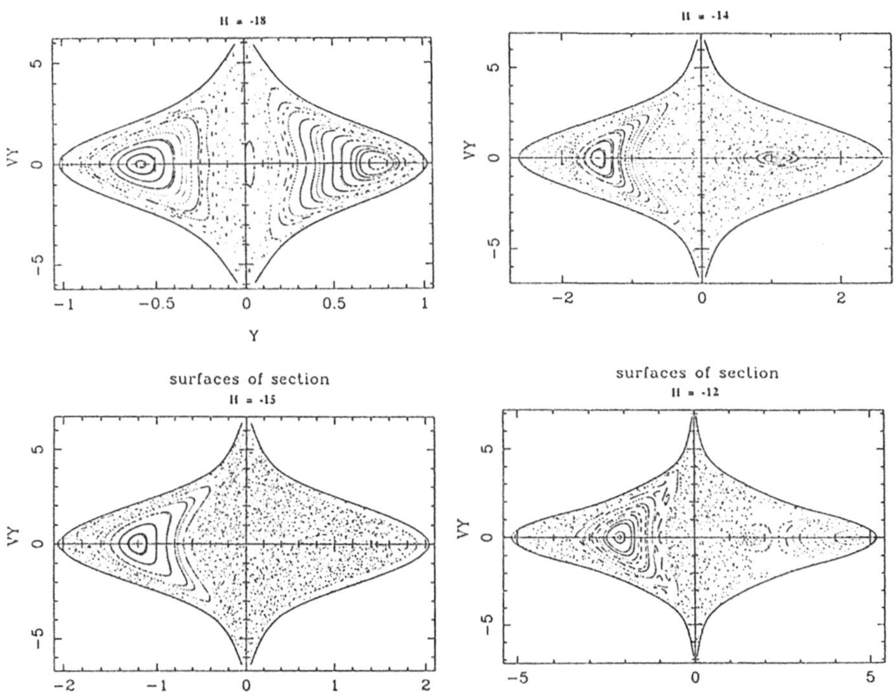

Fig. 5. Surfaces of section for increasing energies in a barred galaxy with 8% central mass, showing the extension of chaotic orbits

One consequence of this feedback cycle is that in evolved bars, there are always just about two ILRs. The behaviour of stellar orbits in presence of a high mass concentration (8%) is displayed in Fig. 5, through surfaces of section for increasing energies. Near the mass concentration, the x_2 family is predominant, becoming round with increasing mass (the potential then becomes axisymmetric in the center). Then comes a region of chaotic orbits, followed near the end of the bar by a region still dominated by the x_1 elongated orbits (with losenge shapes near the 4/1). Once in the chaotic sea, the orbits are bounded only by their limiting energy curve, in the rotating frame, where the potential is $\Phi(r) - 1/2\Omega^2 r^2$, so that outside corotation, there is no bound. Moreover, most orbits between CR and $-4/1$ resonance are stochastic (Contopoulos and Grosbøl 1989). For the ensemble of chaotic orbits inside CR, the corotation region acts as a boundary. This could be the origin of the sharp cut-off of the lens, if we consider that the chaotic orbits due to the destruction of the bar by the central mass concentration contribute in a large part to the lens component.

Test-Particle Simulations. This scenario has been checked through test-particle simulations, including a strong central mass concentration (from 0 to 2.4 % of the total mass), and where a bar is slowly introduced in the potential (Fig. 6). The bar potential is the analytic expression from Kalnajs (1976); we have investigated the behaviour of the stellar component and the gas component, the latter being treated as collisional clouds (Combes and Gerin 1985). In the stellar component, empty regions develop in the bar, between the remaining regular orbits regions, and the bar takes the form of "ansae" for a value of the central mass of a few % (see Fig. 6). In the gas component (and certainly in the young stars formed) collisions scatter particles back in the chaotic region, and the surface density profile looks more like a plateau. At the end of the bar, the gas accumulates in a wide ring, which is not a resonant ring, but corresponds to the region of the last remaining periodic orbits.

***N*-body Simulations.** Self-consistent simulations with stars and gas are shown in Fig. 7. In the 2D run, the central mass concentration is 1% and the total gas mass is 0.5% of the total mass. A very thick oval forms through dissolution of the stellar bar, with the "ansae"-shape. The gas bar is thinner and longer. This half-destroyed bar remains for a $2 \cdot 10^9$ yr time-scale. In the 3D run, the central mass is only 0.5% of the total mass, and the gas mass 1% (i.e. 10^9 $M_\odot$). We can see from the two projections that the bar at the beginning tends to form a box-shape, but the thickening perpendicular to the plane is stopped while the bar is weakening and forming the lens. The luminosity profile of such a disk presents the characteristic of a typical lens (cf. Fig. 7).

Figure 8 shows the amplitude of the bar (P_2, the ratio of the maximum $m = 2$ tangential force to the radial one) and corresponding pattern speed (Ω_2) as a function of time for several N-body runs. First, a purely stellar model without any central concentration can form a quasi-steady bar, with a constant amplitude and pattern speed for more than a Hubble time. In presence of a central concentration, up to 1% in mass, the bar forms in the same way, and reaches about the same amplitude, but it then weakens steadily, while slowing down in speed. The time-scale for decaying by a factor 2 is half a Hubble time. In presence of even a small amount of gas (of the order of 1% of the total mass), the bar can be destroyed much more rapidly, in one tenth of a Hubble time. The

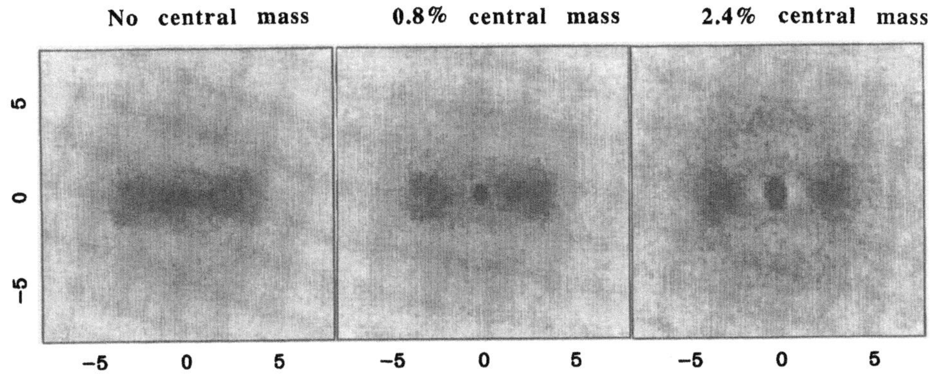

Fig. 6. Results of test particle simulations in a rotating frame of an imposed bar, for three different central mass concentrations: 0, 0.8 and 2.4 % of the total galaxy mass (including a dark halo). The regions in the middle of the bar are depopulated

gas has been driven towards the very center by the gravity torques, and further enhances the central mass.

5 An $m = 1$ Mode

When the gas central concentration is high enough to be gravitationally unstable, it can decouple from the main disk dynamics and form a nuclear bar, rotating with a higher speed than the main bar (Friedli and Martinet 1993). An even more complete decoupling may occur through an $m = 1$ mode (Junqueira and Combes 1996). We have studied the development of such a feature in recent self-consistent simulations with stars and fluid-gas (the beam-scheme hydrodynamical code was adopted, e.g. Sanders and Prendergast 1974). After a transient $m = 2$ spiral, the gas concentration becomes unstable and forms an $m = 1$ spiral mode (Fig. 9). In the density plots, the spiral is conspicuous until 3 kpc radius, and other more transient features, with much lower angular speeds, dominate at larger radii. The Fourier analysis of the total potential indicates a pattern speed of $\Omega_p = 400$ km s^{-1} kpc^{-1} for the central $m = 1$ spiral, which corresponds to a position of the OLR at 3 kpc (Fig. 10). This pattern speed is much higher than the speed at larger radii, there is no coupling between several modes, as was suggested for

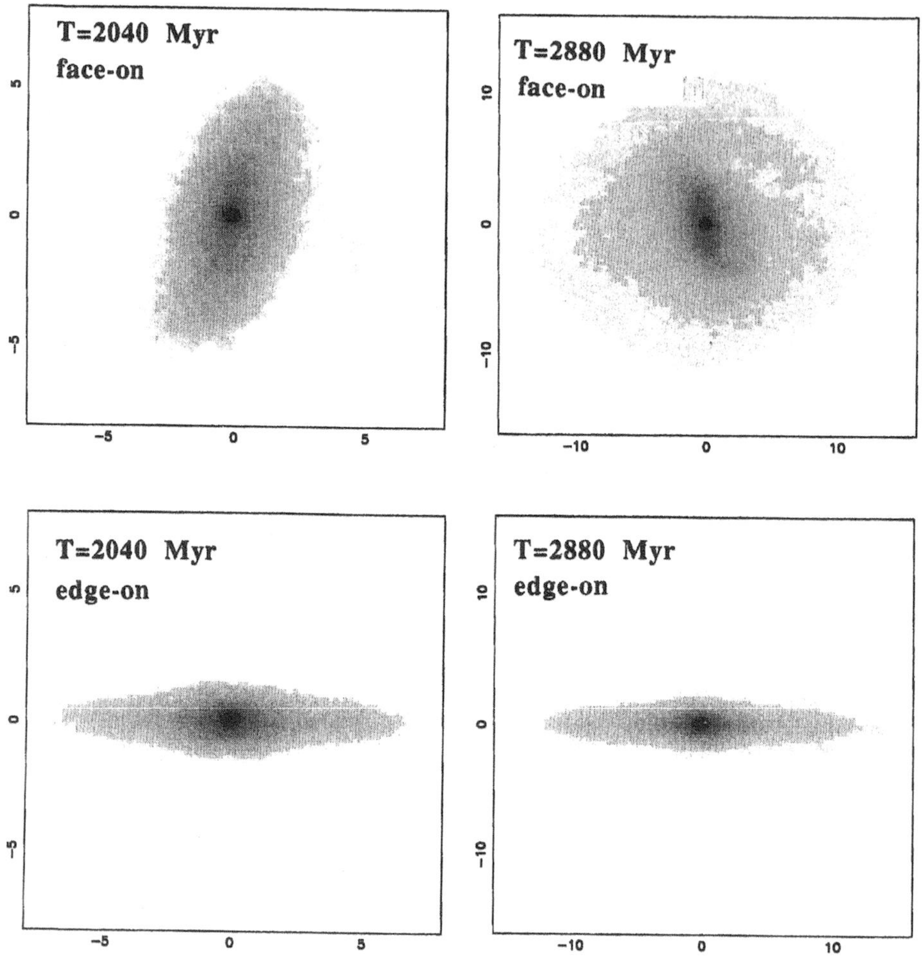

Fig. 7. Weakening of a bar by gas accumulation at the center. **a** 2D N-body simulations, at $T = 5600$ Myr; **b** 3D N-body simulations at $T = 2880$ Myr; **c** Logarithm of the surface density profile parallel and perpendicular to the bar

bars within bars: in the latter case, the corotation of the small bar was the inner resonance of the large one, and there could be a non-linear interaction between the two patterns (Tagger et al. 1987).

The power spectrum analysis for the gas density and the total potential reveals that the influence of the fast-rotating $m = 1$ pattern extends towards the edge (Fig. 11). This can be explained for the potential, since the $m = 1$ perturbation is a periodic move of the center of mass, and is instantaneously felt in the whole disk, but the signature in the gas density is more remarkable. The stellar component does not follow the $m = 1$ spiral with as much contrast, but the off-centering of its center of mass is also conspicuous. The two components

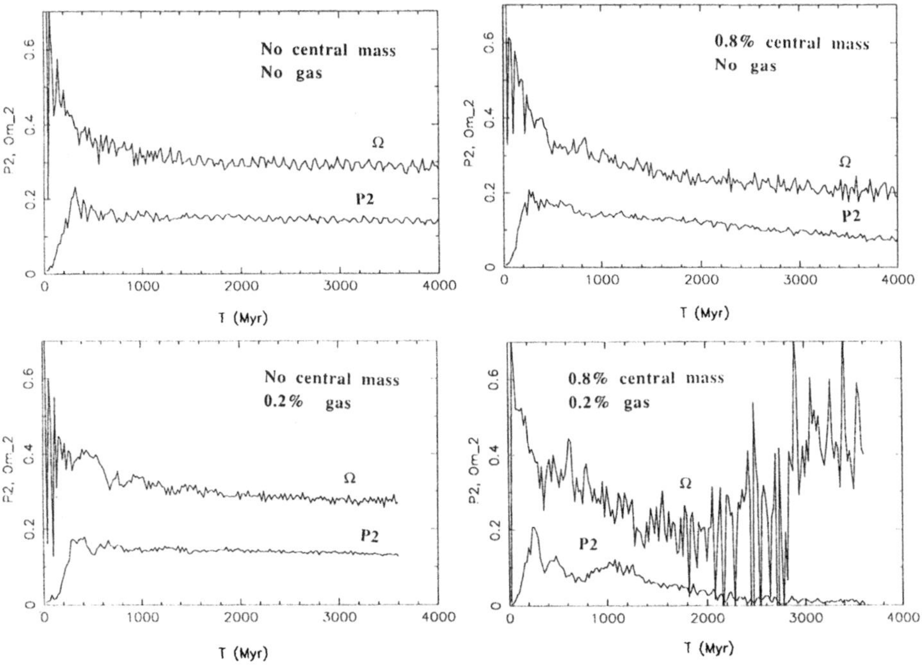

Fig. 8. Amplitude (P_2) and pattern speed (Ω_2) of the bar formed in N-body simulations with or without gas, and with/without a central mass concentration. Only a small amount of gas accumulated in the center by infall is sufficient to weaken the bar

oscillate in phase opposition, while the massive dark halo, represented by an analytical Plummer component, is nailed down to the center.

This $m = 1$ mode is quite similar to what has been studied numerically by Adams et al. (1989) in gaseous disks associated with young stellar objects. They discovered $m = 1$ gravitational instabilities, where the star did not lie at the center of mass of the system. Shu et al. (1990) presented an analytical description of a modal mechanism, the SLING amplification, or "Stimulation by the Long-range Interaction of Newtonian Gravity". This mechanism uses the corotation amplifier, where the birth of positive energy waves outside strengthens negative energy waves inside. In the SLING mechanism, a feedback cycle is provided by four waves outside corotation; long-trailing waves propagate from the OLR inward to the Q-barrier at CR where they refract in short trailing waves. These propagate outwards, cross the OLR, and reflect back at the outer edge of the disk. This is a critical point, the whole amplification mechanism depends on the reflecting character of this outer edge. Then a short leading wave propagates inwards from the edge, through OLR, towards CR, where it refracts again into a long leading wave, which is then reflected at OLR. An essential point here is also the ability of waves not to be absorbed at OLR, that is why this mechanism applies to gaseous disks only. Using a WKBJ analysis, they derived from the dispersion relation, and the condition of a constructive reflection, the required pattern speed for these modes. The instability arises uniquely from the displacement of the central star from the center of mass of the system, which creates an effective forcing potential. In our simulation the same effect is occurring, i.e. the center of mass of the gas and of the stars is displaced from the center of mass of the system and they are displaced in opposite positions, inducing the $m = 1$ wave formation. In other words, the modal mechanism proposed by Shu et al. (1990) may be in action in our experiment.

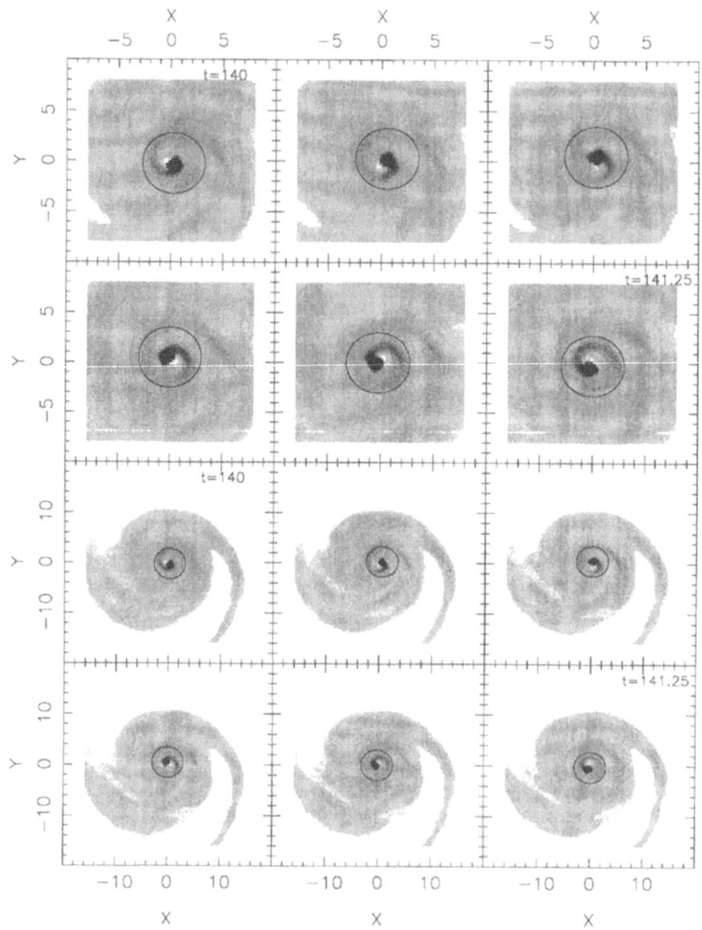

Fig. 9. Shape of the $m = 1$ mode in the gaseous component, rotating at $\Omega_p = 400$ km s^{-1} kpc^{-1}, the circle is at 3 kpc (from Junqueira and Combes 1996)

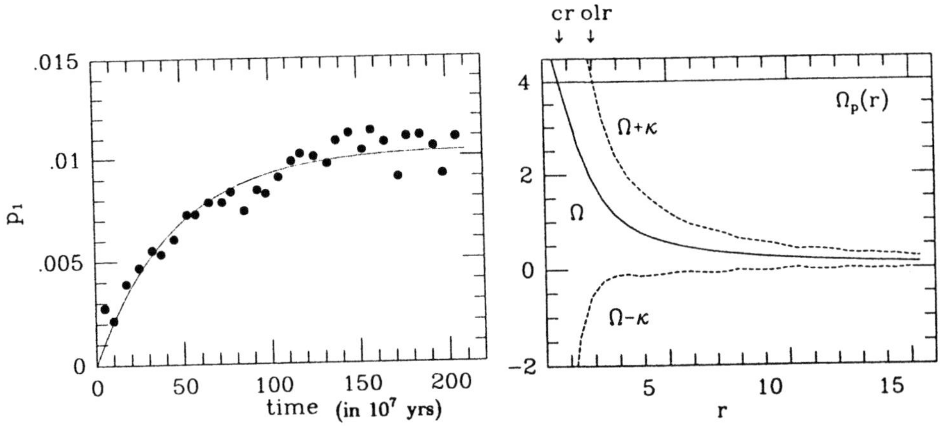

Fig. 10. Amplitude of the $m = 1$ perturbation as a function of time (P_1); pattern speed of the $m = 1$ component superposed on the frequency curves (from Junqueira and Combes 1996)

Observed $m = 1$ and Lop-sidedness. Many galaxies are observed with their nuclei off-centered with respect to neighbouring isophotes, and lopsided morphologies have been known for a long time (Baldwin et al. 1980; Wilson and Baldwin 1985). Asymmetric features are preferentially observed in the distribution of gas in late-type spiral galaxies. In several cases these features can be identified as one-armed spirals ($m = 1$ mode). More frequently, nuclei of galaxies are observed displaced with respect to the gravity center, as in M 33 and M 101 (de Vaucouleurs and Freeman 1970). Our own galaxy appears to experience such an off-centering, at least as far as the gas disk is concerned: about three quarters of the molecular mass of the nuclear disk is at positive longitude, and one quarter at negative longitude (e.g. Bally et al. 1988). Asymmetries are also seen at larger scale in atomic gas, while much of the neutral gas between a few 100 pc and 2 kpc from the nucleus lies in a tilted disk whose plane of symmetry is inclined by 20° with respect to the galactic plane (Burton and Liszt 1978; Liszt and Burton 1978).

Barred spiral galaxies can have their kinematical center displaced from the bar center (Christiansen and Jefferys 1976; Marcelin and Athanassoula 1982). The nucleus of M 31 reveals such an off-centering (Bacon et al. 1994) which has been interpreted in terms of an $m = 1$ perturbation (Tremaine 1995). Miller and Smith (1992) have studied through N-body simulations of disk galaxies a peculiar oscillatory motion of the nucleus with respect to the rest of the axisymmetric galaxy. They interpret the phenomenon as a local instability, or overstability, of the center but not in terms of a normal mode. Their models did not include any gas component. Weinberg (1994) shows that a stellar system can sustain

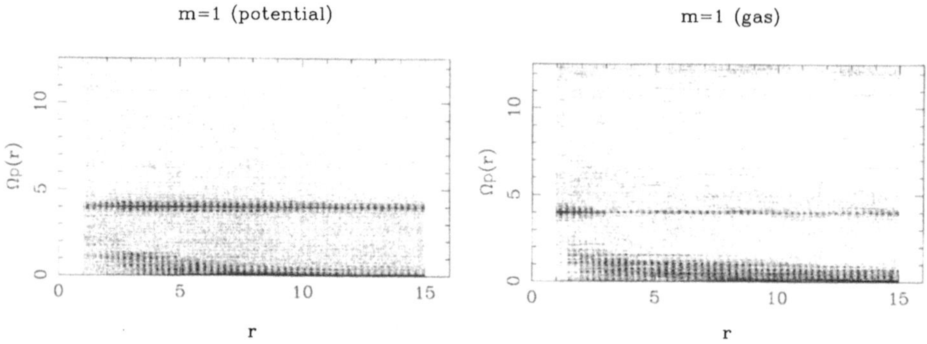

Fig. 11. Results of the power spectrum analysis for the gas density and for the potential. The identified perturbation extends up to the outer parts of the disk (from Junqueira and Combes 1996)

weakly damped $m = 1$ mode for hundreds of crossing times. A fly-by encounter could excite such a mode, and explain off-centering in most spiral galaxies, or the forcing of a halo could maintain a time-dependent potential in the galaxy (Louis and Gerhard 1988).

6 Conclusion

Spiral galaxies evolve on time-scales shorter than a Hubble time, due to the strong density waves, spirals and bars, that develop in their disk. These generate gravity torques and angular momentum transfer, since the wave potential is not exactly in phase with the density. Both components, stars and gas, develop important phase-shifts, but a purely stellar disk is heated by gravitational instabilities, and tends towards a quasi-steady hot state, often possessing a bar (where phase-shifts vanish). Due to the gas dissipation and cooling, spiral instabilities are continuously renewed in the disk, and secular evolution follows. N-body simulations of spiral and bar developments can be used to estimate quantitatively the gravitational "viscosity" time-scales. They are much shorter than the turbulent-viscosity time-scales, and decrease down to the dynamical time-scale at small radii.

The consequent gas inflow towards the center weakens the bar, in producing a central mass concentration. The latter modifies significantly the central potential, and in particular enhances the precession rate $\Omega - \kappa/2$ to a large value. The modification of the frequency curves has the effect of moving the resonances, and the structure of periodic orbits. In particular, two inner Lindblad resonances enter into play, and the existence of perpendicular x_2 orbits begin to weaken the bar in the center. If the gas flow continues, the increase of the central mass will produce large regions where stochastic orbits dominate, corresponding to the intermediate region of the bar. But the weakening of the bar has a regulating effect in decreasing the gravity torques and the subsequent gas flow. A partly destroyed bar could be the origin of lenses.

In parallel, the high gas mass concentration can trigger gravitational instabilities in the very center, which decouples from the outer parts, in developing a nuclear bar or spiral wave. These are rotating with high pattern speeds. These can stop for a few dynamical times the gas flow, since the gas density is more in phase with the potential. The gas coming from the outer parts can then settle in the disk and form stars, re-establishing the balance between bulge and disk mass. Scenarios of alternate growing of bulge and disk, through regulation mechanisms, could explain the dispersion of disk to bulge ratio along the Hubble sequence.

Acknowledgements. It is a great pleasure to thank Per Olof Lindblad, Aage Sandqvist and their colleagues for the organization of a meeting so full of warmth. I am very grateful to my collaborators, Susanne von Linden, Harald Lesch and Selma Junqueira, for authorization to present part of our common work before publication.

References

Adams, F. C., Ruden, S. P., Shu, F. H. (1989): ApJ **347**, 959
Athanassoula, E. (1983): in Proc. IAU Symp. 100 Internal Kinematics and Dynamics of Galaxies, ed. E. Athanassoula, Kluwer, Dordrecht, p. 243
Athanassoula, E. (1992): MNRAS **259**, 345
Athanassoula, E., Bosma, A., Crézé, M., Schwarz, M.P. (1982): A&A **107**, 101
Athanassoula, E., Sellwood, J.A. (1986): MNRAS **221**, 213
Bacon, R., Emsellem, E., Monnet, G., Nieto, J-L. (1994): A&A **281**, 691
Baldwin, J. E., Lynden-Bell, D., Sancisi, R. (1980): MNRAS **193**, 313
Bally, J., Stark, A.A., Wilson, R.W., Henkel, C. (1988), ApJ **324**, 223
Bertin, G., Romeo, A.B. (1988) A&A **195**, 105
Bosma, A. (1983): in Proc. IAU Symp. 100 Internal Kinematics and Dynamics of Galaxies, ed. E. Athanassoula, Kluwer, Dordrecht, p. 253
Burton, W.B., Liszt, H.S (1978): ApJ **225**, 815
Buta, R. (1986): ApJS **61**, 631
Buta, R. (1992): in Morphological and Physical Classification of Galaxies, eds. G. Longo et al., Kluwer, Dordrecht, p. 1
Buta, R., Crocker, D.A. (1991): AJ **102**, 1715
Buta, R., Crocker, D.A. (1993): AJ **105**, 1344
Christiansen, J.H., Jefferys, W.H. (1976): ApJ **205**, 52
Combes, F. (1988): in Galactic and Extragalactic Star Formation, Nato Advanced Studies Series, ed. R.E. Pudritz, M. Fich, Kluwer, Dordrecht, p. 475

Combes, F. (1994): in Mass-Transfer Induced Activity in Galaxies, ed. I. Shlosman, Cambridge University Press, Cambridge, p. 170
Combes, F., Gerin, M. (1985): A&A **150**, 327
Combes, F., Debbasch, F., Friedli, D., Pfenniger, D. (1990a): A&A **233**, 82
Combes, F., Dupraz, C., Gerin, M. (1990b): in Dynamics and Interactions of Galaxies, ed. R. Wielen, Springer, p. 205
Contopoulos G., Gottesman S.T., Hunter J.H., England M.N. (1989): ApJ **343**, 608
Contopoulos, G., Grosbøl, P. (1989): A&AR **1**, 261
Courteau, S., de Jong, R.S., Broeils, A.H. (1996): ApJ **457**, L73
de Vaucouleurs, G., Freeman, K.C. (1970): in Proc. IAU Symp. 38 The Spiral Structure of our Galaxy, eds. W. Becker, G. Contopoulos, Kluwer, Dordrecht, p. 356
Donner, K.J., Thomasson, M. (1994): A&A **290**, 785
Elmegreen, B. G. (1994): ApJ **427**, 384
Frank, J., King, A.R., Raine, D.J. (1985): Accretion Power in Astrophysics, Cambridge Univ. Press, Cambridge
Freeman, K.C. (1975): in Proc. IAU Symp. 69 Dynamics of Stellar Systems, ed. A. Hayli, Kluwer, Dordrecht, p. 367
Friedli, D., Benz, W. (1993): A&A **268**, 65
Friedli, D., Martinet, L. (1993): A&A **277**, 27
Goldreich, P., Tremaine, S. (1979): ApJ **233**, 857
Hasan, H., Norman, C. (1990): ApJ **361**, 69
Hasan, H., Pfenniger, D., Norman, C. (1993): ApJ **409**, 91
Huntley, J. M., Sanders, R. H., Roberts, W. W. (1978): ApJ **221**, 521
Jog, C.J., Solomon, P.M. (1984): ApJ **276**, 114
Junqueira, S., Combes, F. (1996): A&A in press
Kalnajs, A.J. (1971): ApJ **166**, 275
Kalnajs, A.J. (1976): ApJ **205**, 745
Kennicutt, R.C. (1989): ApJ **344**, 685
Kormendy, J. (1977): ApJ **214**, 359
Kormendy, J. (1979): ApJ **227**, 714
Kormendy, J. (1982) in Morphology and Dynamics of Galaxies, 12th Advanced Course of the SSAA, eds. L. Martinet, M. Mayor. Geneva Observatory, Geneva, p. 113
Kormendy, J. (1984): ApJ **286**, 116
Larson, R.B. (1988) in Galactic and Extragalactic Star Formation. Proc. of the NATO, Vol. **232**, 31
Lin, D.N.C., Pringle, J.E. (1987a): MNRAS **225**, 607
Lin, D.N.C., Pringle, J.E. (1987b): ApJ **320**, L87
Liszt, H.S., Burton, W.B. (1978): ApJ **226**, 790
Louis, P.D., Gerhard, O.E. (1988): MNRAS **233**, 337
Lynden-Bell, D., Kalnajs, A.J. (1972): MNRAS **157**, 1
Lynden-Bell, D., Ostriker, J.P. (1967): MNRAS **136**, 293
Marcelin, M., Athanassoula, E. (1982): A&A **105**, 76
Miller, R.H., Smith, B.F. (1992): ApJ **393**, 508
Nishida, M.T., Wakamatsu, K. (1996): AJ in press
Norman, C.A., May, A., van Albada, T.S. (1985): ApJ **296**, 20
Ostriker, J.P., Peebles, P.J.E. (1973): ApJ **186**, 467
Pfenniger, D., Norman, C.A. (1990): ApJ **363**, 391
Pfenniger, D., Combes, F., Martinet, L. (1994): A&A **285**, 79

Phinney, E. S. (1994): in Mass Transfer Induced Activity in Galaxies, ed. I. Shlosman, Cambridge Univ. Press, p. 1
Prendergast, K. (1983): in Proc. IAU Symp. 100 Internal Kinematics and Dynamics of Galaxies, ed. E. Athanassoula, Kluwer, Dordrecht, p. 215
Roberts, W. W. Huntley, J. M., van Albada, G. D. (1979): ApJ **233**, 67
Roberts, W.W., Hausman, M.A. (1984): ApJ **277**, 744
Sanders, R.H., Huntley, J.M. (1976): ApJ **209**, 53
Sanders, R.H., Prendergast, K. H. (1974): ApJ **188**, 489
Sanders, R.H., Tubbs, A. D. (1980): ApJ **235**, 803
Schwarz, M.P. (1981): ApJ **247**, 77
Schwarz, M.P. (1984): MNRAS **209**, 93
Sellwood, J.A., Carlberg, R.G. (1984): ApJ **282**, 61
Sellwood, J.A., Wilkinson, A. (1993): Rep. Prog. Phys., **56**, 173
Shakura, N.I., Sunyaev, R.A. (1973): A&A **24**, 337
Shaw, M.A., Combes, F., Axon, D.J., Wright, G.S. (1993): A&A **273**, 31
Shlosman, I., Noguchi, M. (1993): ApJ **414**, 474.
Shu, F. H., Tremaine, S., Adams, F. C., Ruden, S. P., (1990): ApJ **358**, 495
Tagger, M., Sygnet, J.F., Athanassoula, E., Pellat, R. (1987): ApJ **318**, L43
Toomre, A. (1964): ApJ **139**, 1217
Toomre, A. (1969): ApJ **158**, 899
Toomre, A. (1990): in Dynamics and Interactions of Galaxies, ed. R. Wielen, Springer, p. 292
Tremaine, S. (1995): AJ **110**, 628
van Albada, G. D. (1983): in Proc. IAU Symp. 100 Internal Kinematics and Dynamics of Galaxies, ed. E. Athanassoula, Kluwer, Dordrecht, p. 237
von Linden, S., Lesch, H., Combes, F. (1996): A&A preprint.
Weinberg, M.D. (1994): ApJ **421**, 481
Wilson, A. S., Baldwin, J. A. (1985): ApJ **289**, 124
Zhang, X. (1996): I ApJ **457**, 125, and II preprint

The Barred Galaxy NGC 1530

Peter Teuben, Michael Regan, and Stuart Vogel

Astronomy Department
University of Maryland
College Park, MD 20742, USA

Abstract. The barred galaxy NGC 1530 has been observed in a large number of wavebands, including spectral lines in H I, Hα and CO, giving a nearly complete velocity field across the whole nuclear and bar region. Strong unresolved (4″ or 700 pc) velocity discontinuities are observed across the dustlanes that we identify with large scale shocks observed in hydrodynamic simulations. Although hydrodynamical models are generally used to constrain the mass model, the observations could in turn be used to find a better ISM descriptor.

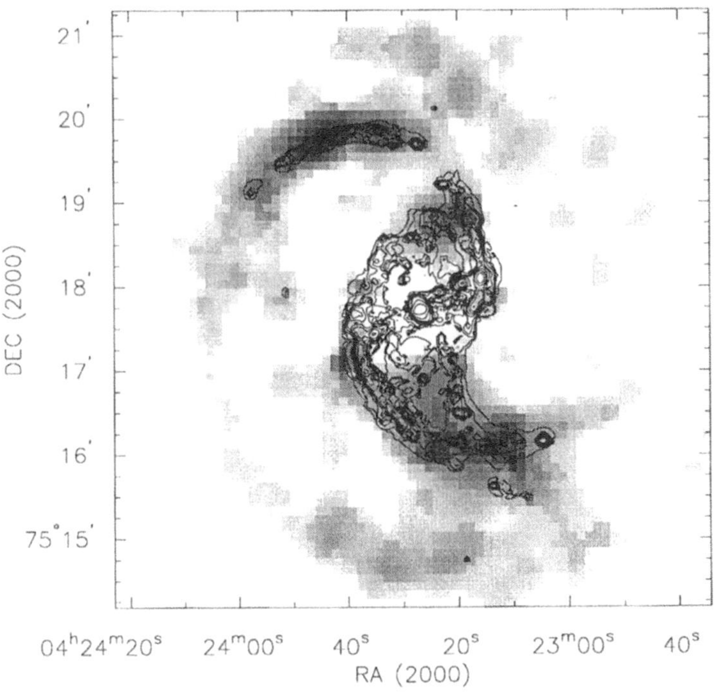

Fig. 1. H I (greyscale) overlayed with contours of Hα emission for NGC 1530

1 Introduction

Kinematic observations of barred galaxies have traditionally been used to constrain their mass models (e.g. Sanders and Tubbs 1980), but with the increased sophistication of hydrodynamic models (FS2, SPH, PPM; see Teuben (1996) for a recent review) modern high resolution spectral data could in turn be used to find a better ISM descriptor instead. In particular one may expect the velocity field across the leading shocks, frequently identified with the dust lanes, to depend sensitively on the simulation method or some of its parameters. We have been using the barred galaxy NGC 1530 to test some of these hypotheses, and in this paper we will summarize the current status of the observations and the beginning of our modeling efforts.

2 Observations

Despite its location ($\delta \approx +75°$) the prominent barred galaxy NGC 1530 has been largely ignored, possibly because of its somewhat large redshift (2450 $\mathrm{km\,s^{-1}}$). Its size makes up for the larger distance though. Assuming a Hubble constant of $H_0 = 75\ \mathrm{km\,s^{-1}}$, the bar, optical and H I disk extend out to a radius of 11 kpc, 30 kpc and 40 kpc resp., a galaxy in the NGC 1365 class.

With the discovery of the large IR flux by IRAS, NGC 1530 has enjoyed a large crowd of observers. Our observations are summarized in Table 1. More details can be found in Regan et al. 1995 (paper I) and Regan et al. 1996a (paper II).

Table 1. Summary of (spectral line) observations for NGC 1530

Instrument	WSRT	BIMA	Palomar 60"	INT	KP
Date(s)	4,9-sep-82	1993, 1994	sep-95	13-jan-91	jan-mar 94
Spectral Line	H I	CO(1-0)	Hα	UBVRI	JHK
Freq/Wavelength	21 cm	105 Ghz	650 nm	300-800nm	1-2μm
Pixel (arcsec)	6	0.5	1.89	0.54	0.95
Beam (arcsec)	13-25	3-5	4	1.3	2
Channels	31	256[(1)]	41[(2)]		
Channel Sep ($\mathrm{km\,s^{-1}}$)	16.6	4	12		
Resolution ($\mathrm{km\,s^{-1}}$)	33.2	8	24		

(1) one correlator section
(2) covering one free spectral range

Although the early WSRT observations showed a lack of H I in the central bar regions, a position velocity diagram along the minor axis already clearly showed that the gas was not streaming on circular orbits. It had the characteristic signature of elliptical orbits oriented along the bar. As can be seen from Fig. 1, the atomic gas is primarily concentrated in the dominant spiral arms and a ring surrounding the bar. Outside the optical radius H I is detected for an additional 180 degrees position angle along the spiral arms, although there are some obvious gaps, for example where the optical arms appear to disappear.

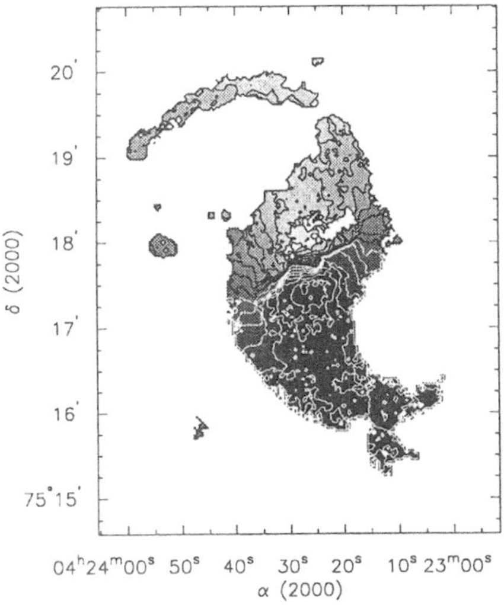

Fig. 2. Velocity field of the ionized gas

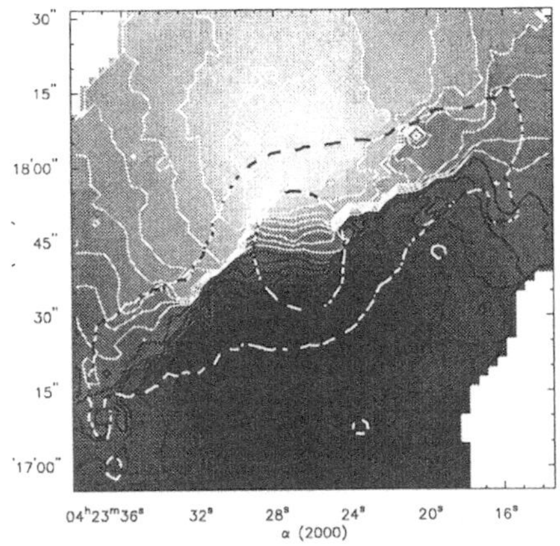

Fig. 3. Same, of the central bar region, with overlayed K-band contours outlining the bar and bulge

Our more recent sensitive Hα Fabry-Perot observations detected ionized gas in both the nuclear and bar region, as well as the spiral arms and the ring surrounding the bar. Diffuse Hα emission in most of the inter-bar region was also detected. Figures 2 and 3 shows the resulting velocity field, as determined from the ionized gas. The outer disk is rather regular, with a slight warp. Quite striking is the abrupt change in the orientation (a "twist") of the velocity field near the edge of the bulge, which can be identified with an inner Lindblad resonance, very similar to some other cases (e.g. NGC 1365: Teuben et al. 1986). A similar change can be seen near the edge of the bar, where corotation must reside. The velocity field is very crowded at the locations of the dustlanes, and at close inspection we find that the profiles are double, if not triple, in many of those locations, suggesting the shock is unresolved at our resolution of 4″ (700 pc). Together with an analysis of the rotation curve (derived from H I and Hα, see Fig. 4), we find a pattern speed of the bar of 20 ± 2 km s^{-1} kpc^{-1}, which places the outer Lindblad resonance at 18 ± 4 kpc (this would approximately coincide with the H I peaks near the major axis).

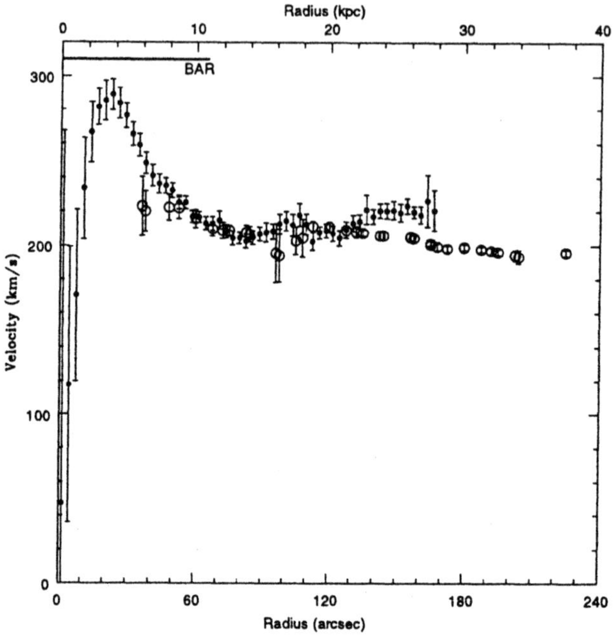

Fig. 4. H I (open symbols) and Hα (closed symbols) rotation curve for fixed inclination (45°) and position angle (188°, see also Table 2)

We have also observed NGC 1530 with the BIMA interferometer in the CO (1–0) line. A single 2′ field centered on the nucleus covers the whole bar. Most of the detected molecular gas has been found near the intersection of the dustlanes and a nuclear dust ring, although Downes et al. (1996) also reports to have detected gas associated with the dust lanes. Using radiative transfer models Regan et al. (1995) showed that the dust scale height in the nuclear ring is about half that of the stellar distribution.

3 Comparison with NGC 1365

It is remarkable how many similarities exist between NGC 1530 and NGC 1365 (see e.g. Jörsäter and van Moorsel 1995):

- both galaxies are large and have large and prominent bars (22 kpc for NGC 1530).
- both have similar distributions of H I (central hole, concentrated in the arms, extent), as well as Hα, compared with the optical (R_{25}). NGC 1530 may be a bit more extended in H I .

- both galaxies exhibit a slight warp in their outer parts, which complicates the interpretation of the rotation curve in terms of a mass model.
- both suffer from a low kinematically derived inclination, compared with the one derived from optical isophotes (but note how these can be biased by the dominant spiral arms).

yet some interesting differences should be noted:

- the rotation curve does not seem to drop, in that sense similar to NGC 1300 and others (Kristen, this volume).
- NGC 1530 has simple bi-symmetric arms, NGC 1365 has multiple, minor, arms, which appear to be reproducable in hydrodynamical simulations (Lindblad et al. 1996).
- NGC 1530 is very isolated, NGC 1365 is an (outlying) member of the Fornax cluster.
- NGC 1530 has a very low surface brightness disk. The "disk" is not well approximated by an exponential disk at all. The lens-like feature inside the ring dominates the light inside the bar region, whereas the very strong spiral arms dominate the light outside. This has also hampered determination of the inclination.

4 Models

Given our well sampled velocity field and the large deviations from circular rotation, the observations can be directly compared to the gas flow in model barred galaxies. We have computed models (Piner et al. 1995) and compared a number of these models with the currently available data. No extensive survey has been done to date, neither have we compared different model techniques in detail with the observations (Teuben et al., in prep). A representative model is shown in Fig. 5, where the gas density and radial velocity field are shown for the projection parameters derived for NGC 1530. It is remarkable how well the major features are reproduced in the models. It should be noted that for a given mass model the allowed range in pattern speed is fairly small, in order to reproduce the major change in features near the inner Lindblad resonance near the edge of the bulge. For example, in the lower right panel the bar is spinning a little fast, such that corotation is exactly at the edge of the bar, and clearly the ILR has disappeared, allowing the gas to flow into the nuclear regions almost unhindered, and the model velocity field is very different from the observed one (cf. Fig. 3). On the other extreme the upper left panel shows a slightly slower rotating bar: both the rather large radius of the ILR as well as the smaller twist of the iso-velocity contours are indicative that this pattern speed is too low.

Acknowledgements. We wish to thank our collaborators Jim Stone and Glenn Piner for the modeling, and Thijs van der Hulst for taking the UBVRI data. This work was supported in part by NSF grant AST 9314847. PT thanks the Nobel Foundation for making it possible to attend this symposium.

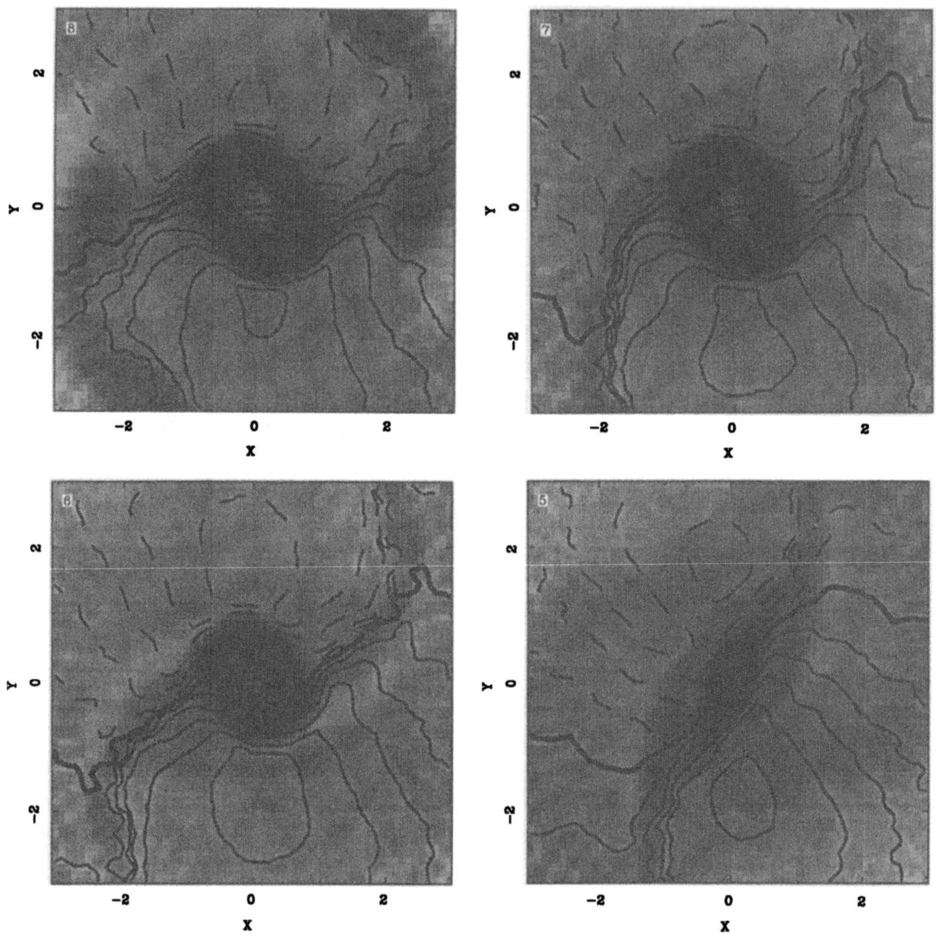

Fig. 5. Series of models, with varying pattern speed of the bar, at the observed orientation and resolution of NGC 1530. The corotation radius (in arbitrary kpc units) is labeled in the top left corner of each panel, in kpc, where the bar ends at 5 kpc. Velocity contours are in from -250 to 250 km s^{-1}, in steps of 50 km s^{-1}

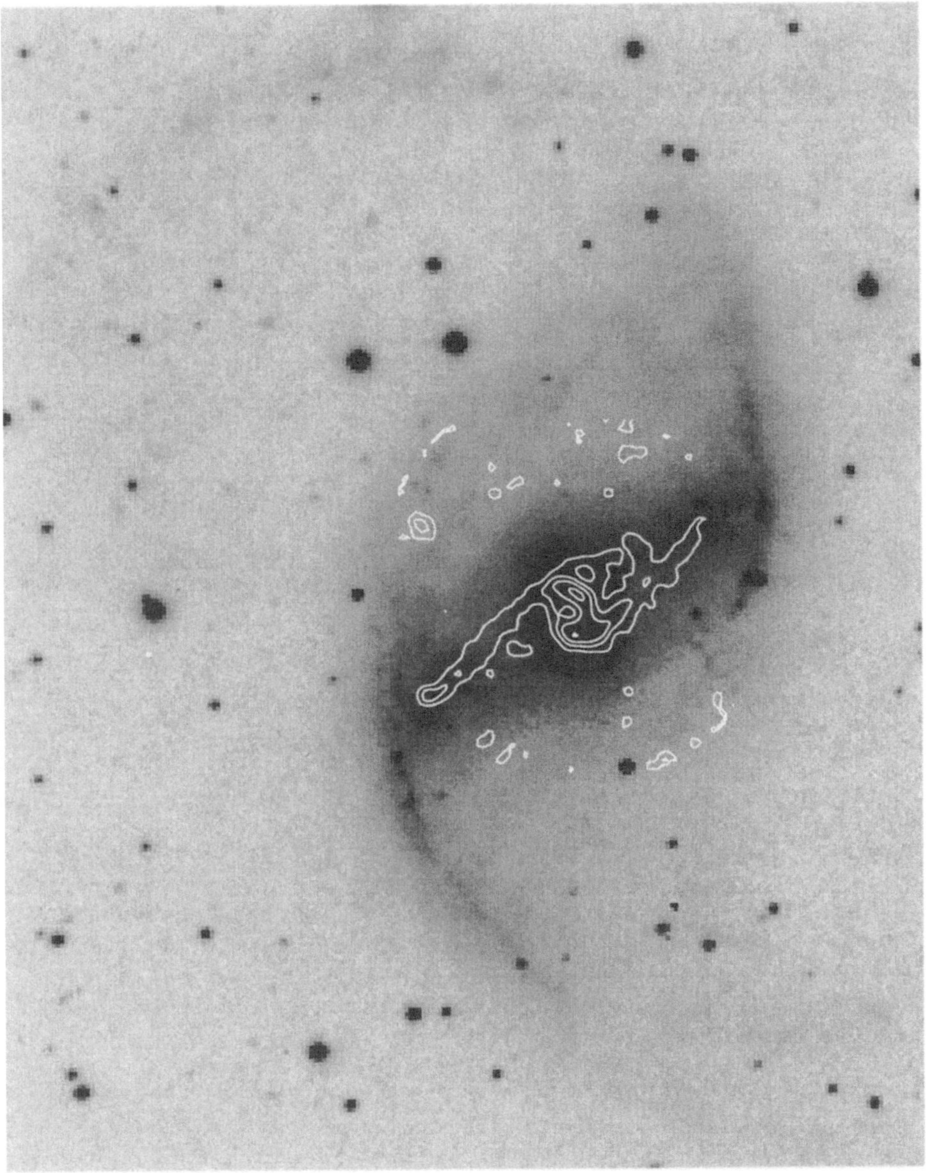

Fig. 6. I band image of NGC 1530 with overlayed I−K color contours that best outline the dust lanes

Table 2. Basic and Derived parameters for NGC 1530.

K-band peak:		
R.A. (J2000)	$4^h23^m26^s\pm7''$	1
Decl. (J2000)	$75°17'43''\pm 8''$	1
Optical size (D_{25})	4.9′	2
Systemic velocity (LSR)	2457±3 km s^{-1}	3
Adopted distance (H_0=75)	35 Mpc	
Linear scale	165 pc/″	
Position Angle	5°	4
Inclination	57°	4
Bar length	100″	1
Bar position angle	121°	1
Systemic Velocity	2447.5±2 km s^{-1}	
Position Angle (receding)	188±1°	
Inclination	45±5°	
Dynamical Center:		
R.A. (J2000)	$4^h23^m26^s\pm8''$	
Decl. (J2000)	$75°17'45''\pm8''$	
Pattern Speed	20±2 km s^{-1}	
Outer Lindblad Resonance	18± kpc	

1. Regan et al. 1995
2. Young et al. 1989
3. Stavely-Smith and Davies 1987
4. de Vaucouleurs et al (RC3) 1993

References

Downes, D., Reynaud, D., Solomon, P.M., and Radford, S.J.E. (1996): preprint

Jörsäter, S., van Moorsel, G.A. (1995): AJ **110**, 2037.

Lindblad, P.A.B., Lindblad, P.O., Athanassoula, E. (1996): in Proc. IAU Coll. 157 Barred Galaxies, eds R. Buta, D.A. Crocker, B.G. Elmegreen, PASPC **91**, p. 413

Piner, B.G., Stone, J., Teuben, P.J. (1995): ApJ **449**, 508

Regan, M.W., Teuben, P.J., Vogel, S.N., van der Hulst (1996a): submitted (paper II)

Regan, M.W., Vogel, S.N. (1996): in preparation

Regan, M.W., Vogel, S.N., Teuben, P.J. (1995): ApJ **449**, 576 (paper I)

Sanders, R.H., Tubbs, A.D. (1980): ApJ **235**, 803

Teuben, P.J., Sanders, R.H., Atherton, P.D. van Albada, G.D. (1986): MNRAS **223**, 1

Teuben, P.J. (1996): in Proc. IAU Coll. 157 Barred Galaxies, eds R. Buta, D.A. Crocker, B.G. Elmegreen, PASPC **91**, p. 299

A Circumnuclear Molecular Torus in NGC 1365

Aa. Sandqvist

Stockholm Observatory, S-133 36 Saltsjöbaden, Sweden

Abstract. The central region of NGC 1365 has been mapped in the $J = 3 - 2$ CO emission line with the 15-m SEST, which has a HPBW of 15″ at the frequency of this transition. The observing grid has a 5″-spacing in the inner and a 10″-spacing in the outer region. A circumnuclear molecular torus with a radius of about 5″ is the dominant feature. Molecular emission is also seen coming from various dust streamers in the bar of the galaxy. The velocity field of the molecular region agrees well with predictions of models of gas streaming in the bar and nuclear region.

1 Introduction

NGC 1365 is a prominent barred spiral galaxy in the Fornax cluster with a heliocentric velocity of +1630 km s^{-1}. The galaxy displays a wide range of phenomena indicating activity – including a Seyfert 1.5 type nucleus with strong, broad and narrow Hα lines and ejection of hot gas from the nucleus in the plane of the galaxy (Veron et al. 1981; Jörsäter et al. 1984; Jörsäter and Lindblad 1989). The nucleus is a strong infrared source as seen by IRAS (Ghosh et al. 1993), and observations with the Einsetein satellite have shown the nuclear region to be a source of soft X-rays (Maccacaro et al. 1982).

Over the past decade we have been studying the central region of NGC 1365 in great detail using the VLA, the NRAO 12-m millimeter wave telescope and the 15-m SEST (Sandqvist et al. 1982, 1988, 1995). The VLA radio continuum observations at 2, 6 and 20 cm, with resolutions of 0.″25 × 0.″10 and 2″ × 1″, have revealed a 5″ radio jet with a steep spectral index(−0.9) emanating from the Seyfert nucleus in a southeastern direction along the apparent minor axis of the galaxy. The radio jet is aligned with the axis of a conical shell of hot ioinized gas in accelerated outflow seen in the [O III] emission line (P.O. Lindblad, these proceedings). A circumnuclear radio ring, containing a number of non-thermal radio sources, has angular dimensions of 8″×20″ with a major axis position angle of 30°. Three of the radio sources are of similar character having a rather flat non-thermal spectrum (0 to −0.4) and contain components still unresolved with a linear resolution of 10 pc. They might be related to the very bright compact H II regions (also only partially resolved at 10-pc resolution) seen in our observations with the HST (P.O. Lindblad, these proceedings). The radio sources lie near the edge of a rapidly rotating nuclear disk of ionized matter with a radius of 7″ (Lindblad 1978), but there is no one-to-one correspondence of the radio sources with the H II hot spots in the same region. We have suggested that the radio components and optical hot spots are manifestations of starburst activity. The

Seyfert nucleus has been detected as a weak radio source with a spectral index of -0.9. Extended radio emission has also been delected along the bar near the prominent dust lanes, where large-scale galactic shocks have been shown to be present.

Molecular gas has previously been mapped in the bar and central region of the Seyfert galaxy NGC 1365 with the 15-m SEST using the $J = 1-0$ and $2-1$ CO lines with resolutions of 44″ and 25″, respectively (Sandqvist et al. 1995). The CO molecular gas is strongly concentrated to the nucleus, where the CO integrated line intensity has a maximum, and the global CO distribution falls off roughly exponentially with the distance from the centre of the galaxy. There is some CO alignmnent with the dust lanes in the bar and some weak emission has been detected in the western spiral arm near the end of the bar at the position of a major H I concentration observed by Jörsäter and van Moorsel (1995).

The central CO luminosity corresponds to a molecular hydrogen mass of $6.3 \times 10^9\,M_\odot$ in the central region of NGC 1365 within a projected radius of 2.2 kpc. At an assumed distance of 20 Mpc, 1″ corresponds to 100 pc. The global molecular hydrogen gas mass is $20 \times 10^9\,M_\odot$, which is similar to the total amount of neutral atomic hydrogen, $15 \times 10^9\,M_\odot$, found by Jörsäter and van Moorsel (1995) using the VLA. The distribution of the H I is, however, radically different from that of the CO. Whereas the molecular mass is concentrated to the nucleus and bar region, the H I is predominantly located in the spiral arm regions. In particular, the H I distribution shows a hole in the central region which coincidies with the CO emission. This indicates that the gas is predominantly molecular in the centre and the inner bar regions.

A few months ago, we carried out new observations of the central region of NGC 1365, predominantly in the $J = 3-2$ CO line, but also in other molecular line transitions in the millimetre wave region. Since the $J = 3-2$ CO line is excited in regions of higher excitation and density than the $J = 2-1$ and $1-0$ lines, it is a good probe of the molecular gas as it passes through the shocks in the inner bar. The higher resolution offered by the $J = 3-2$ line observations also enables a better comparison with predictions of molecular gas kinematic transport inward along the bar to the central star burst region, expected from theoretical models of gas flow in bars of galaxies (see e.g. P.A.B. Lindblad, these proceedings).

2 SEST Observations

Observations of NGC 1365 were carried out in August 1995 using the 15-m Swedish ESO-Submillimetre Telescope (SEST) on La Silla in Chile. During the first four nights of the observing run, the atmospheric conditions were excellent, permitting submillimetre observations of the $J = 3-2$ CO line. The weather then deteriorated somewhat, forcing a change to the millimetre wavelength region. The telescope properties at the observed frequencies are presented in Table 1.

A dual beam switch mode, with a beam separation of 11′.6, was used placing the source alternatively in the two beams to eliminate asymmetries in the

Table 1. SEST Parameters

Frequency (GHz)	Molecule	Transition	HPBW ($''$)	η_{mb}	$T_{receiver}$ (K)	Channel Resolution (km s^{-1})
88.6	HCN	$(1-0)$	57	0.92	100	22
89.2	HCO^+	$(1-0)$	57	0.92	100	22
110	^{13}CO	$(1-0)$	48	0.82	110	19
141	H_2CO	$(2_{1,2}-1_{1,1})$	36	0.68	130	15
147	CS	$(3-2)$	34	0.66	135	14
346	^{12}CO	$(3-2)$	15	0.26	425	6

signal paths. Three different SIS receivers were used in conjunction with two low-resolution acousto-optical spectrometers, each with a total bandwidth of about 1 GHz (channel resolution of 1.4 MHz). The average receiver temperatures ($T_{receiver}$) are given in Table 1, as are the channel velocity resolutions after smoothing operations. Only linear baseline subtractions were performed on the profiles. All profile temperatures have been converted to main beam brightness temperatures (T_{mb}) by dividing the antenna temperatures (T_A^*) by the respective main beam efficiencies (η_{mb}). The velocities are heliocentric radial velocities.

Four new molecular species have been detected in NGC 1365, namely HCN, HCO^+, H_2CO and CS. These four profiles are presented in Fig. 1 together with that of $J=2-1$ ^{13}CO. All five species are tracers of high density gas. The integrated main beam brightness line intensities, $\int T_{mb}dV$, are HCN: 4.91, HCO^+: 4.25, ^{13}CO: 7.61, H_2CO: 0.93 and CS: 2.21 K km s^{-1}. The T_{mb} rms noise level of the profiles are of the order of 0.002 K. Submillimetre $J=3-2$ CO line profiles observed towards three positions in the centre of NGC 1365 are also presented in Fig. 1; the equatorial offsets from the optical nucleus for these observations are given in square brackets. The total integrated CO line intensities, $\int T_{mb}dV$, are 276 and 277 K km s^{-1} at the southwest and northeast maxima, respectively; at the centre, this value is 236 K km s^{-1}.

The mapping in the $J=3-2$ CO line was done over an approximately $120'' \times 60''$ region, centered on the optical nucleus ($\alpha(1950.0) = 3^h31^m41^s.80, \delta(1950.0) = -36°18'26''.6$) and covering the bar region. A total number of 133 positions were observed. A grid spacing of $10''$ was used for the outer parts of this region. For the inner part (approximately $70'' \times 40''$) a grid spacing of $5''$ was used, i.e. a sampling rate of three points per HPBW. Great care was applied to frequent pointing checks. This included using the central profile of NGC 1365 as a pointing check, since an error of a few arcseconds would be immediately noticeable in the relative amplitudes of the two main components in the central profile. Furthermore, observations of the inner part were made only during night time, after midnight and before sunrise, which is the time of maximum atmospheric stability. In addition, observations were only made at elevations between 50° and 80° in order to minimize beam distortion and maximize aperture efficiency. A profile map including all the $J=3-2$ CO observations is presented in Fig. 2.

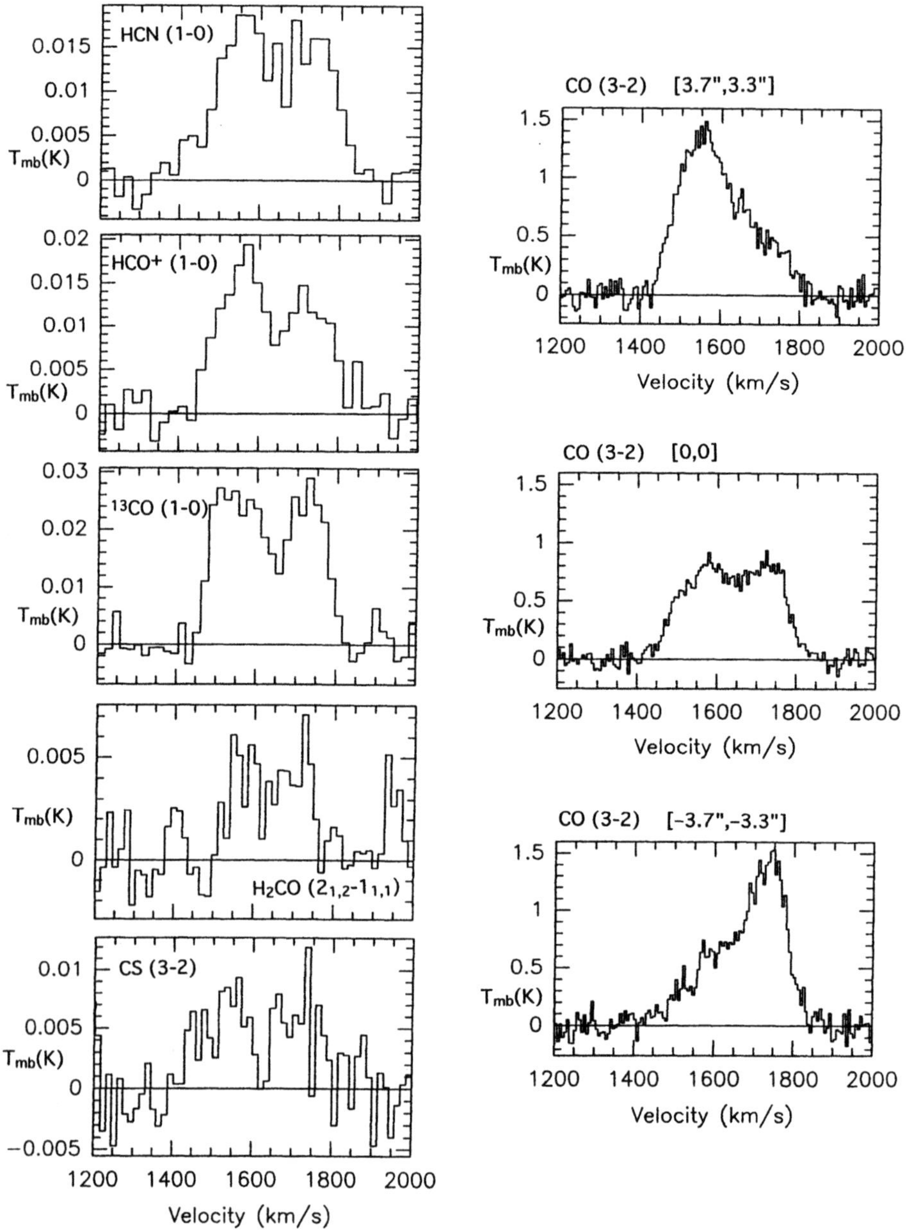

Fig. 1. Left: Millimetre molecular line profiles observed towards the central position in NGC 1365. **Right**: Submillimetre $J = 3 - 2$ ^{12}CO molecular line profiles observed towards three positions in the centre of NGC 1365

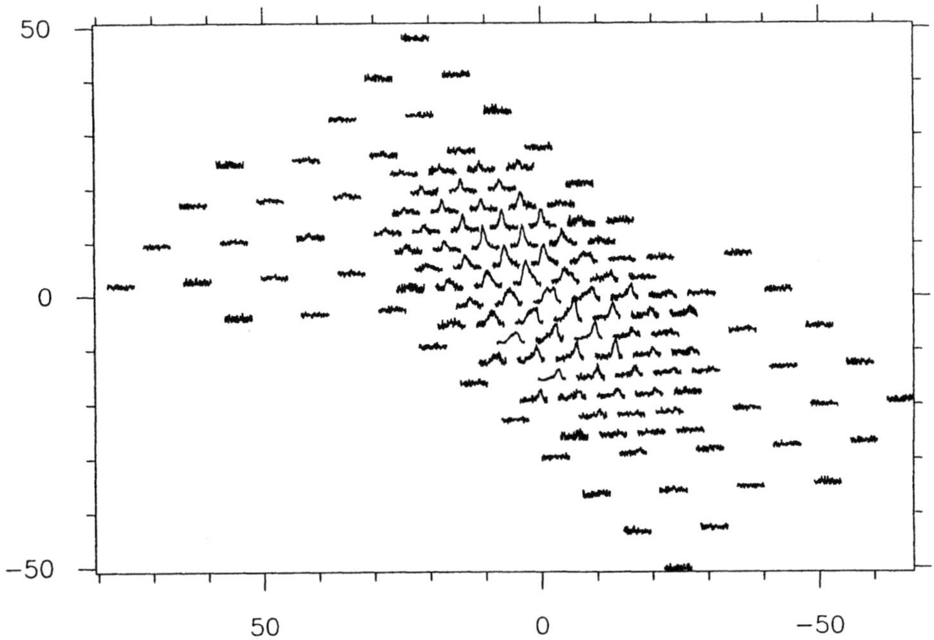

Fig. 2. $J = 3 - 2$ CO profile map of NGC 1365. The equatorial offsets are in units of arcseconds and are measured from the optical nucleus

3 The CO Distribution and Velocity Structure

The interrelation between the emission regions of the CO molecular line, the dust lanes and the H II hot spot regions in the central region of NGC 1365 is presented in Fig. 3. The optical image is a $B-$Gunn$z$ colour index map, which emphasizes the dust lanes as white areas and regions with hot stars and H II regions as dark areas. The CO is presented here by the $J = 3 - 2$ total integrated line intensity $\int T_{\rm mb} dV$-distribution.

There is clear correspondence between the extended CO emission, as represented by the outermost contours, and the dust lanes at the preceding edges of the bar. Even the curved dust feature near the western end of the bar has a corresponding distinctly curved CO component, which can be seen in the lowest contour level. Other dust streamers also contain observable CO.

The most interesting phenomenon, however, is the doubly-peaked CO structure seen near the optical nucleus, with a local minimum right at the nucleus. This structure and its alignment along the major axis of the galaxy is suggestive of a circumnuclear molecular torus with a radius of 5″ (500 pc). From the torus,

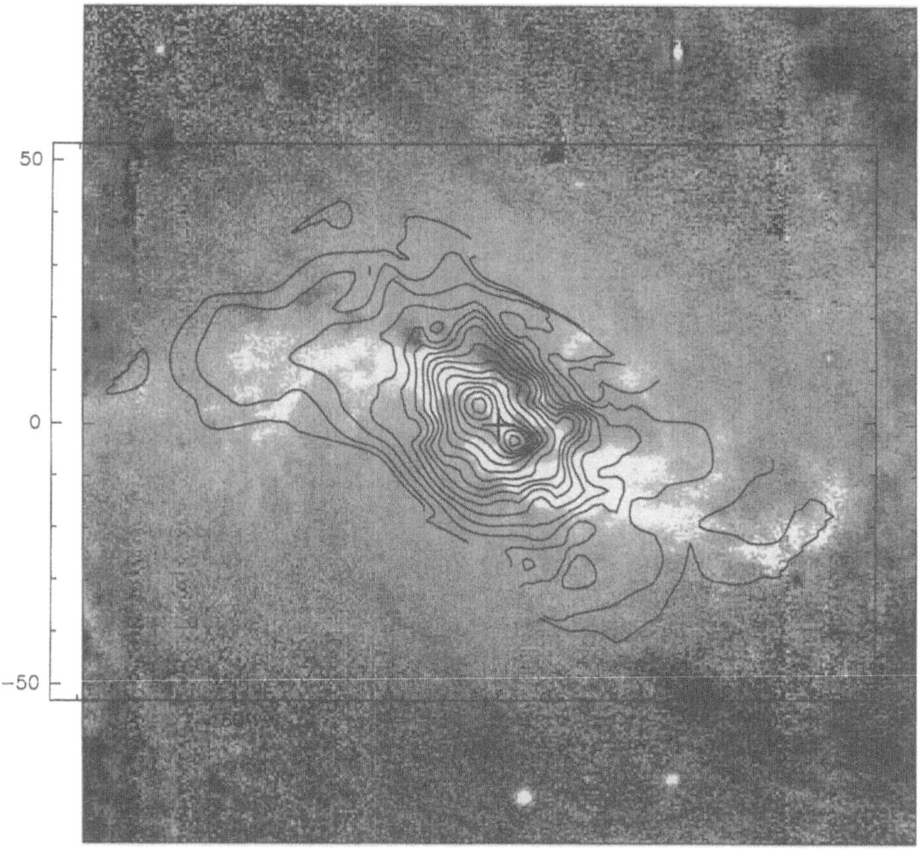

Fig. 3. The $\int T_{\rm mb} dV$-distribution of the $J = 3-2$ CO emission line (the two lowest contour values are 7.7 and 19.2 K km s^{-1}, thereafter the contour interval is 19.2 K km s^{-1}) superimposed on a $B -$ Gunn$z$ colour index map showing dust lanes (light areas) as well as regions of hot stars and H II-regions (dark areas). The equatorial offsets are in units of arcseconds and are measured from the optical nucleus which is marked by a cross

there are CO extensions leading out into the two dominant eastern and western dust lanes.

The overall central CO velocity field and a position-velocity map along the major axis of the galaxy are displayed in Fig. 4. The velocities in the isovelocity diagram have been calculated by taking moments of the line profiles. The velocity gradient across the molecular torus has its maximum value along the major axis and its linear character in this region may reflect rotation of the torus. A change of peak-temperature velocity of 190 km s^{-1} is found over the 10″ between the two torus maxima, which corresponds to 1 kpc on a linear scale. In regions about

25″ from the nucleus (e.g. near equatorial offsets of (−20″, +5″)) the isovelocity contours are very close together. This may be the velocity signature of orbit crowding of the molecular gas as it crosses the Inner Lindblad Resonance (IRL). The radius of the IRL in NGC 1365 has been determined to be 27″ by P.A.B. Lindblad (these proceedings).

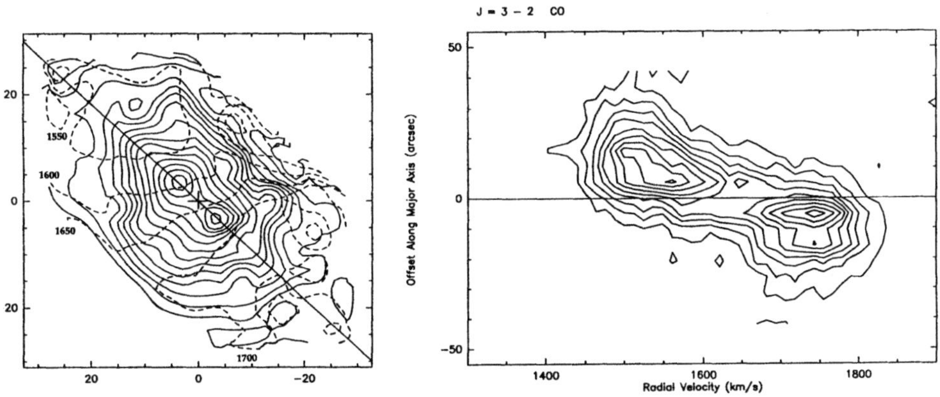

Fig. 4. Left: The central $\int T_{\rm mb} dV$ $J = 3-2$ CO line contours from Fig. 3 (solid lines) and CO isovelocity contours (dashed lines) with values indicated in km s^{-1}. The diagonal straight line indicates the major axis and the cross the optical nucleus of NGC 1365. **Right**: $J = 3-2 T_{\rm mb}$ CO position-velocity map along the major axis of NGC 1365. The lowest contour value and the contour interval are 0.15 K

4 The Molecular Torus and Other Nuclear Structure

A convenient presentation of the molecular torus and its interrelation with other (circum)nuclear structures in NGC 1365 is presented in Fig. 5. At the core of the galaxy there is the weak, steeply nonthermal radio source from which emanates the 5″-long radio jet, also steeply nonthermal. The jet is aligned along the axis of the conical shell of hot ionized [O III] gas, and both are aligned along the minor axis of the galaxy, out of the plane of the galaxy. The next structure is the molecular torus with a radius of 5″ (500 pc) which is aligned along the major axis, in the plane of the galaxy. On the outer edge of the torus, there is the radio ring with the unresolved radio sources, and outside this component the optical hot spots at a radius of about 7″ (700 pc). In Fig. 5, these hot spots are identified by their [O III] emission, north and west of the radio ring components. Finally at a radius of 25″ (2.5 kpc) there is the IRL molecular gas (identified by its orbit-crowding velocity effects), and beyond this the gas and dust streamers reaching out through the bar of the galaxy.

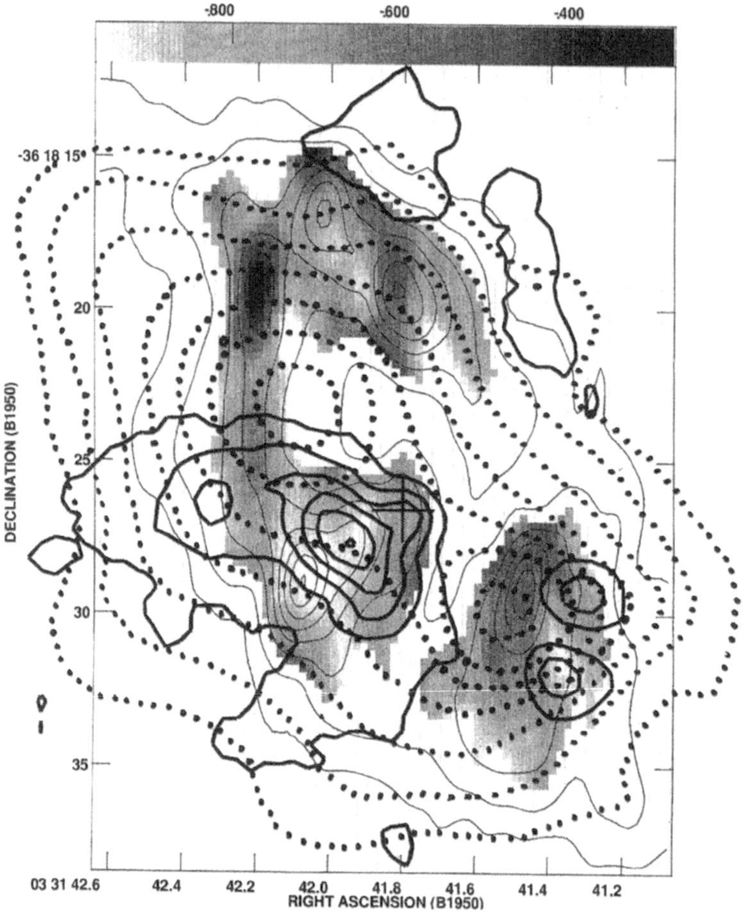

Fig. 5. The CO molecular torus (dots) superimposed on a 20-cm radio continuum map (thin solid contours) and [O III] emission (thick solid contours). The gray scale indicates the spectral index of the radio ring emission (white: −1.0, black: −0.3). The cross marks position of the optical nucleus

References

Ghosh, S.K., Verma, R., Rengarajan, T., Das, B., Saraiya, H. (1993): ApJS **86**, 401

Jörsäter, S., Lindblad, P.O. (1989): in ESO Conf. Workshop Proc. 32 Extranuclear Activity in Galaxies, eds. E.J.A. Meurs, R.A.E. Fosbury, p. 39

Jörsäter, S., Lindblad, P.O., Boksenberg, A. (1985): A&A **140**, 288

Jörsäter, S., van Moorsel, G. (1995): AJ **110**, 2037

Maccacaro, T., Perola, G.C., Elvis M. (1982): ApJ **257**, 47

Sandqvist, Aa., Elfhag, T., Jörsäter, S. (1988): A&A **201**, 223

Sandqvist, Aa., Jörsäter, S., Lindblad, P.O. (1982): A&A **110**, 336

Sandqvist, Aa., Jörsäter, S., Lindblad, P. O. (1995): A&A **295**, 585

Dynamics of Inner Galactic Disks: The Striking Case of M 100

Isaac Shlosman

Department of Physics and Astronomy, University of Kentucky
Lexington, KY 40506-0055, U.S.A.

Abstract. We investigate gas dynamics in the presence of a *double* inner Lindblad resonance within a barred disk galaxy. Using an example of a prominent spiral, M 100, we reproduce the basic central morphology, including four dominant regions of star formation corresponding to the compression maxima in the gas. These active star forming sites delineate an inner boundary (so-called nuclear ring) of a rather broad oval detected in the near-infrared. We find that inclusion of self-gravitational effects in the gas is necessary in order to understand its behavior in the vicinity of the resonances and its subsequent evolution. The self-gravity of the gas is also crucial to estimate the effect of a massive nuclear ring on periodic orbits in the stellar bar.

1 Nuclear Starburst Galaxies

The role of active galaxies within the framework of galactic evolution is far from clear. In particular, it is still unknown if central activity encompasses a small percentage of galaxies or if it is a normal evolutionary stage. It also remains undetermined whether two notable types of such activity, nuclear starbursts and Seyfert nuclei, are 'genetically' linked.

A subgroup of disk galaxies shows intense star forming activity within the central few hundred parsecs, in the so-called nuclear rings. In the visual and ultraviolet wavelength ranges and at a high spatial resolution, these rings frequently appear to be patchy and incomplete, and/or consist of a pair of tightly wound spiral arms (Buta and Crocker 1993). In the near-infrared (NIR), they are regular and weakly elliptical (Knapen et al. 1995a; Shaw et al. 1995). Molecular gas distribution based on CO emission reveals a complex morphology at and interior to the rings (review by Kenney 1996). The origin of these rings is related to the inner Lindblad resonance (ILR), i.e. the resonance between the planar stellar orbits and the perturbing force of a stellar bar or of an oval distortion. Although self-gravity in the gas was ignored in earlier numerical simulations of disk galaxies, these simulations clearly showed that gas is focused into nearly circular orbits interior to the *outer* ILR (OILR; e.g. Schwarz 1984; Combes and Gerin 1985). As the gas accumulates in this region, no further evolution occurs and it was suggested that the gas is converted into stars with a high efficiency (Elmegreen 1994). More sophisticated 2D modeling involving self-gravity in the gas supported this picture (Shaw et al. 1993), indicating at the same time that increasing surface density in the gas may lead to its fragmentation and destruction of the gaseous ring (Wada and Habe 1992). Effects of fragmentation are lessened if energy deposition by massive stars is taken into account by inducing turbulence in the gas (Heller and Shlosman 1994).

Recent observations of molecular gas distribution in the centers of a number of barred galaxies provide some indirect evidence (based on stellar rotation curves) that a double ILR may in fact reside there (Kenney et al. 1992; Knapen et al. 1995a). Although gas flows in barred galaxies have been modeled for about two decades now, gas behavior in the vicinity of a double ILR is not well understood. Due to the low resolution of numerical schemes and to a common belief that the *inner* ILR (IILR) is probably located too close to the center to be observationally resolved, most efforts have been devoted to the study of gas dynamics between the OILR and outer Lindblad resonance (OLR; e.g. Athanassoula 1992). For these reasons, numerical schemes usually fail to catch the nuclear ring phenomenon.

Here we investigate a self-consistent gas evolution in the central resonance region of a moderately barred galaxy. Effects of star formation are incorporated at some level. For convenience, we choose the M 100-like total mass distribution which was claimed to possess a double ILR (Arsenault et al. 1988; Knapen et al. 1995a). Furthermore, we highlight the effect that massive nuclear rings have on the dominant stellar orbits in the galactic disk, i.e. the back-reaction of the stellar component to the self-gravitating gas accumulating between the ILRs. A full account can be found in Knapen et al. (1995b) and Heller and Shlosman (1996).

2 Twisting of NIR Isophotes in M 100

M 100, the brightest (barred) spiral galaxy in Virgo, displays all the virtues of a nuclear starburst and is inclined at $30° \pm 3°$. Surprisingly, the UV/optical starburst ring formed by a tight pair of spirals, lies around $10''$ from the center ($1''$ corresponds to 83 pc at a distance of 17.1 Mpc), whereas its NIR (2.2μm) counterpart has a substantially larger radius of $\sim 18''$ and a considerable width of $\sim 16''$ (Knapen et al. 1995a). It is weakly oval (minimal ellipticity ~ 0.13) and its semimajor axis leads the stellar bar by about $60° - 70°$. Most unusual is the observed gradual twist of the NIR isophotes, from the position angle (P.A.) of the ring towards the P.A. of the stellar bar, both *exterior* and *interior* to the ring (see Fig. 1). In other words, the large-scale bar is dissected by an oval ringlike zone oblique to the bar. The existence of this inner bar-like feature was confirmed independently by high-resolution CO observations of the velocity field (Rand et al. 1996). Molecular gas within the central $10''$ participates in high velocity $\sim 100\,\mathrm{km\,s^{-1}}$ non-circular motions along the P.A. of the stellar bar.

Stellar orbits are expected to change their orientation abruptly by $90°$ at each resonance in the disk, which can be understood within the framework of forced oscillations. These orbits are oriented along the stellar bar between the OILR and the corotation radius (x_1 family), they are perpendicular to the bar between the IILR and OILR (x_2 family), and are again aligned with the bar

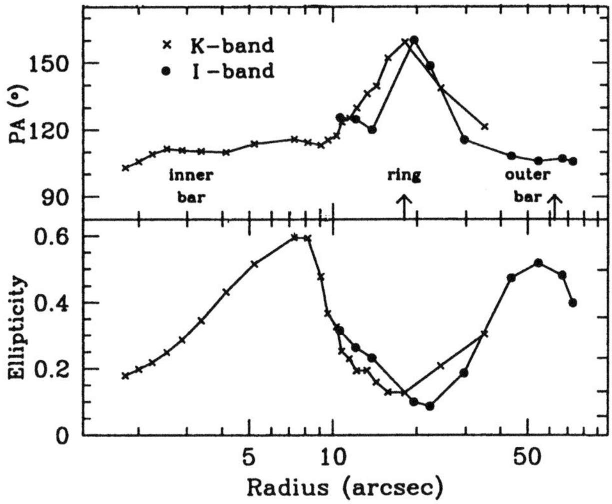

Fig.1. The K and I isophotes in M 100: position angle (*top*) and ellipticity (*bottom*) (From Knapen et al. 1995b)

between the center and the IILR. NIR isophotes in a quiescent galaxy are believed to follow the overall mass distribution, i.e. they arise from an old stellar population represented by K and M giants, (e.g. Frogel 1985). However, *gradual* skewing of isophotes in M 100 is indicative of gas behavior across the resonances rather than the behavior of stars. There are two possible ways to explain this twisting of NIR isophotes between 10″ to 30″ in M 100. First, there can be a sufficient contribution from massive young stars, K and M supergiants (Knapen et al. 1995a). These stars are expected to be found in the star bursting regions, and, being dynamically young (age less than 10^7 yrs), they should follow gas rather than stellar orbits. In such a case, the NIR light will *not* follow the mass distribution. Knapen et al. (1995b) found that this is the most plausible explanation for two NIR 'hot spots' in M 100, and similar conclusions have been reached for active star forming regions in NGC 1309 (Rhoads 1996). Alternatively, gas gravity can drag some of the old population stars in the stellar bar towards the x_2 orbits in the ring (Shaw et al. 1993). As we show in Sect. 4, the growing accumulation of gas between the ILRs is capable of affecting the main periodic orbits in the bar (especially the x_1 orbits outside the OILR) in such a way as to support a gradual twist of *stellar* orbits, from being aligned with the bar to becoming almost perpendicular to it.

3 Simulations of Stellar and Gas Dynamics in M 100

To further understand the circumnuclear morphology in M 100 and to confirm that it is compatible with the presense of a double ILR there, we have tailored our numerical simulations to a M 100-*like* mass distribution. Stars and gas, embedded in halo and bulge potentials, were evolved by means of a 3D hybrid

SPH/N-body code (Heller and Shlosman 1994). Models without and with star formation have been constructed, Q1 and Q2 respectively. In the Q2 model, the gas is considered to undergo 'star formation' if it is Jeans unstable, if it participates in a locally converging flow, and if its density exceeds 20 $M_\odot\,pc^{-3}$. We assume that only massive OB stars form, which deposit 10^{51} ergs per 10^6 yr per star in the gas by means of their line-driven winds, as well as deposit 10^{51} ergs per 10^4 yr as supernovae, leaving no remnants. This energy is assumed to be instantly radiated away by the gas, with only 5% being retained and converted into turbulent motion. One unit of time, $[\tau]$, corresponds to 3.75×10^7 yr.

Initially, an axisymmetric model of a galactic disk was constructed which was dynamically unstable and developed a stellar bar in the process of evolution. The resulting bar strength was moderate, $q \sim 0.3$, where q is defined as a maximum ratio of the sum of $m = 2, 4, 6$, and 8 Fourier components of the nonaxisymmetric gravitational force to the $m = 0$ axisymmetric component. Based on the (axisymmetric) rotation curve of the model after $\sim$ 3 rotation periods, a double ILR could be found in the central region, IILR at $\sim$ 240 pc and OILR at $\sim$ 1.4 kpc. However, the epicyclic approximation is invalid because of the relatively strong bar. Instead, we have analyzed the dominant families of periodic orbits in the model potential. It is customary to extend the ILRs into the nonlinear regime based on the radial extent of the x_2 family of orbits. The corresponding limiting semimajor orbital axes in Q1 and Q2 have been found at $\sim$ 500 pc and $\sim$ 1.3 kpc, and adopted as the nonlinear IILR and OILR, respectively. Hence, the resonance region between the ILRs was reduced substantially compared to the linear regime.

After an initial transient in the Q1 model, the gas formed a pair of trailing shocks along the leading edge of the stellar bar, in agreement with 2D numerical simulations (e.g. Athanassoula 1992). Deeper in the potential well, gas dynamics in the vicinity of ILRs was dominated by a pair of tightly wound trailing shocks and a pair of leading shocks (see Fig. 2, at $\tau = 15$). Such a shock system in the vicinity of a double ILR is predicted on the base of a simple epicyclic approximation which is not adequate at this stage. In particular, we note that two systems of shocks, trailing and leading, interact non-linearly through a pronounced cuspy feature (caustic; Fig. 2, $\tau \gtrsim 17$). To the extent that this shock system delineates spiral arms, we observe a transient pseudo-ring made out of a pair of tightly wound spirals between the ILRs. In fact, two separate gas circulations (oval 'rings') form between the ILRs due to the action of the shocks. The outer gaseous ring evolves as to (almost) align itself with the minor axis of the stellar bar, while the inner gaseous ring (almost) aligns itself with the major axis of the bar.

This evolution comes about as a response to the gravitational torques by the bar and in order to minimize them. However, the effects of self-gravity in the gas are crucial in order to understand its dynamics. The self-gravity acts in a manner similar to surface tension: the outer gaseous ring settles down on lower energy x_2 orbits, away from the OILR. At the same time, the inner ring shrinks across the IILR and settles on the x_1 orbits. This behavior is depicted in Fig. 2, at around

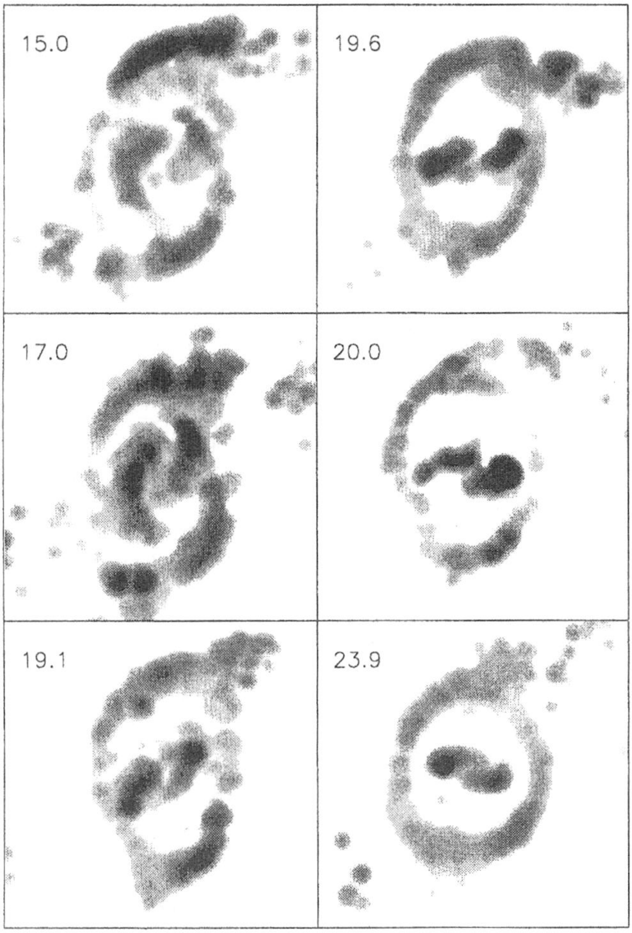

Fig.2. Logarithmic gray-scale map of shock dissipation inside the OILR in M 100 (Q1 model, without star formation). Time is given in the upper left corners, $[\tau = 1] = 3.75 \times 10^7$ yr. Each frame is 2.6 kpc across. The gas flows counterclockwise, the stellar bar is horizontal. (From Knapen et al. 1995b)

$\tau \sim 17 - 24$. It is imperative to mention that the gas is *not* locked between the ILRs, but merely experiences a temporary slowdown there. The outer ring is constantly perturbed by a number of density inhomogeneities and 'rains' onto the inner ring which contains orbits deeper in the potential well. (In the presence of star formation, as we discuss below, these perturbations are caused by the turbulent motions in the gas excited by stellar winds and supernovae.) As the inner ring becomes more massive, it is subject to gravitational fragmentation. Neglecting the effects of star formation at this stage is not justified.

Technically speaking, star formation in the Q2 model has introduced 'turbulent' motion in the gas and induced mixing between material with a different angular momentum and energy. As a result, gaseous circulations between the ILRs widen notably and merge, and it is more appropriate to speak of a gaseous disk which extends inwards, from the outermost x_2 orbits, much of the way to the center. This disk is an oval, and its semimajor axis is positioned almost at right angles to the bar.

As expected, the shock system outlines the regions of intense star formation in the Q2 model (see Fig. 3). However, there is no one-to-one correspondence between the shocked and gravitationally unstable gas, as the shock and star formation maps show. At the same time, four major sites of star formation in the resonance region persist during most of the simulation time — all corresponding to the maxima of dissipation in the gas. Two elongated star forming regions are found downstream from the place where the inflow along the outer shocks crosses the bar's minor axis and encounters the gas circulation on the x_2 orbits. Azimuthal smearing of star forming regions is a direct consequence of the time scale for Jeans instability becoming an appreciable fraction of the orbital time scale so close to the rotation axis. Another pair of star forming regions is located around the IILR ($\tau \sim 10 - 25$, in Fig. 3), slightly ahead of the bar's major axis and where the cuspy feature is seen in the Q1 model. This prevailing morphology dominated by four star forming regions (at and just outside the IILR) appears to be robust during most of the simulation time. It is compatible with the loci of star formation in the U, V, and Hα images of the inner 2 kpc in M 100 (Knapen et al. 1995a). No star formation correlates with the position of the OILR.

The star formation rate in the Q2 model reaches its peak around $\tau \sim 25 - 28$, exhibiting burst behavior with a typical time scale of $\sim 10^7$ yr. Around $\tau \sim 28$, the mass inflow rate across the IILR peaks strongly, indicating a catastrophic loss of angular momentum by the gas ($\tau = 28.4$, Fig. 3). The subsequent evolution of this dynamical runaway in the self-gravitating gas was discussed by Heller and Shlosman (1994) in the absence of star formation. We view this process as the initial phase of decoupling of the gaseous bar from the large-scale stellar bar, as envisioned in the 'bars within bars' scenario (Shlosman, Frank and Begelman 1989). It is not clear how much of the underlying old population participates in this process (and if it does at all). The amount of gas at the onset of instability, $\sim$few$\times 10^9\ \mathrm{M}_\odot$, is $\sim 10\% - 20\%$ of the mass *interior* to the runaway region, which is more than an order of magnitude less than quoted by Ho, Filippenko and Sargent (1996). This amount of gas is available even in the early-type disks.

In the more realistic case, advanced here in the Q2 model, the outcome of evolution depends in a sensitive way on the efficiency of star formation, the state of the interstellar medium, and its ability to retain energy deposited by massive stars. These questions are potentially relevant for our understanding of fueling the nonthermal activity in Seyfert nuclei and must be addressed in future work. We conclude, that a characteristic time for the gas to 'filter' through the resonance region is $\lesssim 10^9$ yr. This can be taken as a rough estimate of a nuclear ring's lifetime.

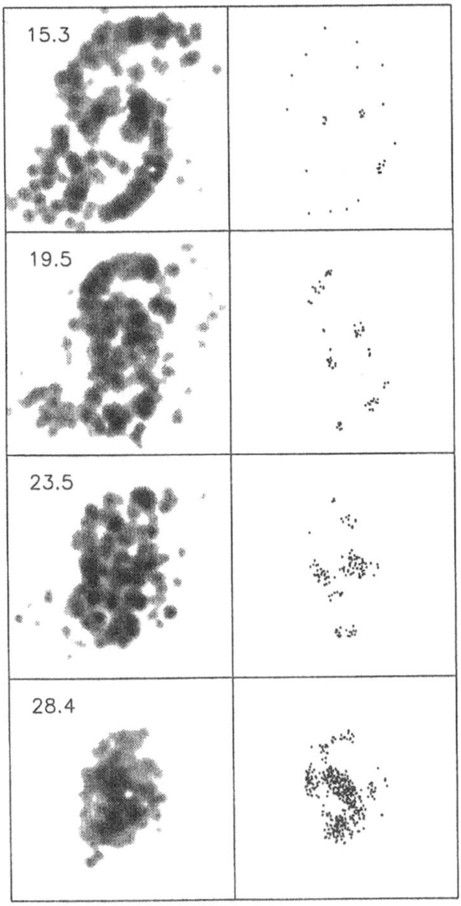

Fig.3. Left: Logarithmic gray-scale map of shock dissipation inside the OILR in M 100 (model with star formation). Time is given in the upper left corners, $[\tau = 1] = 3.75\times10^7$ yr. Each frame is 2.6 kpc across. The gas flows counterclockwise, the stellar bar is horizontal. **Right**: Star formation map corresponding to the region shown on the left. (From Knapen et al. 1995b)

4 Massive Nuclear Rings: Affecting Periodic Orbits

The above numerical simulation of gas dynamics in the presence of a double ILR demonstrates explicitly that the radial gas inflow is not stopped between the resonances but only slows down there — a kind of 'self-mulching lawn mower' effect. This results in gas accumulation in the form of a massive elliptical 'ring'. Molecular mass of as much as a few$\times10^9\,M_\odot$ stored in the ring is not out of the question. Besides enhanced star formation caused by favorable conditions in the region, such a ring will have gravitational effects on the stellar component in the disk. Here, we are mainly interested in how the main periodic orbits are affected, and what consequences it may have on the gas circulation in the bar.

For simplicity we use a 3D analytical model of a galaxy consisting of a disk, bulge, halo and stellar bar. Furthermore, in an attempt to simulate the effect of a massive nuclear ring within the central kpc, we make assumptions about its shape and the orientation of its major axis with respect to the stellar bar. Results of a 3D orbit analysis for such gravitational potentials are presented elsewhere (Heller and Shlosman 1996). Here we discuss only necessary details.

The most pronounced change in the periodic orbits, when a circular ring (or elliptical ring whose major axis coincides with that of the bar) is added, is that the extent of the ILR resonance region is increased with the ring's mass, weakening the stellar bar. In addition, the x_1 orbits of different energy intersect in the vicinity of the ring (as do x_2 orbits within the ring). This has a two-fold effect on the gas: the phase space available to x_2 has increased, but at the same time, orbits near the ring became intersecting and unable to hold gas, amplifying dissipation there. Thus, we expect that the gas will fall through the IILR after an initial stage of accumulation, exactly as observed in the numerical simulations.

An additional and qualitatively different effect is obtained when the ring is mildly elliptical and its major axis is oblique to the bar, leading it in the direction of galactic rotation. This configuration is the one typically observed in nuclear starburst galaxies and appears as a long-lived transient in our numerical simulations. The behavior of x_1 orbits can be qualitatively understood in this case as a response to the perturbing forces of the stellar bar and of the oblique ring. Both forces have the same driving frequency but are phase-shifted. A straightforward application of an epicyclic approximation to the motion of a viscous 'fluid' reveals the rich variety of possible responses in the gas to this driving force. A representative case, calculated using fully nonlinear orbit analysis when the ring leads the stellar bar by 60°, is shown in Fig. 4. The change in the position angle of the x_1 with distance to the ring is rather dramatic, starting at large radii with a slowly growing phase shift which reaches a maximum of 34° in the leading direction, followed by a rapid decline to 0° just interior to the ring and then continuing to −11° in the trailing direction deep into the bulge.

5 Implications: Dynamics of the Circumnuclear Region

Our dynamical study of gas flow across a double ILR resonance in the circumnuclear region of a barred galaxy provides some insight into relevant processes which accompany this flow. First, self-gravitating effects in the gas accumulating between the resonances are crucial in understanding its subsequent evolution (besides the star formation). In particular, self-gravity acts as a 'surface tension' and the gas moves deeper into the potential well, away from the OILR and across the IILR. Star formation is clearly peaked around the IILR, partly because the gravitational torque's sign changes across this resonance causing additional compression in the gas and creating conditions favorable to Jeans instability there.

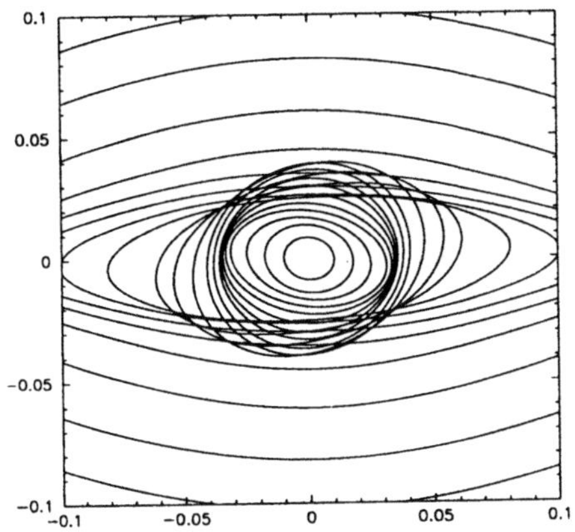

Fig.4. Twisting of x_1 periodic orbits supporting the stellar bar in the presence of a massive $10^9\,M_\odot$ oblique elliptical ring ($e = 0.4$). Ring's potential is softened with $\epsilon = 100$ pc. The stellar bar is horizontal. Rotation is counterclockwise and corotation is at 5 kpc. The ring's semimajor axis ($[r = 0.04] = 400$ pc) is offset to the bar by 60° in the leading direction. The frame is 2 kpc on a side. (From Heller and Shlosman 1996)

Second, we find a complicated but basically a ring-like morphology for the distribution of star forming regions in the nuclear starbursts. Dominant regions of star formation lie downstream from two main compression sites of the gas ('twin peaks'), when it crosses the minor axis of the stellar bar. This is somewhat outside the IILR. Additional pair of star forming regions are found at the IILR, on the major axis of the bar and slightly leading it, and correspond to the twin 'hot spots' found in M 100.

Third, there is no prominent star formation associated with the OILR. This explains why the UV/optical and Hα observations identify the star forming ring in M 100 *inside* its NIR ring. Unless the star formation, already at this stage, is very efficient in consuming the gas at the IILR, the gas 'breaks through' and falls towards the center, an event that is accompanied by a prominent burst of star formation. This dynamical runaway of self-gravitating gas *inside* the IILR depends on details of star formation and physics of the interstellar medium, and it is outside the scope of this study. We only comment that the characteristic time scale of this runaway is $\sim$few$\times 10^7$ yr, much shorter than the time it takes for the gas to 'filter' through the resonance region, $\lesssim 10^9$ yr.

Fourth, we find that massive nuclear rings are capable of perturbing the gravitational potential in the circumnuclear regions, thus affecting the main periodic orbits there. The phase space allowed to the orbits aligned with the minor axis of the bar (x_2 family) is substantially increased and the orbits aligned with the bar (x_1 family) are significantly distorted. Orbits with different values of the Jacobi integral are found to intersect, meaning that gas cannot be retained there and will move inwards across the IILR. In the most interesting case of an elliptical ring oblique to the bar, x_1 orbits are gradually twisted, in a way similar to the skewing of NIR isophotes observed in M 100. So these orbits can trap both gas (accompanied by K and M supergiants) and old population stars. It may be possible, in a such a case, to estimate the 'degree of asymmetry' in the nuclear ring (i.e. azimuthal mass distribution) based on the observed change in the ellipticity and position angle of NIR isophotes with radius.

To summarize, the gas seems to be the prime dynamic agent in the circumnuclear regions of at least some disk galaxies, although its mass is only a fraction of the mass in the stellar component there. Further observations of molecular gas kinematics in active and normal galaxies will provide clues to understanding their central activity.

Acknowledgements. It is a pleasure to thank organizers of this stimulating meeting for financial assistance. I am indebted to John Beckman, Roelof de Jong, Clayton Heller, Johan Knapen and Reynier Peletier for collaboration on some of the research described above.

References

Arsenault, R., Boulesteix, J., Georgelin, Y., Roy, J.-R. (1988): A&A **200**, 29
Athanassoula, E. (1992): MNRAS **259**, 345
Buta, R., Crocker, D.A. (1993): AJ **105**, 1344
Combes, F., Gerin, M. (1985): A&A **150**, 327
Elmegreen, B.G. (1994): ApJ **425**, L73
Frogel, J.A. (1985): ApJ **298**, 528
Heller, C.H., Shlosman, I. (1994): ApJ **424**, 84
Heller, C.H., Shlosman, I. (1996): ApJ, submitted
Ho, L.C., Filippenko, A.V., Sargent, W.L.W. (1996): in Proc. IAU Colloq. 157 Barred Galaxies, eds. R. Buta et al., Kluwer, Dordrecht, in press
Kenney, J.D.P. (1996): The Interstellar Medium in Galaxies, ed. J.M. van der Hulst, Kluwer, Dordrecht, in press
Kenney, J.D.P., Wilson, C.D., Scoville, N.Z., Devereux, N.A., Young, J.S. (1992): ApJ **395**, L79
Knapen, J.H., Beckman, J.E., Heller, C.H., Shlosman, I., de Jong, R.S. (1995b): ApJ **454**, 623
Knapen, J.H., Beckman, J.E., Shlosman, I., Peletier, R.F., Heller, C.H., de Jong, R.S. (1995a): ApJ **443**, L43
Rand et al. (1996): in preparation
Rhoads, J.E. (1996): ApJ, submitted
Schwarz, M.P. (1984): MNRAS **209**, 93
Shaw, M., Axon, D., Probst, R., Gatley, I. (1995): MNRAS **274**, 369
Shaw, M., Combes, F., Axon, D.J., Wright, G.S. (1993): A&A **273**, 31
Shlosman, I., Frank, J., Begelman, M.C. (1989): Nat **338**, 45
Wada, K., Habe, A. (1992): MNRAS **259**, 82

Dynamical Substructures in Two Nearby Galaxy Nuclei

Roland Bacon[1] and Eric Emsellem[1,2]

[1] Centre de Recherche Astronomique de Lyon, Observatoire de Lyon, F-69561 St Genis-Laval cedex, France
[2] Sterrewacht Leiden, Postbus 9513, 2300 RA Leiden, The Netherlands

Abstract. Using the integral field spectrograph TIGER at CFHT, we have observed a few nearby galaxy nuclei suspected to harbor a supermassive black hole. The two remarkable objects presented here, M 31 and M 104, display peculiar kinematics: a rapidly rotating cold nuclear disk and puzzling photometric and kinematical asymetries in the case of M 104 and M 31 respectively. Axisymmetric models confirm the presence of supermassive black holes. However at the attained subparsec resolution, we are far from understanding the present observations of M 31's nucleus.

1 Introduction

The improved resolution of HST has shown that some galaxies exhibit nuclear structures such as disks or bars. Using the new integral field spectrograph TIGER at CFHT, we have investigated the kinematics of a few nearby galaxy nuclei in detail.

2 M 31

Because of its proximity and its well defined nucleus, M 31 is an ideal target to study in detail the kinematics of the nucleus of a giant galaxy. Although Lallemand already showed in 1960 that the nucleus is rotating rapidly, the first extensive kinematical studies were conducted by Kormendy (1988) and Dressler and Richstone (1988). They both showed that simple dynamical models required a large central mass concentration to fit the observational data, the best candidate for such a mass concentration being a supermassive black hole of $\sim 10^7$ $M_\odot$. The fact that the center of rotation was not coincident with the maximum of light was noticed in both studies, but not much commented.

More recently, the attention focussed again on M 31's nucleus when Lauer et al. (1993) and Bacon et al. (1994) found new and puzzling indications of its complex structure.

2.1 The Double Photometric Structure

The asymmetric appearance of the nucleus of M 31 was known since the Stratoscope II observations presented by Light et al. (1972). Lauer et al. (1993), using the spatial resolution of the HST[1], resolved the nucleus into two components[2]: P1 (the brighest peak) and P2 which is nearly coincident with the center of the outer isophotes of the nucleus. This double structure is seen from the near UV to the IR (Nieto et al. 1986; Mould et al. 1989; Rich et al. 1995) and is unlikely to be due to dust absorption. However, King et al. (1995) showed that P2 is brighter than P1 in the far UV.

2.2 The Double Kinematical Structure

The nucleus of M 31 was observed with the IFS TIGER at CFHT in November 1990 and September 1991. For the first time, these observations provided two-dimensional fields of the stellar velocity and velocity dispersion of the nucleus. These maps revealed that its kinematics are much more complex than originally expected: the center of rotation nearly corresponds to P2, but the velocity dispersion peaks at a point roughly symmetric from P1 with respect to P2. Although most of these results were outlined in previously released long-slit data, the merit of the IFS observations was to focus the attention on the peculiarities (asymmetries) of the kinematics of M 31's nucleus. Such results are much more difficult to uncover using long-slit spectrography, since the slit is generally aligned with one of the main photometric axes (major or minor axis), which are *a priori* assumed to be aligned with the kinematical ones. Observational results and a preliminary theoretical analysis are presented in Bacon et al. (1994).

2.3 Some Possible Models

These results have motivated a few theoretical studies. The proposed models can be split in two main classes: those considering P1 as an intruding object, and those assuming that the asymmetries reflect some intrinsic properties of the nucleus.

An Eccentric Disk. Tremaine (1995) proposed that the nucleus of M 31 could be an eccentric disk of stars travelling on nearly keplerian orbits around a central black hole. As shown by Tremaine, such a crowding effect can qualitatively reproduce the observed asymetries. However, there are a few difficulties with this model: (i) the eccentric disk is intrinsically very thin and the ringlets have to be arbitrarily thickened to account for the axis ratio of the nucleus, (ii) the modelled rotation curve must be strongly asymmetric with respect to P2, which is in contradiction with the observations.

---

[1] Deconvolved pre-Costar observations.

[2] Bacon et al. (1994) reached the same conclusions almost simultaneously using high resolution images obtained at the CFHT.

A Cold Stellar Cluster. Lauer et al. (1993), following Dressler and Richstone (1988) and Mould et al. (1989), have proposed that P1 could be an accreted globular cluster or the nucleus of a dwarf galaxy. However, these authors pointed out that the timescale of such an event is short, since dynamical friction would force the additional stellar system to decay rapidly ($< 10^5$ yr).

The smooth two-dimensional TIGER velocity field does not show any significant peculiarities at the location of P1. This indicates that the stellar cluster has a radial velocity close to the mean projected velocity of the nucleus near P1, and suggests that it is located within the nucleus.

Recently, Miller and Smith (1995) have conducted simulations where a stellar cluster orbits in a quasi-homogeneous nucleus. They show that a stellar cluster can survive longer than expected in such an environment. However, if the nucleus is axisymmetric, which is the case in the Miller and Smith's simulation, the observed velocity and velocity dispersion gradients imply a dense object of a few $\sim 10^7$ $M_\odot$ at P2, which renders the nucleus far from homogeneous.

Emsellem and Combes (1996) have examined the evolution of a dense stellar cluster falling into the nucleus by means of N body simulations. In their models, the nucleus is considered as a structure kinematically independent from the bulge. This assumption was motivated by the photometric and kinematics observations of the nucleus (Bacon et al. 1994), and allows the bulge to be modelled as a simple fixed potential. At $t = 0$, the nucleus is assumed to be a thick Toomre disk represented by more than 130 000 particles, and a central dark mass in the range of $10^7 - 10^8$ $M_\odot$. The dense stellar cluster, simulated as a plummer sphere of 1.7 10^6 $M_\odot$ ($\sim$ 15 000 particles), is launched in the equatorial plane of the nucleus. The resolution (grid size) of the simulation is 0.15 pc.

The simulations show that the main effect is not due, as previously believed, to the orbital decay but to the tidal forces: the stellar cluster is rapidly disrupted and spread into a ring-like structure. During the disruption, the transfer of angular momentum between the nucleus and the cluster is rather small ($< 10\%$).

When viewed nearly edge-on, this model reproduces the main observed asymmetries, including the kinematical ones. The main drawback of the stellar cluster hypothesis is the very short lifetime of such an event. As mentioned above, it is unlikely that P1 is spatially very distant from the nucleus and only seen near the center because of a projection effect. The timescale for the disruption of the cluster is strongly dependent on the black hole mass. Therefore, if indeed the presence of a central dark mass of a few 10^7 $M_\odot$ is confirmed, it is then likely that we observe P1 just before its complete tidal disruption. In the absence of such a central density, the stellar cluster could survive significantly longer.

A Nuclear Bar. Gerhard (1986) suggested that the observed photometry and dynamics could be reproduced by a nuclear stellar bar without the need of a supermassive black hole. In Gerhard's model, the bar is supposed to share the same equatorial plane as the outer disk, the observed photometric twist between the nucleus and disk minor axis being due to the projection of a triaxial ellipsoid. Bacon et al. (1994) showed that the inclination of the bar imposed by this model predicts a velocity field which would appear twisted with respect to the photometric minor axis (see their Fig. 20): this is in contradiction with the observed velocity field.

However the apparent kinematical decoupling of the nucleus from the disk and bulge could be real, and we can relax the condition of alignment between the nucleus and the disk reference axes. Recent numerical simulations performed by Combes and Emsellem (1996) show that it is then possible to reproduce the observed photometric and dynamical properties, without invoking the presence of a SBH.

3 M 104

The Sombrero galaxy (M 104, NGC 4594) is a nearby, bright and Sa galaxy and a candidate for harboring a central massive black hole (Kormendy 1988) of $\sim 10^9$ $M_\odot$.

3.1 A Rapidly Rotating Inner Disk

Kormendy (1988) detected a very rapidly rotating stellar component in the inner 10 arcseconds. With the help of ground-based photometry, he suggested that it corresponded to a central bright disk, dominating the surface brightness in the center. Subsequent studies seemed to confirm this hypothesis: Wagner et al. (1989) detected an asymmetric projected velocity distribution along the photometric major axis, indicative of the superposition of a hot slowly rotating component ("the bulge") and a cold rapidly rotating one ("the inner disk"); Emsellem et al. (1994) carried out a disk/bulge decomposition, using high resolution CFHT photometry, which indeed revealed the presence of a very flattened ($\epsilon > 0.9$) bright inner disk extending out to $\sim$ 15 arcseconds.

The Sombrero galaxy was again observed at the CFHT in April 1992, using the IFS TIGER. This allowed Emsellem et al. (1996) to obtain the full two-dimensional map of the Line Of Sight Velocity Distributions (LOSVDs) in the inner 6 arcseconds. The contribution of the inner disk can easily be traced from the asymmetries of the obtained LOSVDs: there is a smooth transition in their shapes as we move away from the major axis (Fig.1). Emsellem et al. (1996) then used the photometric model of M 104 (Emsellem 1995) to constrain the disk-to-bulge surface brightness ratio, and disentangle the kinematical contribution of the inner disk and bulge to the LOSVDs. The two-dimensional velocity field of the inner disk is (as expected) rather flattened, exhibits a very strong central gradient and a maximum of $\sim$ 300 km s^{-1} at about 5 arcseconds along the major axis. This is indeed much higher that the mean velocity of 235 km s^{-1} observed at the same point (at a resolution of $\sim$ 1 arcsec FWHM). Emsellem and Qian (1996) have recently derived a full two-integral distribution function which nicely fits these features, and show that the LOSVDs should exhibit a highly contrasted peak due to the inner disk as spatial and spectral resolution increase. They also predict the presence of a supermassive black hole of 2 10^9 $M_\odot$.

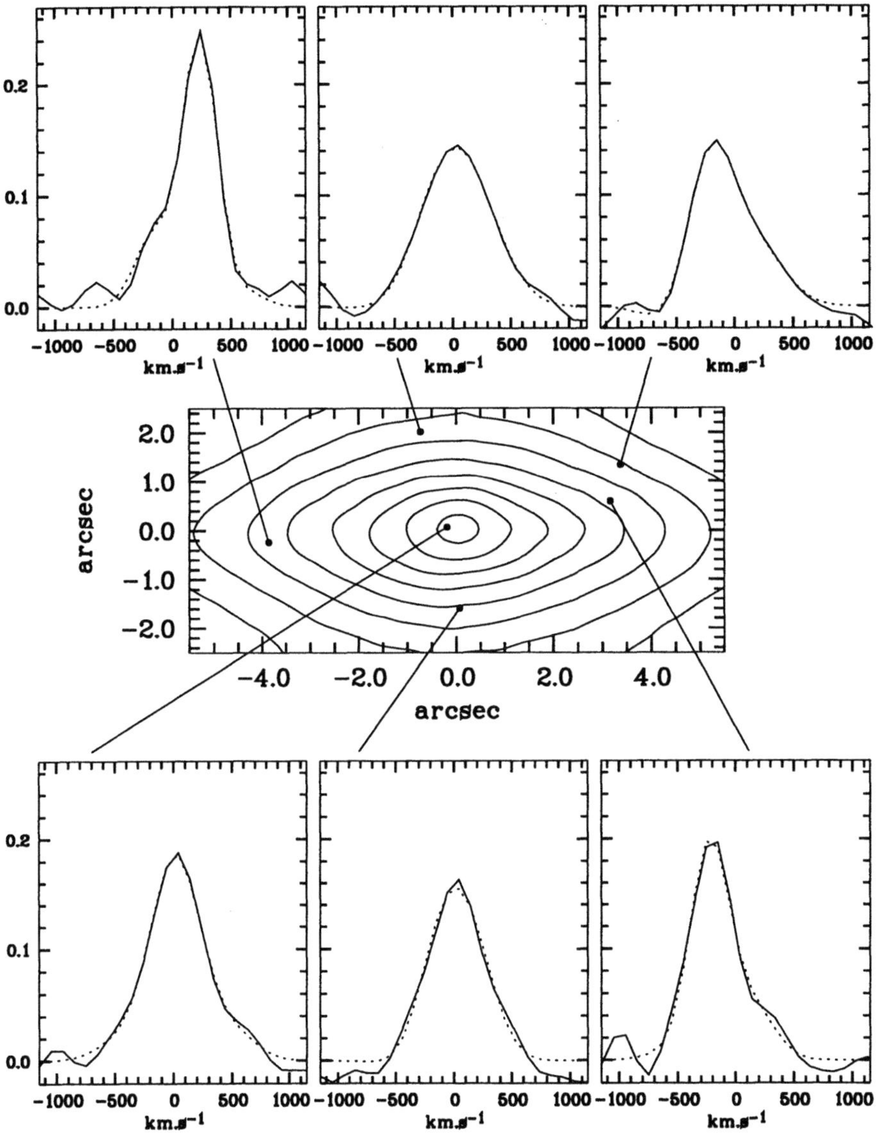

Fig. 1. Examples of M 104 line-of-sight velocity distribution profiles

4 Conclusion

We have obtained 2D kinematics of two nearby galaxy nuclei suspected to harbor a supermassive black hole. In M 104 we find a rapidly rotating cold nuclear disk, while in M 31 we observe strong asymmetries which are still not well understood. Axisymmetric models of these two objects imply large concentrations of mass, presumably supermassive black holes of respectively 2 10^9 and 7 10^7 $M_\odot$.

In M 31, the best resolved case, some models have been proposed to explain the observed asymmetries, but none is really satisfying. Clearly new theoretical developments *and* high spatial resolution integral field spectrography are needed to understand what is going on in the nucleus of our neighbouring giant galaxy.

The latter could be provided by the integral field spectrograph OASIS, which is being developed at the Observatoire de Lyon. This instrument, which will succeed the TIGER prototype early in 1997, has been specially designed for the CFHT adaptive bonnette. It will thus provide high spatial resolution ($\sim$ 0.2 arcsec FWHM) together with improved 2D capabilities (1500 spectra obtained simultaneously).

References

Bacon R., Emsellem E., Monnet G., Nieto J.L. (1994): A&A **281**, 691

Bacon R., Adam G., Courtès G., Dubet D., Dubois J.P., Emsellem E., Ferruit P., Georgelin Y., Monnet G., Pécontal E., Rousset A., Sayède F. (1995): A&AS **113**, 347

Combes F., Emsellem E. (1996): A&A, in preparation

Dressler A., Richstone D.O. (1988): ApJ **324**, 701

Emsellem E. (1995): A&A **303**, 673

Emsellem E., Bacon R., Monnet G., Poulain P. (1996): A&A, submitted

Emsellem E., Combes F. (1996): A&A, in preparation

Emsellem E., Monnet G., Bacon R., Nieto J.-L. (1994): A&A **285**, 739

Gerhard O.E. (1986): MNRAS **219**, 373

King I.R., Stanford S.A., Crane P. (1995): AJ **109**, 164

Kormendy J. (1988): ApJ **325**, 128

Kormendy J. (1988): ApJ **335**, 40

Lallemand A., Duschene M., Walker M.F. (1960): PASP **72**, 76

Lauer T.R., Faber S.M., Groth E.J., Shaya E.J., Campbell B., Code A., Currie D.G., Baum W.A., Ewald S.P., Hester J.J., Holtzman J.A., Kristian J., Light R.M., Lynds C.R., O'Neil E.J., Westphal J.A. (1993): AJ **106**, 1436

Light E.S., Danielson, R.E., Schwarzschild, M. (1974): ApJ **194**, 257

Miller R.H., Smith B.F. (1995): preprint

Mould J., Graham J., Matthews K., Soifer B.T., Phinney E. S. (1989): ApJ **339**, L21

Nieto J.L., Macchetto F.D., Perryman M.A.C., Serego Alighieri S., Lelievre G. (1986): A&A **165**, 189

Rich R.M. Mighell K.J. (1995): preprint

Tremaine S. (1995): CITA preprint 95-4

Wagner S. J., Dettmar, R.-J., Bender, R. (1989): A&A **215**, 243

The Spheroidal Component of Seyfert Galaxies

Charles H. Nelson[1], Mark Whittle[2], and John W. MacKenty[1]

[1] Space Telescope Science Institute, 3700 San Martin Drive, Baltimore, MD 21218, USA
[2] Astronomy Dept. University of Virginia, Box 3818 Charlottesville, VA 22903, USA

Abstract. We use measurements of the stellar and gaseous kinematics for a large sample of Seyfert galaxies to examine properties of the host galaxy and their relationship to the active nucleus. We find that Seyferts are offset from the Faber-Jackson relation for normal galaxies having brighter bulges at a given velocity dispersion than normal galaxies. This indicates that Seyferts have lower mean mass-to-light ratios than normal galaxies and therefore younger stellar populations. Comparing gas and stars, we find that the kinematics of ionized gas in Seyferts are largely due to gravitational motion in the host galaxy potential. Additional acceleration of emission line gas can occur in objects with kiloparsec-scale linear radio sources and in interacting or morphologically disturbed galaxies. We also find correlations between the emission line and radio luminosity and the stellar velocity dispersion suggesting a link between the host galaxy potential and the strength of the NLR emissions.

We also discuss some results from Hubble Space Telescope snapshots of a sample of 52 Markarian Seyfert galaxies. A number of these show small scale-bars and double nuclei. Also the nuclei of type 1 – 1.5 Seyfert galaxies are dominated by strong point sources, while those of Seyfert 2 galaxies tend to be resolved and resemble normal bulge luminosity profiles. This suggests that the nuclear continuum observed in Seyfert 2 galaxies is extended, covering several tens of parsecs or more, in agreement with unified models of active galactic nuclei.

1 Introduction

Activity in galaxies is a phenomenon of the nucleus — the deepest part of the gravitational potential. It is natural, therefore, to consider the possibility that the host galaxy plays a critical role in the formation and development of the active nucleus. The parameter which perhaps best scales with the depth of the gravitational potential is the nuclear stellar velocity dispersion, σ_*. Until recently relatively few measurements of σ_* have been available for active galaxies. We have therefore obtained stellar kinematic measurements for a large sample of Seyferts to address a number of fundamental questions regarding the host galaxy and its relationship to the active nucleus. Are the dynamics of Seyfert bulges different from those of normal galaxies? What physical processes accelerate emission line gas in the narrow line region? Are these processes related to the host galaxy or to the active nucleus? What links exist between the emission from the active nucleus and the host potential?

2 Background

Several studies of Seyfert galaxies have discussed the possibility of a virial origin for the kinematics of ionized gas in the narrow line region (NLR). For example, Wilson and Heckman (1985) plotted the FWHM of the [O III] λ5007 emission line against σ_* for a sample of Seyferts and LINERs using data obtained from the literature. They found roughly equal gas and stellar velocity widths with considerable scatter. Whittle (1992a,b) using *indirect* virial parameters V_{max}, the maximum of the rotation amplitude of the galaxy (obtained from rotation curves and H I profiles), and M_{bul}, the bulge magnitude, found that for most Seyferts the kinematics of emission line gas have a gravitational origin. However, Seyferts which harbor nuclear kiloparsec-scale linear radio sources have emission lines considerably broader than expected for normal gravitational motion. In these objects there is an additional acceleration mechanism which is most likely related to the interaction of the line emitting gas with the radio plasma ejected from the nucleus. There have also been indications that the nuclear radio luminosity scales with the optical luminosity of the host galaxy (Meurs and Wilson 1984; Edelson 1987; Whittle 1992b).

3 Sample and Observations

To examine these issues further we have obtained stellar and gaseous kinematic measurements for a sample of 85 objects: 73 Seyferts, 9 LINERs, 3 normal galaxies (Nelson and Whittle 1995). This increases the number of stellar velocity dispersion measurements available for Seyferts by roughly a factor of 3. Where possible we have combined our measurements with previously published values (e.g. Whitmore et al. 1985; Terlevich et al. 1990) to yield a best value for σ_*. We generally find good agreement between these measurements and our own with no systematic trends (see Nelson and Whittle 1995 for a comparison).

The primary difficulty in measuring the stellar kinematics in active galaxies is obtaining high signal-to-noise spectroscopy of the stellar absorption lines. This is most difficult in objects with strong featureless continua (Seyfert 1 galaxies) which dilute the absorption lines. In the near-IR, however, the strength of the nuclear continuum emission relative to the host galaxy starlight is considerably reduced. We have therefore obtained spectroscopy in two wavelength ranges. The first, in the near-infrared, is centered on the Ca II triplet absorption lines. The second, in the visual, is centered on the Mg*b* absorption lines and includes the [O III] λ5007 emission line. Systemic velocities and velocity dispersions were obtained using the cross-correlation method (Tonry and Davis 1979) and the widths of emission lines were also measured.

4 The Faber-Jackson Relation for Seyfert Galaxies

Seyfert and normal galaxies can be compared using relationships between virial parameters. In this paper I will describe our results for the Faber-Jackson relation usually stated as $L \propto \sigma_*^n$, where L is the luminosity of the spheroidal component, and the exponent is typically $n \simeq 3-4$. For the Seyferts we have applied a correction for the luminosity of the active nucleus based on the fluxes of the emission lines and estimates of the strength of the featureless continuum. In addition, since surface photometry separating bulge and disk light is not available for our sample we use the relationship between Hubble type and bulge-to-total ratio obtained by Simien and de Vaucouleurs (1986) to convert total magnitudes to bulge magnitudes. The results are shown Fig. 1.

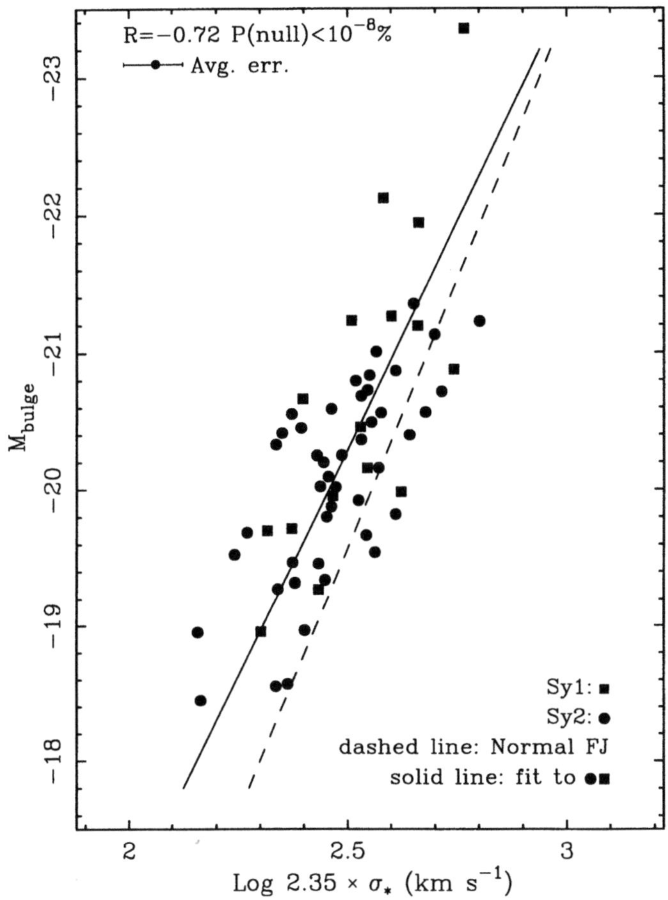

Fig. 1. The Faber-Jackson relation for Seyfert galaxies. Note that the velocity dispersion is plotted as log $2.35 \times \sigma_*$

We find a strong correlation for the Seyferts similar to that found for normal galaxies with a best fit (solid line) yielding a value for the exponent $n \simeq 3$. The tightness of the correlation indicates that our σ_* values are reliable. More importantly, it suggests that Seyferts have essentially normal stellar kinematics. However, the dashed line shows the Faber-Jackson relation for normal galaxies determined by Whittle (1992b) using published data for normal ellipticals and spiral bulges. Notice that the Seyferts are offset from this relation having lower σ_* at a given value of $M_{\rm bul}$ ($\Delta \log\sigma_* \simeq -0.1$) or alternatively having brighter bulges ($\Delta M_{bul} \simeq -0.6$ mag) for a given σ_*. We believe this offset is real since we consider our σ_* values to be accurate, the mean correction for AGN light is small ($\langle \Delta m_{\rm AGN} \rangle = 0.04$) and the Hubble types used for the bulge corrections would have to be systematically 2 stages too early (Sa is really Sb) to account for the effect (see Nelson and Whittle 1996, hereafter NW96).

Why should Seyferts be offset from the Faber-Jackson relation for normal galaxies? One possibility is that Seyferts have massive black holes which alter the nuclear stellar kinematics. However, the observed shift is in the wrong direction since we would expect higher values of σ_* in objects with massive black holes. Furthermore, our apertures are too large ($\sim 1 - 2''$ corresponding to several hundred pc) to detect the kinematic influence of even a very large black hole.

Spheroidal stellar systems follow a planar relationship in the parameter space defined by σ_*, $M_{\rm bul}$ and $\langle \mu \rangle$, the mean surface brightness, known as the fundamental plane. The Faber-Jackson relation is a projection of this onto the σ_*-$M_{\rm bul}$ plane. Thus it is possible that Seyferts, although offset from the Faber-Jackson relation, lie on the fundamental plane but have systematically different surface brightnesses. If true, Seyferts would have fainter surface brightnesses than normal galaxies by ~ 1 mag. We consider this possibility unlikely since Seyferts are generally considered to be systems with prominent bulges, although surface photometry of Seyferts explicitly separating bulge, disk and nuclear emission is clearly important.

The most likely explanation of the Faber-Jackson offset is that Seyferts have lower mean mass-to-light ratios, indicating younger stellar populations than those found in the bulges of normal galaxies. Comparing the Faber-Jackson residuals for the Seyfert and normal galaxies we find that the Seyferts, although shifted, are actually less scattered (NW96). In fact if we interpret these residuals solely in terms of differences in M/L we find that Seyferts do not have excessively low M/L but rather seem to avoid objects with high M/L. Thus, those systems which have only small amounts of interstellar gas in their nuclei and are unable to sustain even moderate star formation rates make poor candidates for Seyfert galaxies. We have also found that Seyferts follow the same relationship between rotation amplitude, $V_{\rm max}$, and σ_* as normal galaxies suggesting normal stellar kinematics and that it is the photometric parameter, $M_{\rm bul}$, and not the kinematic one, σ_*, that is primarily responsible for the offset (NW96).

5 Stellar and Gaseous Kinematics in Seyfert Galaxies

We now turn to a comparison of the stellar and gaseous kinematics in Seyfert galaxies. In Fig. 2 we plot the FWHM of the [O III] λ5007 emission line against

$2.35 \times \sigma_* = \mathrm{FWHM}_*$. The Seyferts are loosely grouped around the $X = Y$ (dashed) line with a moderately strong correlation and a number of objects showing rather high emission line widths relative to their stellar widths. Thus for most objects it appears that $\mathrm{FWHM}_{[\mathrm{OIII}]} \simeq \mathrm{FWHM}_*$. This suggests that the primary influence on the NLR kinematics is the host galaxy potential and confirms Whittle's (1992a,b) previous results using indirect virial parameters.

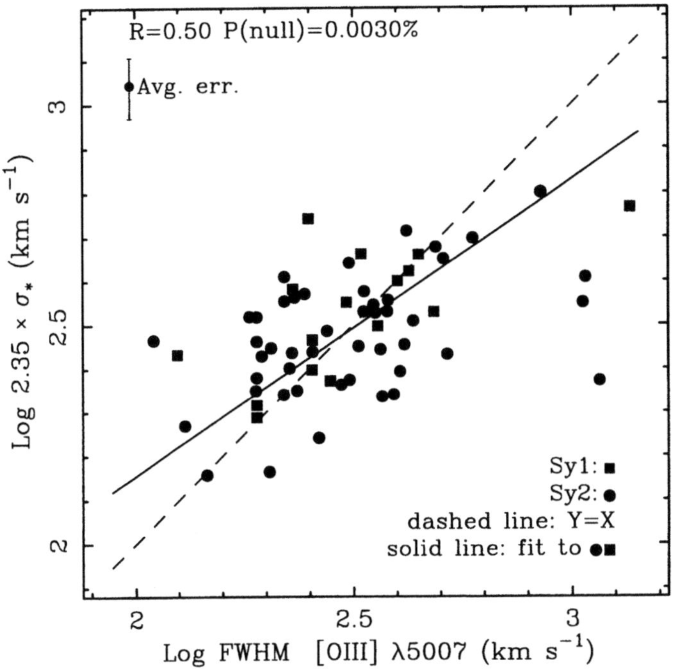

Fig. 2. $\mathrm{FWHM}_{[\mathrm{OIII}]}$ vs. FWHM_* is shown. The solid line is a fit to the data and the dashed line shows $X = Y$

We note however that the scatter in this diagram is larger than the measurement error indicating that a significant portion of it is real. (The mean error on σ_* is indicated in the upper left corner of Fig. 2). Thus we can look for secondary influences on the gas kinematics by searching for trends in the scatter about the mean relation.

We first consider trends with radio morphology. Four of the five objects with the largest emission line widths (NGC 1068, NGC 1275, Mkn 78, Mkn 3) are known to have linear radio sources. The fifth (Mkn 622) may also be a linear radio source because of its peculiar rectangular shaped [O III] profile which is considerably different from its much narrower Hβ line. Flagging other objects with linear radio sources also shows a tendency for the more luminous of these to lie on the high emission line width side of the fit to the remaining objects (NW96). A similar deviation for linear radio sources was found by Whittle (1992a,b). These results suggest that interaction between the radio plasma and the ionized gas provides additional acceleration and broadens the emission lines.

We also find that interacting and morphologically disturbed objects tend to have higher emission line widths at a given σ_* than the remaining galaxies (NW96). Thus, galaxy interactions can also perturb the NLR velocity field. Barred galaxies show only a weak tendency to have broader lines at a given velocity dispersion and instead are more scattered about the mean relation. Similar results were found by Whittle (1992a,b).

6 Active Properties and the Gravitational Potential

We now consider links between the host galaxy and properties more directly associated with the active nucleus. We begin by noting that Seyfert type was not a distinguishing parameter in any of the previous analysis. If the observed differences in Seyfert 1 and Seyfert 2 galaxies are due to differences in the orientation of an obscuring molecular torus, as postulated in unified models of AGN, then we conclude that the kinematics of stars and gas show no preference for motion in the plane of the torus.

In Fig. 3 we plot the radio luminosity at 1415 MHz, L_{1415}, versus σ_* showing a moderately strong correlation. Objects flagged with + symbols are linear radio sources and have higher radio luminosity for a given value of σ_*. A similar correlation exits between the luminosity of the [O III] λ5007 line and σ_*.

These relationships demonstrate that the strength of the emission from the NLR scales with the depth of the host galaxy potential. The simplest explanation is that galaxies with more massive bulges have more powerful central engines perhaps as a result of more massive nuclear black holes. However, it is also possible that physical conditions in the NLR are set by the bulge potential. For example, the pressure in the hot ISM of the bulge may well depend on the depth of the potential. Since the emissivity of radio plasma depends on the pressure, more luminous radio sources may be produced in galaxies with more massive bulges all other things being equal. We have also found that this correlation extends to the cores of radio galaxies (NW96) suggesting a continuity in the galactic scale radio properties of radio quiet (e.g. Seyferts) and radio loud AGN.

7 HST Snapshot Imaging of Markarian Seyferts

We also discuss some results from an imaging survey of 52 Seyfert and 50 non-Seyfert Markarian galaxies obtained with the Hubble Space Telescope (HST). These observations were taken using WF/PC-1 in a near-infrared bandpass (the F785LP filter) which is virtually free of emission lines. Therefore, these images are almost entirely continuum emission from starlight and nuclear processes. The goals of this project are to study the small-scale structure in Seyfert galaxies and to examine the host galaxy bulge. We find that several galaxies possess stellar bars less than a kiloparsec in length and several others have double nuclei (MacKenty et al. 1994; MacKenty et al. 1996).

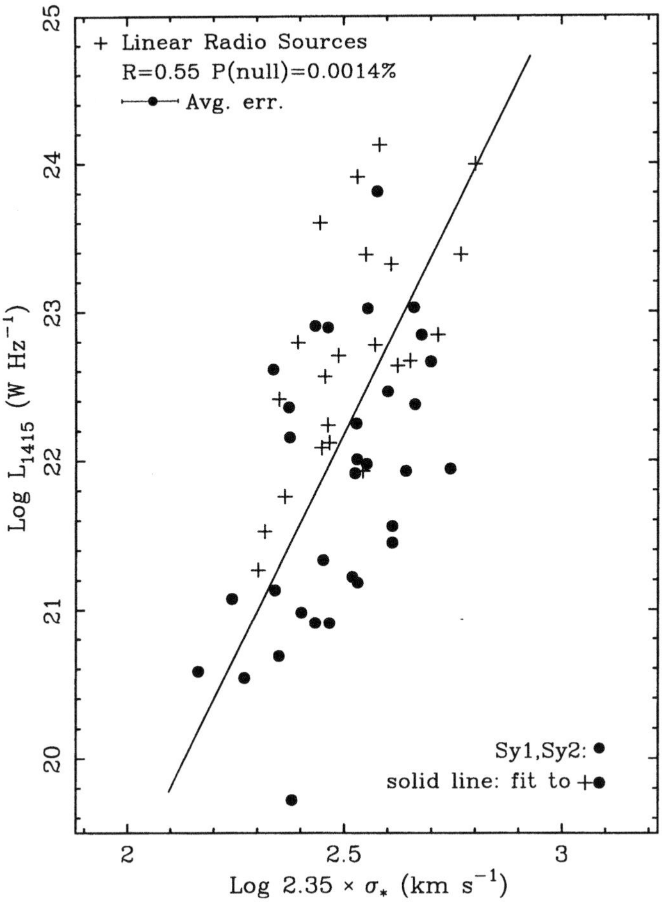

Fig. 3. Radio luminosity at 1415 MHz is plotted against σ_*. We find a moderately strong correlation and a significant tendency for objects with linear radio sources to have relatively higher [O III] luminosity at a given σ_*

We have also found a striking difference in the appearance of Seyfert 1 and Seyfert 2 nuclei in this sample (Nelson et al. 1996). The nuclei of Seyfert 1 galaxies are dominated by a strong point source consistent with being the nuclear continuum source viewed directly. Seyfert 2 galaxies by contrast are weaker, often resolved and have luminosity profiles typical of normal bulges. Since dominant point sources are not detected in the Seyfert 2s the featureless continuum emission seen in these objects must arise in an extended source perhaps as large as 50 – 100 parsecs across. In unified models of AGN the continuum seen in Seyfert 2 galaxies is seen only as scattered light "reflected" over the edge of the obscuring molecular torus (see e.g. Antonucci 1993) and is therefore highly polarized. However, Tran (1995) has found that a substantial amount of nuclear continuum is unpolarized and thus viewed directly. Another possibility is that since the profiles of Seyfert and non-Seyfert Markarian galaxies are similar, there may be a substantial contribution from circumnuclear star formation in the type 2 Seyferts. All of these possible contributors to the continuum emission in Seyfert 2 galaxies are expected to be extended sources in agreement with our observations.

8 Summary

In order to study the relationship between the host galaxy and the active nucleus in Seyfert galaxies we have obtained stellar and gaseous kinematic measurements for a large sample. Our results show that Seyfert galaxies follow a tight relation between σ_* and $M_{\rm bul}$ similar to the Faber-Jackson relation although offset from the one found for normal galaxies. This indicates that Seyfert bulges have lower mean M/L than those in normal galaxies suggesting younger stellar populations. We have also found that the gravitational potential is the primary influence on the NLR gas kinematics. Secondary influences are related to the effects of kpc scale linear radio sources and galaxy interactions. Correlations of emission line and radio luminosities with σ_* show that the strength of the NLR emission is also linked to the mass of the bulge.

Results from a WF/PC-1 imaging survey of Markarian Seyfert galaxies are also reported. We find that a number of these objects contain double-nuclei and small scale bars. Also differences in the appearance of Seyfert 1 and Seyfert 2 nuclei indicate that the featureless continua in the type 2s are likely to be extended sources in agreement with expectations from unified models of AGN.

References

Antonucci, R. (1993): ARAA **21**, 473

Edelson, R. A. (1987): ApJ **313**, 651

MacKenty, J. M., Simkin, S. M., Griffiths, R. E., Ulvestad, J. S., Wilson, A. S (1994): ApJ **435**, 71

MacKenty, J. M., Nelson, C. H., Simkin, S. M., Griffiths, R. E., (1996): in preparation

Meurs, E. J. A., Wilson, A. S. (1984): A&A **136**, 206

Nelson, C. H., MacKenty, J. M., Simkin, S. M., Griffiths, R. E., (1996): ApJ in press

Nelson, C. H., Whittle, M. (1995): ApJS **99**, 67

Nelson, C. H., Whittle, M. (1996): ApJ, in press (NW96)

Simien, F., de Vaucouleurs, G. (1986): ApJ **302**, 564

Terlevich, E., Diaz, A. I., Terlevich, R. (1990): MNRAS **242**, 271

Tonry, J., Davis, M. (1979): AJ, **84** 1511

Tran, H. D. (1995): ApJ **440**, 597

Whitmore, B. C., McElroy, D. B., Tonry, J. L. (1985): ApJS **59**, 1

Whittle, M. (1992a): ApJ **387**, 109

Whittle, M. (1992b): ApJ **387**, 121

Wilson, A. S., Heckman, T. M. (1985): in Astrophysics of Active Galaxies and Quasi-Stellar Objects, ed. J. S. Miller, California: University Science, p. 39

The Pattern Speed of the Galactic Bulge

Agris J. Kalnajs

Mount Stromlo and Siding Spring Observatories, Private Bag, Weston Creek PO, 2611, Australia

Abstract. Bar formation and the buckling of hot disks are two independent processes. Together they can produce rotating triaxial systems which resemble the Galactic bulge. The triaxial system behaves like the bar mode of a disk: giving it angular momentum makes it rounder, while removing angular momentum makes it more triaxial. The relative position of the Sun can be determined by matching the COBE 3.5 μm light distribution along $l = 0°$, and the velocity dispersion ratios in Baade's window. From the velocity dispersion in Baade's window, and an assumed distance of 8 kpc to the Galactic center we obtain a mass of $2.7 \times 10^{10} M_\odot$ and a pattern speed of 39 $\mathrm{km\,s^{-1}\,kpc^{-1}}$.

1 Introduction

The bulge of a disk galaxy may contain only a modest fraction of the mass, but because of its location and compact size it plays a very important role in the dynamics of the system.

In this paper I will discuss a family of rotating triaxial bulge models which I accidentally discovered while performing numerical simulations of the buckling of hot disks. There is a strong suspicion that each member of this family is just a small or finite amplitude bar mode of an oblate rotating system.

Rigid bulges have been used, in both numerical and analytic discussions of disk dynamics, to successfully tame the bar and other disk instabilities. The axisymmetric force field from the rigid bulge stiffens the central part of the disk by increasing the orbital frequencies and this makes it less responsive.

Real bulges will also stiffen the central part of the disk, but because they are not rigid they can be deformed by external fields. The deformation depends on the strength and the pattern speed of the field. A particularly interesting situation arises when the bulge has a bar mode. When such a discrete mode is perturbed in a periodic fashion it behaves like a harmonic oscillator. Recall that such an oscillator responds sympathetically when forced at a frequency less than its own natural frequency, but oppositely when the forcing frequency has been raised above the latter. Similarly here any discrete bar mode can be expected to interact strongly with external fields whose pattern speeds are close to that of the bar, augmenting those which are faster, and opposing those which are slower.

I have conjectured that the pattern speed of the Galactic disk is high enough to place us just outside the outer Lindblad resonance (Kalnajs 1991), which makes it close to 46 $\mathrm{km\,s^{-1}\,kpc^{-1}}$. If the pattern speed of the bulge is close to but below this value, one can expect a cooperative interaction between the disk and the bulge, whereas if the bulge preferred a pattern speed higher than this, the two will oppose each other. Such opposition would dampen the prospect for a long-lived grand design spiral structure in the Galactic disk.

2 Models

My interest in three-dimensional disk dynamics was kindled by Sellwood's report on the buckling instabilities of bars (Sellwood 1991). These buckling instabilities were first observed in numerical simulations of warm disks which initially developed planar bars (Combes et al. 1990; Raha et al 1991). Combes et al. (1990) argued that the subsequent thickening of the bar was related to planar orbital instabilities associated with the bar potential, whereas Raha et al. (1991) argued more convincingly that the thickening was due to a buckling instability (Toomre 1966; Merritt and Sellwood 1994).

I had already constructed a family of flat self-gravitating finite disk models. These differentially rotating models have a phase space distribution which is a function of $\Psi = E - \Omega J + (J/R_0)^2/2$, where E is the specific energy and J is the specific angular momentum. The distribution function $F(\Psi)$ is constant if $\Psi_- < \Psi < \Psi_+$ and is zero elsewhere. One can easily integrate $F(\Psi)$ over the velocities and obtain the surface density as a function of radius and potential. The harder part is finding the density and potential pair which satisfies the resulting integral equation. Because these models were designed to provide starting conditions for numerical studies of particle correlations, the solution of Poisson's equation incorporates gravity softening. Here R_0 is a free parameter, while Ω is determined by the condition that the mean circular velocity should equal the circular velocity at the edge.

The presence of the quadratic term in J makes the models rotate differentially in the mean, and regulates the amount of random kinetic energy. In units where the gravitational constant, G, the total mass, M, and the radius of the edge, R, are all 1, one can find models with R_0 ranging from 0.7 to ∞.

Table 1 gives a brief summary of the disk models. The amount of shear present can be judged from the circular angular velocities at the center, Ω_c, and the edge, Ω_e. The Ostriker-Peebles parameter, t, is a dimensionless measure of the amount of organised motion. The ratio of Ω_c and the *mean* angular velocity at the center, Ω, is another measure. There are two values for each parameter, the first is for no gravity softening, and the second for the softening parameter $a = 0.0125$ appropriate for the model discussed at length in the following sections.

The unusual choice of the functional form of the distribution function greatly simplifies the analytical stability calculations, which should be performed if one wishes to demonstrate the stability of a subset of these models. But that involves hard work. It is much easier to run a few numerical simulations, and these show that the models are unstable and form bars at the low end of R_0, and appear to be stable when R_0 is large. The critical value of R_0 which separates the stable and unstable models is difficult to determine using only a few thousand particles,

Table 1. Disk model parameters

R_0^2	t		Ω		Ω_c		Ω_e	
1	0.1826	0.1892	2.2038	2.2051	4.3732	4.1802	1.1019	1.1025
2	0.1480	0.1520	1.6724	1.6723	3.9072	3.7700	1.1150	1.1149
4	0.1293	0.1323	1.4048	1.4042	3.6459	3.5328	1.1238	1.1234
∞	0.1093	0.1114	1.1354	1.1344	3.3609	3.2706	1.1354	1.1344

because as one approaches instability the fluctuations become large. It appears to be ≈ 2.

With the help of these models it was easy to convince oneself that the bar formation and the buckling of a disk are two independent processes, as the following reasoning illustrates.

The left panel of Fig. 1 shows the time history of the square roots of the principal moments of inertia of four unstable $R_0 = 1$ models. (The planar moments have been tapered to minimise the contribution from the high energy tail created by the bar instability). The two-dimensional equilibrium was perturbed by giving each of the 8 000 particles a small displacement in z coordinate, chosen at random from the interval $[-0.0025, 0.0025]$. The initial planar density fluctuations are much larger and this gives the bar instability a head start. The onset of the buckling can be advanced by larger initial z displacements, or a smaller value of gravitational softening, a, which in these cases was 0.025 .

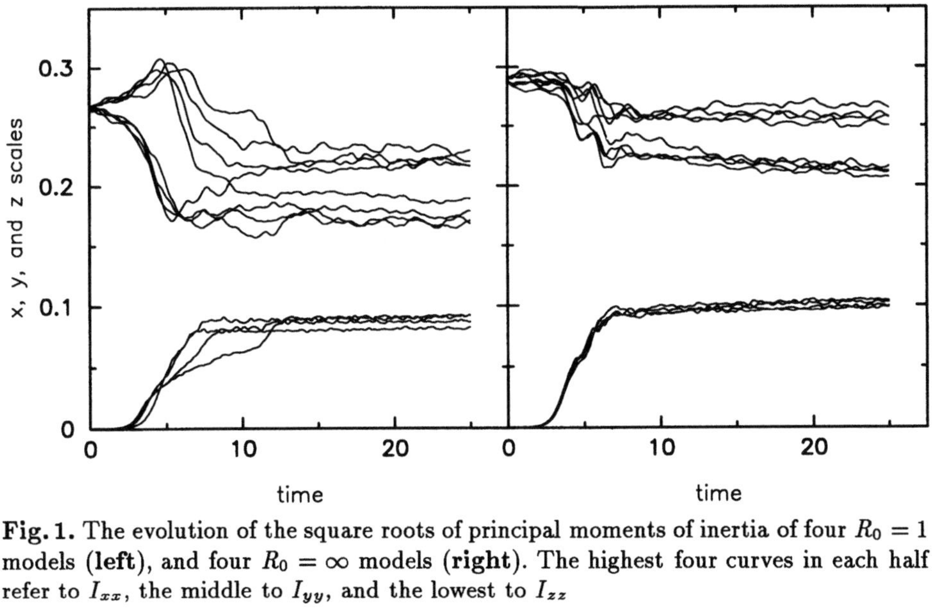

Fig. 1. The evolution of the square roots of principal moments of inertia of four $R_0 = 1$ models (**left**), and four $R_0 = \infty$ models (**right**). The highest four curves in each half refer to I_{xx}, the middle to I_{yy}, and the lowest to I_{zz}

Because the bar developed before the buckling, one might be forgiven for suspecting a causal connection between the bar formation and the buckling. However this should be dispelled by the right panel of Fig. 1 which shows the corresponding histories of four stable $R_0 = \infty$ models. The small initial jump in the bar amplitude is probably due to the establishment of particle-particle correlations. Here it is pretty clear that the final triaxial figure is caused by the finite amplitude buckling, whose shape is seldom axially symmetric.

3 Triaxial Figures

Our main interest lies in the rotating triaxial end products produced by the violent buckling instabilities and not in the history of their formation. It is most unlikely that real bulges are formed by the above process, since it requires implausible initial conditions. However that is the way I built them – a more clever person might have constructed the same objects by means of linear programming, or from an inspired guess of the equilibrium state.

The $R_0^2 = 2$ disk models usually produce somewhat more pronounced triaxial objects than those seen in the $R_0 = 1$ and $R_0 = \infty$ experiments described above. There may also be some lingering doubt about the effect of the gravity softening on the final z thickness, since it is only four times larger than the $a = 0.025$ used in the above experiments. These reasons motivated the choice of a $R_0^2 = 2$ disk model with $a = 0.0125$ as the subject for a more detailed study.

Model 222, named after the random number seed used to produce it, was started just like the previous experiments. It began to buckle sooner because a was halved. By $t = 25$ it had evolved into a triaxial object rotating with a pattern speed $\Omega_{\mathrm{p}} = 1.140$. At this point the single experiment split into five parallel ones. During the interval $25 < t < 40$ an external field was applied. The shape of this field closely approximated that of the $m = 2$ component from the triaxial density at $t = 25$, but the phase was advanced by $\pi/2$ radians and this shift was maintained. The field was turned on and off in a cosbel fashion. The duration of each transition was 3 time units. The relative amplitudes of the five fields were $-5, -1, 0, 1, 5$, and these torques produced -19.5%, -3.4%, 0%, 2.8%, and 6.7% changes in the total angular momentum.

Figure 2 shows the time evolution of the square roots of I_{zz} and the azimuthally averaged I_{xx}. The initial expansion perpendicular to the plane is accompanied by a contraction along the plane. The central density rises, and the circular rotation rate in the center increases by a factor of three. The shape of the axisymmetric part of the potential in the plane closely resembles a Plummer potential with a scale length of 0.2075 . Later the torques produce small changes in the equatorial scale, which shrinks as the angular momentum decreases.

Figure 3 is more interesting. It shows the time evolution of the bar-like aspect of Model 222 as measured by $(I_{xx} - I_{yy})/(I_{xx} + I_{yy})$. The bar becomes more pronounced as angular momentum is removed from the system. A 6.7% increase in the angular momentum is sufficient to make the system round.

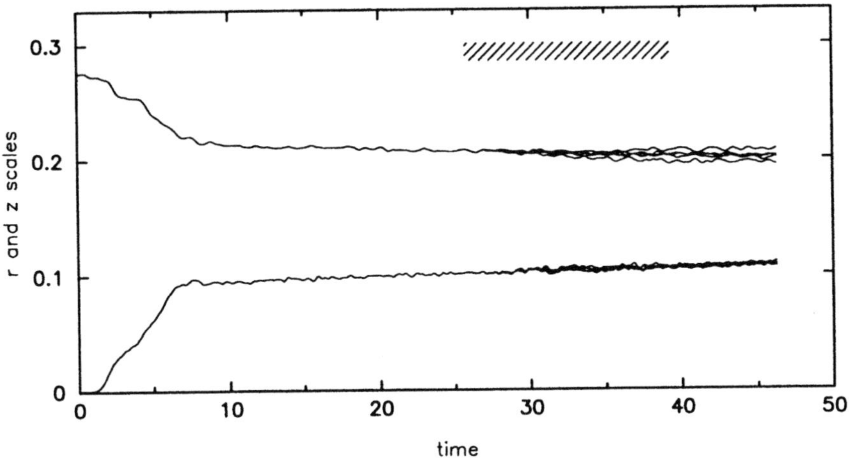

Fig. 2. The evolution of the square roots of $(I_{xx} + I_{yy})/2$ (**top**), and I_{zz} (**bottom**) of the $R_0^2 = 2$ model. The hatched region shows the effective duration of the torques

A more graphic representation of the end results of the three experiments which produced the changes of 6.7%, 0%, and −19.5% is shown in Fig. 4.

After the initial rapid changes the system settles down into a slowly evolving phase, and not a strict equilibrium. For example, the middle curve in Fig. 3 shows a linear decrease of the bar amplitude which corresponds to a damping time of 120, or 22 bar revolutions. At this stage it is not clear how much of this decrease and the gentle drifts seen in Fig. 2 are due to the roughness of the 8 000 particle simulation and how much would survive in the large N limit.

For the record: the forces are calculated by direct summation and the leapfrog integration time step is 0.00325 .

4 A Model of the Galactic Bulge

For some time I considered the triaxial objects as mere curiosities. I began to take them more seriously after discovering that the projected densities resembled the bulge light distribution measured by the COBE satellite (Dwek et al. 1995).

One can determine the solar radius in model units by matching the shape of the model $l = 0°$ latitude light distribution to the 3.5 μm $l = 0°$ latitude scan measured by Dwek et al. (1995). Figure 5 shows that both the bulge light and model surface density fall off exponentially away from the plane and that a good match of the exponential scale length can be achieved when the Sun is placed at 1.75 units from the center. The fit is not sensitive to the orientation of the bar.

The model fails to reproduce the central peak. The missing light can be accounted for by adding a miniature version of the model – one that is shrunk by a factor of eight in linear dimension and has the orientation shown in the middle panel of Fig. 4. This addition contains 3.3% of the bulge light. There is

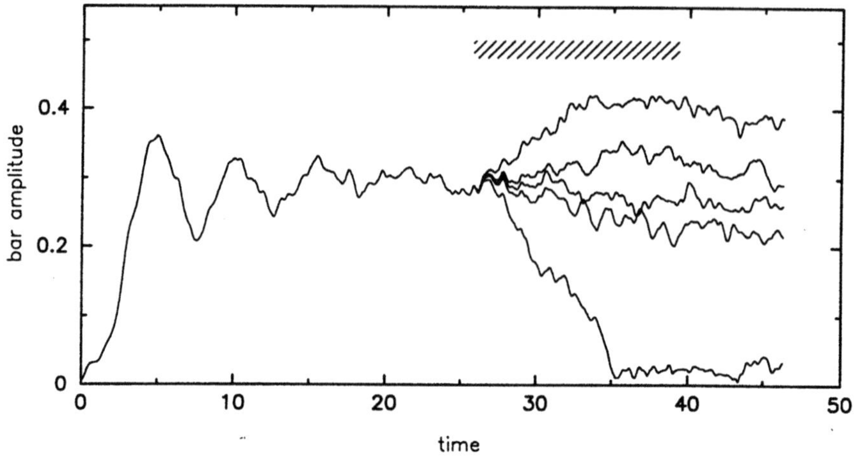

Fig. 3. The evolution of the bar amplitudes $(I_{xx}-I_{yy})/(I_{xx}+I_{yy})$ of the $R_0^2 = 2$ model. The highest final curve resulted from the 19.5% loss of angular momentum, whereas the lowest curve arose from the 6.7% gain. The hatched region shows the effective duration of the torques

good evidence, derived from the motions of OH/IR stars (Lindqvist et al. 1992), that the central light peak is associated with a similar increase in mass.

The velocity scale and an orientation for the bar can be obtained by matching the observed velocity dispersions in Baade's window, ($l = 1°$, $b = -3.9°$). The three components of velocity dispersion of bulge stars are: $\sigma_r = 110 \pm 10$, $\sigma_l = 116 \pm 9$, and $\sigma_b = 105 \pm 8$ kms/sec (Terndrup et al. 1995). Table 2 below shows the values of the dispersion velocities and the mean radial velocity, v_r, predicted by the model. The velocity scaling is obtained by setting the radial dispersion equal to 110 km/sec. Good agreement is achieved when the long axis of the bar is parallel to a direction which lies in the range $315° < l < 330°$ and $b = 0°$. Good agreement can also be achieved when the direction lies in the range $30° < l < 45°$. But this range would put the near side of the bulge on the wrong side of the center.

The apparent axial ratio of the bulge, measured at the 25% of peak level, is 0.50 at the $l = 315°$ orientation, and 0.58 at $l = 330°$. These values can be adjusted by changing the intrinsic axial ratios of the model.

If we assume that the Sun is at a distance of 8 kpc from the center, and use a value of 159 $\mathrm{km\,s^{-1}}$ for the velocity scaling, the mass of the bulge becomes $2.69 \times 10^{10} \mathrm{M_\odot}$, and the pattern speed 39 $\mathrm{km\,s^{-1}\,kpc^{-1}}$.

An independent determination of the velocity scaling and orientation has been obtained by Beaulieu (1996) who used the radial velocities and the l, b distribution of 97 planetary nebulae from a new survey covering the region $-10° < b < -5°$ and $-20° < l < 20°$. The two results are in agreement.

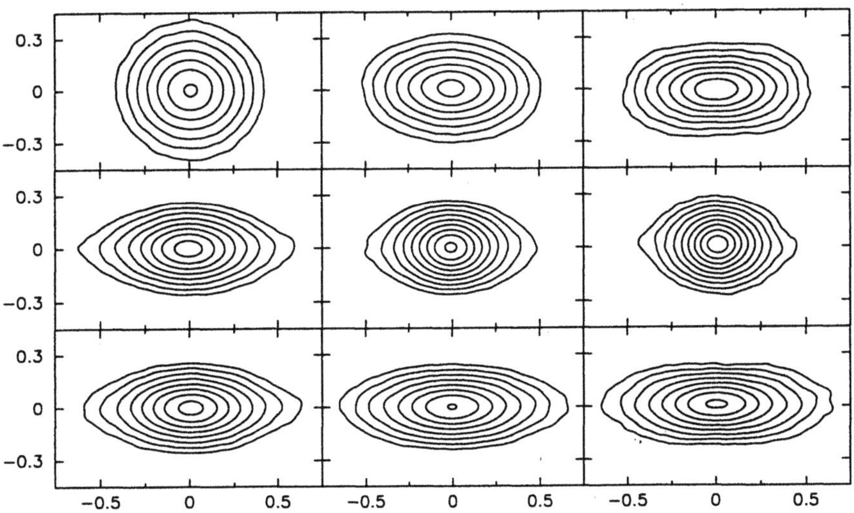

Fig. 4. Three triaxial figures viewed along their principal axes. Top row shows the view along the z-axis, middle row along the x-axis, and bottom row along the y-axis. The figure which gained most angular momentum is shown in the left column, and the one which lost most is on the right. The angular momentum of the central figure was unchanged. The outermost isophote is the same in all panels, and the levels increase in −.5 magnitude steps

Table 2. Model predictions for Baade's Window

θ	*scale*	v_r	σ_r	σ_l	σ_b
270°	168	11.82	110.0	139.5	107.3
285°	168	13.02	110.0	139.9	107.6
300°	162	15.98	110.0	129.8	102.8
315°	161	13.65	110.0	127.6	102.0
330°	155	16.15	110.0	117.7	97.7
345°	152	17.14	110.0	112.6	93.4
360°	153	16.18	110.0	111.1	92.2

5 The Missing Pieces

5.1 The Disk

The rotation curve produced by the bulge alone has fallen to 56% of its peak value at the position of the Sun, and therefore a substantial disk is needed to keep it level. The dynamics of the bulge is not very sensitive to the axisymmetrical part of the force arising from such a disk. For example, the slow addition of

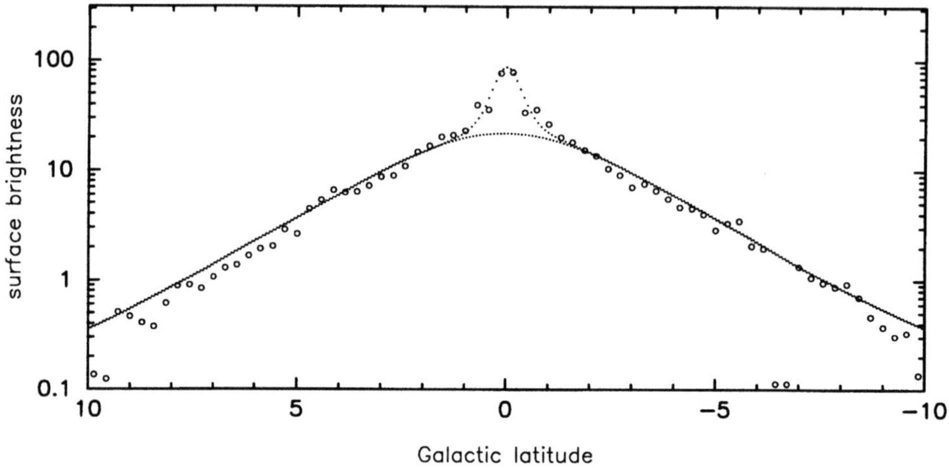

Fig. 5. Comparison of the COBE/DIRBE 3.5 μm light distribution (circles) with model (lower dotted curve), and model plus a scaled version of the model (upper dotted curve)

the force from a rigid Plummer sphere with scale length of 2 and a mass of 8, produces a rotation curve which is nearly flat, and raises $\Omega_{\rm p}$ by 6% to 1.203. The more interesting interaction with the non-axisymmetric part of the stellar and gaseous part of the disk which can give rise to torques, is still to be examined.

5.2 The Central Core and Damping

That small 3.3% of excess light not accounted for by the model cannot be ignored. When a small Plummer sphere with a scale length of 0.025 and containing 3% of the mass is slowly added, $\Omega_{\rm p}$ rises to 1.358 and the bar amplitude begins to decline linearly with time. The damping time is 45, or 10 bar revolutions.

The explanation for this seemingly dramatic change lies in the sympathetic response of the bulge to the added central mass. A sphere of radius 0.15 would on the average contain 1751 bare bulge particles. The added Plummer sphere contributes the equivalent of another 230, which in turn attract a further 521 bulge particles. The characteristic size of the attracted cloud is determined by the structure of bulge, which means that the density changes will occur where they matter. The addition of an extra 751 particles to the 1751 existing ones is no longer a 3% effect. The effects of the added Plummer sphere and the resultant density redistribution on $\Omega - \kappa/2$ are shown in Fig. 6.

The response of a spherical system to an imposed central mass point was first calculated by Gilbert (1970). He showed that, in the limit of vanishingly small mass, the effective mass of the imposed point will be increased by a factor of 2.75 by what he called polarisation. For our rotating, flattened bulge the factors are 3.26, 3.51, and 3.92 for 3%, 2%, and 1% mass, which extrapolates to 4.40 for a vanishingly small mass.

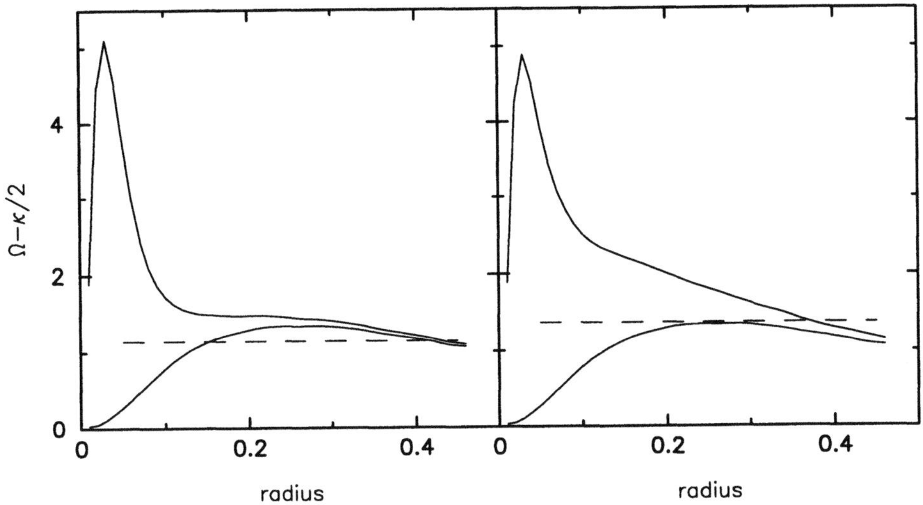

Fig. 6. The Lindblad precession frequencies, $\Omega - \kappa/2$, of the bare bulge (lower solid curves). Expected change produced by the force from a small Plummer sphere containing 3% of the mass (upper curve, left), and actual change (upper curve, right). The dashed lines show pattern speeds of the bare bulge (left), and bulge plus core (right)

The Lindblad precession frequency, $\Omega - \kappa/2$, has a well defined meaning for planar orbits moving in an axially symmetric potential. In a Plummer potential the precession rate of an orbit with a fixed angular momentum decreases with increasing radial action. Therefore the precession rate of any orbit will not exceed the peak of the circular $\Omega - \kappa/2$ curve. Because the Ω_p of the bare bulge lies below the peak, there must be some resonant planar orbits which lose angular momentum and in so doing damp the wave. The meaning of $\Omega - \kappa/2$ becomes blurred when the orbits are allowed to move out of the plane. But one suspects that the damping is related to a similar resonant interaction, and the circular $\Omega - \kappa/2$ curve appears to be a good *indicator* of the number of potential angular momentum losers.

When the damping is small, a small negative torque can arrest it. But a negative torque produces secular changes which slow down the pattern speed and raises the peak of the $\Omega - \kappa/2$ curve. In the largest negative torque experiment, summarised by the top curve in Fig. 3, Ω_p fell to 58% of its initial value and the peak rose by 15%.

A better way to counteract the damping is to raise Ω_p.

6 Conclusion

The initial experiments which produced the rotating triaxial objects provided a simple answer to the question of why some bulges are triaxial: they are modes, which once excited, will last for a long time. The small amount of damping which was present did not seem significant and could have been caused by the roughness of the 8 000 particle simulations. Longer integrations and the observation that the addition of a small amount of mass in the center increased the damping and decreased the lifetime of the triaxial shape to 1.3×10^9 years, spoiled that tidy answer.

This failure may be a blessing in disguise. It is well known that there is a relation between the bulge and disk mass distributions in the sense that the peak circular velocity produced by the bulge seldom exceeds that of the disk by more than about 10%. The 216 $\mathrm{km\,s^{-1}}$ peak at 1.37 kpc produced by Model 222 (which rises to 231 at 1.04 kpc when the extra 3% core is added) follows this tradition. The relation between the disk and bulge mass distributions reinforces the notion that the two components must interact via non-axisymmetric forces.

Such an interaction has immediate benefits for the bulge: negative torques can make it more triaxial and any induced bar structure in the disk will raise the pattern speed. It remains to be seen whether these promised benefits are sufficient to sustain a long-lived grand design.

References

Beaulieu, S. (1996): Ph. D. thesis, Australian National University
Combes, F., Debbasch, F., Friedli, D., Pfenniger, D. (1990): A&A **233**, 82
Dwek, E., Arendt, R.G, Hauser, M.G., Kelsall, T., Lisse, C.M., Moseley, S.H., Silverberg, R.F., Sodrosky, T.J., Weiland, J.L. (1995): ApJ **445**, 716
Gilbert, I.H., (1970): ApJ **159**, 239
Kalnajs, A.J. (1991): in Dynamics of Disc Galaxies, ed. B.Sundelius (Göteborgs University), p. 323.
Raha, N., Sellwood, J.A., James, R.A, Kahn, F.D. (1991): Nat **352**, 411
Sellwood, J.A. (1991): in Dynamics of Disc Galaxies, ed. B.Sundelius (Göteborgs University), p. 123.
Lindqvist, M., Habing, H.J., Winnberg, A. (1992): A&A **259**, 118
Merritt, D., Sellwood, J.A., (1994): ApJ **425**, 551
Terndrup, D.M., Sadler, E.M., Rich, R.M. (1995): AJ **110**, 1774
Toomre, A. (1966): in Geophysical Fluid Dynamics, Notes on the 1966 Summer Study Program at the Woods Hole Oceanographic Institution, ref. no. 66-46, p. 111

The Central Parsec of the Milky Way: Star Formation and Central Dark Mass

Reinhard Genzel

Max-Planck-Institut für extraterrestrische Physik, Giessenbachstraße, D-85748 Garching, Germany

Abstract. Recent high resolution near-infrared imaging and spectroscopy give detailed information about the structure, evolution and mass distribution in the nucleus of the Milky Way. The central parsec is powered by a cluster of luminous helium rich, blue supergiants. The most likely scenario for the formation of these massive stars is a burst of star formation a few million years ago. Radial velocity measurements for about 300 early and late type stars between 0.1 and 5 pc radius from the dynamic center now strongly favor the existence of a central dark mass of 2.5 to 3.3×10^6 $M_\odot$ within 0.1 pc of the dynamic center. This central dark mass cannot be a cluster of neutron stars. It is either a compact cluster of stellar black holes or, most likely, a single massive black hole.

1 Introduction

The nucleus of the Milky Way (adopted distance 8.5 kpc) is one hundred times closer to the Earth than the nearest large external galaxy and more than a thousand times closer than the nearest active galactic nuclei. We can therefore study physical processes happening in our own Galactic Center at a level of detail that will never be reached in the more distant, but usually also more spectacular systems. What powers these nuclei and how do they evolve? What are the properties of the nuclear stellar clusters? Is star formation happening there? Do dormant massive black holes reside in their cores? In the present paper I will describe the status of our present knowledge about these key questions. For a more extensive discussion I refer to Genzel, Hollenbach and Townes (1994).

The nuclear mass is dominated by stars, except probably in the innermost parsec. The density of stars increases with decreasing radius R from the dynamic center approximately as $1/R^2$ and attains a value of $\sim 4 \times 10^6$ $M_\odot$ per pc^3 in the central few tenths of a parsec. Infrared observations on a scale of 100 pc to 1 kpc show that these stars appear to be distributed in a rotating bar (Blitz and Spergel 1991). The gravitational torque of this bar may also explain the non-circular motions of interstellar gas clouds found by radio spectroscopy (Binney et al. 1991). The non-circular motions in turn may trigger gas infall into the nucleus. There is increasing evidence from gamma-ray spectroscopy of the 1.8 MeV ^{26}Al line (Diehl et al. 1993) and from infrared stellar spectrophotometry (e.g. Lebofsky and Rieke 1987; Cotera et al. 1994; Figer 1995) that (massive) star formation has occured throughout the Galactic Center region no longer than 10 million years ago.

Also on a scale of ~100 pc several variable, spectacular hard X-ray and gamma-ray sources have been found (e.g. Skinner 1993). They may represent stellar black holes or neutron stars accreting gas from a companion or from nearby dense gas clouds. Throughout the central few hundred parsecs giant molecular clouds (~10^6 $M_\odot$) are found whose gas density ($n(H_2)$ ~10^4 to 10^6 cm^{-3}) and temperature (40 to 200 K) are significantly greater than those of the clouds in the Galactic disk (e.g. Güsten 1989). The dynamics of this central molecular cloud layer is characterized by large internal random motions and unusual streaming velocities that can be partially explained by the presence of the central bar potential mentioned above. On a scale of a few parsecs from the dynamic center there is a system of dense orbiting molecular filaments approximately arranged in form of a circum-nuclear 'disk' (Genzel et al. 1985; Jackson et al. 1993). The circum-nuclear disk is probably fed by gas infall from dense molecular clouds at 10 pc. Internal to the circum-nuclear disk one finds a number of ionized streamers (the "mini-spiral") orbiting the center (Lo and Claussen 1983).

Magnetic fields as large as ~1 mGauss appear to permeate the central 50 pc and are aligned approximately perpendicular to the Galactic plane (e.g. Morris 1993; Sofue 1994). Where they interact with neutral gas clouds remarkable filaments of nonthermal radio synchrotron emission are seen (Yusef-Zadeh, Morris and Chance 1984). The central radio source, Sgr A, can be separated into a thermal source, Sgr A West and a non-thermal source, Sgr A East. Sgr A East may be evidence for one or several supernovae that have exploded in the central 10 parsecs within the last 10^5 years (Mezger et al. 1989).

While the velocities of gas and stars are approximately constant outside of a few parsec, the velocities are observed to increase within the inner core (e.g. Genzel and Townes 1987). The first evidence for this increase in gas velocities came from mid-infrared spectroscopy of [Ne II] by Wollman (1976) and Lacy et al. (1979, 1980). These authors and others following interpreted the ~250 $km\,s^{-1}$ gas velocities as signalling a concentration of non-stellar mass in the Galactic Center, possibly a few million solar mass black hole at the dynamic center (Lacy et al. 1982; Serabyn and Lacy 1985). However, gas is affected by magnetic, frictional and wind forces, in addition to gravity so that stellar velocities are required to unambiguously determine the mass distribution.

At the dynamic center is a compact radio source, Sgr A* which is close to, but not coincident with a group of bright near-infrared sources (IRS 16) of blue color (e.g. Backer 1994). Since its discovery 25 years ago Sgr A* has been the most probable candidate for a central black hole.

2 What Powers the Central Parsec?

The observed broad band emission of the central few parsecs is dominated by intense mid- and far-infrared emission from 50 to 100 K dust grains originating in the circum-nuclear disk and in a cloud ridge associated with the ionized 'mini-spiral'. Taking into account the radiation not intercepted by the circum-nuclear gas the total UV and visible luminosity of the central parsec has been estimated to lie between 1 and 3×10^7 $L_\odot$ (Davidson et al. 1992). The Lyman continuum flux is about 2 to 3×10^{50} s^{-1} as determined from the thermal radio continuum (Lacy et al. 1980). Infrared spectroscopy of fine structure lines sampling a wide range of excitation stages implies that the effective temperature of the UV radiation field in the central parsec is only about 30 000 to 35 000 K. The line ratios also suggest a heavy element abundance of about twice that in the Sun (Lacy et al. 1980).

What powers this low excitation H II region and what are the properties of the central star cluster? Through the advent of sensitive, large format infrared detector arrays and speckle imaging it has become possible in the last few years to image the central parsec at the resolution to the diffraction limit of 4 m class telescopes ($\sim$0.1″ or 0.04 pc at 2 μm, Eckart et al. 1992, 1993, 1995). The best current images resolve the near-infrared emission of the central parsec into about 700 stars with K-band (2.2 μm) magnitudes $\leq$16. Thus all red and most blue supergiants, all red giants/AGB stars later than K5 and all main sequence stars earlier than B0.5 should be detected in those images. The central IRS 16 complex located within 1″ of the compact radio source Sgr A* consists of about two dozen single (or perhaps multiple) stars (see also Simons et al. 1990; Simon et al. 1990). From the number distribution of the near-infrared sources it appears that the centroid of the stellar cluster is more likely on Sgr A* than on the IRS 16 complex and that the core radius of the $K < 15$ stellar number density distribution is about 0.2 to 0.4 pc (Eckart et al. 1993, 1995; Genzel et al. 1996). If the stars with $K < 15$ are representative of the overall mass distribution of the cluster (an assumption that appears very plausible based on recent spectroscopic identification of the stars, Genzel et al. 1996) this core radius together with the mass of stars estimated to lie within a few parsecs indicates that the stellar density in the core about 3 to 8×10^6 $M_\odot$ pc^{-3}. Recent imaging spectroscopy with the new MPE 3D spectrometer shows that within the core radius bright late type stars (supergiants and the brightest AGB stars) are absent but that the core is surrounded by a ring of red supergiants/AGB stars (Genzel et al. 1996). Following earlier discussions of Lacy et al. (1982), Phinney (1989) and Sellgren et al. (1990), Genzel et al. (1996) interpret this finding in terms of destruction of the brightest (and hence largest) late type giant stars by collisions with main sequence stars. Assuming that such collisions in fact permanently destroy the outer atmosphere of the giants (see Davies et al. 1991) and that collisional destruction becomes observable whenever the collision time is less or equal than the lifetime of the red giant/supergiant phase, the observed lack of stars brighter than $K < 10$ also implies a core stellar density of about 4×10^6 $M_\odot$ pc^{-3}, in excellent agreement with the density estimated from the number counts.

Another important ingredient of the near-infrared story has been the discovery of a He I/Brγ near-infrared emission line star (the AF-star, Forrest et al. 1987; Allen et al. 1990), followed by the discovery of an entire cluster of about 25 such stars in the central parsec and centered on the IRS 16/IRS 13 complex (Krabbe et al. 1991, 1995). Several of the brightest members of the

IRS 16 complex are He I stars, as is the nearby bright source IRS 13 (Eckart et al. 1995; Krabbe et al. 1995; Libonate et al. 1995; Blum et al. 1995b; Tamblyn et al. 1996). The IRS 16 He I "broad line region" discovered a decade ago by Hall et al. (1982) and Geballe et al. (1984) thus is now identified as a group of luminous mass losing, He-rich stars. Non-local thermodynamic equilibrium (NLTE) stellar atmosphere modeling of the observed emission characteristics of the AF-star (Najarro et al. 1994) confirms and quantifies earlier proposals (Allen et al. 1990; Krabbe et al. 1991) that the AF-star is a WN9/Ofpe star. WN9/Ofpe stars are a rare class of luminous blue supergiants related to luminous blue variables (LBVs), WNL Wolf-Rayet stars and Of/ON supergiants (Allen et al. 1990; Krabbe et al. 1991; Najarro et al. 1994; Libonate et al. 1994; Blum et al. 1995b; Tamblyn et al. 1996). These stars very likely represent the post-main sequence phase (including perhaps the last part of the main sequence) of massive stars (20 to 120 $M_{\odot}$) before they explode as supernovae. The AF-star has a luminosity of about 3×10^5 $L_{\odot}$, effective temperature near 20 000 K and main-sequence mass between 25 and 40 $M_{\odot}$ (Najarro et al. 1994). The surface He/H abundance ratio is near unity and the mass loss rate is 6×10^{-5} $M_{\odot}$ yr^{-1} at a velocity of 700 km/s (Najarro et al. 1994). Based on the most recent 3D spectroscopy and modelling the brightest He I stars (IRS 16NE,C,SW, IRS 13) also have effective temperatures between 20 000 and 30 000 K, are Helium-rich and are about 5 to 10 times more luminous than the AF star (Krabbe et al. 1995). Their progenitor O stars likely had masses near 100 $M_{\odot}$. In addition several stars display C III/C IV/N III emission lines, characteristic for late WC and WN Wolf-Rayet stars (Blum et al. 1995a; Krabbe et al. 1995; Genzel et al. 1996). Combining the contributions from all its members, the He I-star cluster can plausibly account for essentially all of the bolometric and Lyman-continuum luminosities of the central parsec (Krabbe et al. 1995). The He I star cluster also provides in excess of 10^{38} erg s^{-1} in mechanical wind luminosity which may have a significant impact on the gas dynamics in the central parsec (Genzel, Hollenbach and Townes 1994). Krabbe et al. (1995) fit the properties of the massive early type stars in the central parsec by a model of a star formation burst between 9 and 3×10^6 years ago in which a few hundred OB stars and perhaps a total of a few thousand stars were formed. This conclusion is in excellent agreement with earlier proposals by Lacy, Townes and Hollenbach (1982), Rieke and Lebofsky (1982) and Allen and Sanders (1986). In the model of Krabbe et al. the He I stars are the most massive cluster members that in the mean time have evolved off the main sequence and the central parsec is now in the late, wind-dominated phase of the burst. The starburst model accounts naturally for the low excitation of the Sgr A West H II region. Although there is also evidence for some very young, embedded OB stars the present star formation activity appears to be significantly less than during the peak of the burst. The present gas density in the central parsec is too low for gravitational collapse of gas clouds to stars in the presence of the strong tidal forces (Morris 1993). Perhaps the burst was triggered by infall of a dense gas cloud less than 10 million years ago, a scenario that is supported by an overall counter-rotation (in the sense of Galactic rotation) of the He I star cluster (Genzel et al. 1996).

Based on earlier theoretical work by Lee (1987), Eckart et al. (1993) have proposed sequential merging by collisions as an alternative to the starburst scenario. A recent Fokker-Planck calculation of an evolving Galactic Center type, dense cluster with merging shows, however, that merging can account for only ~10–20 $M_\odot$ stars and no $\geq$ 30 $M_\odot$ stars (Lee 1994). The basic reason is that in the calculations a sufficiently dense stellar core (density 10^7 $M_\odot$ pc^{-3} or greater) cannot be maintained for a long enough time to build up very many massive stars. Morris (1993) has suggested that the He I stars are not classical blue supergiants at all but transitory objects that have been created in collisions between (~10 $M_\odot$) stellar black holes and solar-mass red giants. Both accounts of the He I stars just cited are very specific to the high density environment of the central parsec. However, a number of stars similar to the Sgr A He I stars have now been found in several clusters 2 to 13′ away from the central, high density Sgr A region (Okuda et al. 1990; Moneti et al. 1991; Cotera et al. 1994; Harris et al. 1994; Figer 1995). In the case of the Morris scenario (1993) one probably would also expect a much larger X-ray emission than is observed. These facts and the requirement of having to account for ~100 $M_\odot$ stars and the presence of heavy element nucleosynthesis products discussed above in my opinion now strongly favors the star formation model over the other scenarios.

There are less than a dozen red supergiants ($L \geq 10^4$ $L_\odot$) in the central starburst zone. Even after correction for collisional destruction of late type stars mentioned above this suggests that there was relatively little star formation prior to 10–15 million years ago. In comparison there are a much great number of late type stars with luminosities 10^3 to 10^4 $L_\odot$, both inside and outside (Haller and Rieke 1989) the central parsec. These medium luminosity stars are likely asymptotic giant branch stars of moderate mass (2 to 7 $M_\odot$). They may signify another starburst episode that happened ~ 10^8 years ago (Haller and Rieke 1989; Krabbe et al. 1995).

3 Is Sgr A* a Massive Black Hole?

The next key issue that I want to discuss is the evidence for a central massive black hole. Ever since the original discovery of the nonthermal compact radio source Sgr A* at the core of the nuclear star cluster (Ekers and Lynden-Bell 1971; Downes and Martin 1971; Balick and Brown 1974) that source has been the primary black hole candidate, in analogy to compact nuclear radio sources in other nearby normal galaxies (Lynden-Bell and Rees 1971). In fact ever more detailed radio observations have confirmed the unique nature of Sgr A* in the Galaxy. Recent very long baseline interferometry (VLBI) observations at 7 mm show its size to be less than a few AU (Backer 1994; Krichbaum et al. 1994). Its proper motion relative to a background quasar is now known to be less than about 38 $km\,s^{-1}$ (Backer 1994), at least 6 times smaller than the (2D) velocity dispersion of the stars. Hence Sgr A* must have a mass in excess of about 150

$M_\odot$. The source shows a mm/submm excess above the flat cm-spectral energy distribution (Zylka et al. 1992) probably indicative of the presence of a very compact ($\sim 10^{12}$ cm) radio core of stellar dimensions.

Yet observations at shorter wavelengths indicate nothing particularly impressive toward the radio position Sgr A*. The high resolution maps of Eckart et al. (1995) for the first time show that Sgr A* is located near the centroid of a T-shaped concentration of ~10 compact near-infrared sources (=Sgr A*(IR)). These sources are likely stars and one might speculate whether they represent a central stellar cusp around Sgr A*. Sgr A*(IR) also does not show intrinsic variability on scales of minutes or years, or significant line emission (Eckart et al. 1995). Depending on spectral type any possible infrared counterpart of Sgr A* has a luminosity between a few 10^2 and 10^4 $L_\odot$. (Variable) hard X-ray emission is commonly considered a key signature of black holes. However, in contrast to the fairly bright infrared emission the present 1 to 30 keV X-ray luminosity of Sgr A West and Sgr A* is less than a few hundred $L_\odot$ (Skinner 1993; Goldwurm et al. 1994). Recent observations with ASCA suggests that Sgr A*'s X-ray luminosity may have been larger in the past few hundred years (a few 10^5 $L_\odot$, Koyama et al. 1996) but still orders of magnitude smaller than the Eddington rate of a million solar mass black hole (Sunyaev et al. 1993).

The evidence for a (dark) central mass concentration in the Galactic Center thus is based entirely on the gas and stellar dynamics. As mentioned in the Introduction, evidence for a central mass concentration based on gas dynamics had already been growing in the 1980s but had not been considered compelling by most researchers in the field. However, ever better stellar velocities have become available during the past 8 years, fully vindicating the earlier measurements of gas velocities and substantially strengthening the evidence for a compact central dark mass in the Galactic Center (Rieke and Rieke 1988; McGinn et al. 1989; Sellgren et al. 1990; Lindqvist et al. 1992; Krabbe et al. 1995; Haller et al. 1996; Genzel et al. 1996).

The most recent determinations by Sellgren et al. (1990), Krabbe et al. (1995), Haller et al. (1996) and Genzel et al. (1996) now are all in good agreement and show a very significant increase of stellar radial velocity dispersion from about 55 $\mathrm{km\,s^{-1}}$ at 5 pc to about 180 $\mathrm{km\,s^{-1}}$ at 0.15 pc. From $\sim 1''$ resolution 3D spectroscopy Genzel et al. (1996) have obtained velocities for 222 early and late type stars between $1''$ and $22''$ distance from Sgr A*. After deprojection of the observed projected velocity dispersions and stellar surface densities Genzel et al. (1996) carried out a Jeans equation analysis. Assuming an isotropic stellar velocity field, the new 3D data in combination with the other stellar measurements mentioned above require a combination of a $M/L_{2\,\mu\mathrm{m}} \sim 2$ stellar cluster, and in addition a 2.5 to 3.3×10^6 $M_\odot$ dark mass. The dark compact mass is required at 6 to 8 sigma significance. It can be reduced but not fully removed even if highly anisotropic velocity fields are considered. For comparison, Haller et al. (1996) conclude that there must be a central mass of just under 2×10^6 $M_\odot$ and the most recent gas dynamics estimates find a central mass between 2 and 4×10^6 $M_\odot$ (Serabyn et al. 1988; Lacy et al. 1991; Herbst et al. 1993). The dark mass is not resolved (R(core) < 0.07 pc), has a $M/L_{2\,\mu m}$ ratio of at least 100 and a density of $\geq 10^9$ $M_\odot$ pc^{-3} (Genzel et al. 1996).

An experiment is now well underway to measure the proper motions of stars between 0.3″ and 10″ from Sgr A* from repeated high resolution near-infrared imaging with the SHARP camera on the ESO NTT (Eckart and Genzel, in preparation). This experiment should soon give a clearcut answer on the anisotropy of the stellar orbits.

As the dark mass has a core radius at least 5 times smaller and a core density at least 250 times greater than that of the visible (old) stellar cluster (average stellar mass ~0.7 $M_\odot$) it does not seem plausible that it consists of solar mass remnants (neutron stars or white dwarfs). Calculations by Chernoff and Weinberg (1990) indicate that such large density ratios between similar mass components cannot be attained even in core collapsed globular clusters. The dark mass concentration could either be a single massive black hole, or a very compact cluster of stellar mass (~10 $M_\odot$) black holes (Morris 1993; Lee 1995) should such a cluster be stable. The most likely configuration is probably a single massive black hole.

If Sgr A* is indeed a million solar mass black hole, the riddle is why it is presently so inactive. It is very interesting that the Galactic Center shares this 'luminosity deficiency' or 'blackness' problem with essentially all nearby nuclei for which there is substantial evidence for dark central masses (Kormendy and Richstone 1995), including the presently most convincing case, the 'mega' H_2O maser source NGC 4258 (Myoshi et al. 1995). It is possible that the tidal disruption and accretion of stars by the hole (happening in the Galactic Center at a rate of ~10^{-4} to 10^{-3} yr^{-1}) occurs very efficiently albeit at low duty cycle (Rees 1988). Accretion of interstellar gas streamers by the hole may be prevented by the need to overcome the angular momentum problem, coupled with the outward force of the stellar winds as discussed above. Finally, the wind gas itself may be accreted largely spherically, with very low radiation efficiency (Melia 1992). A final and very interesting possibility is that most of the energy of the accreting material is advected into the hole and not radiated (Narayan et al. 1995). Nevertheless current models of black hole accretion have to be stretched to be comensurate with Sgr A* being an underfed million solar mass black hole (Ozernoy and Genzel 1996).

Acknowledgements. I thank P.O. Lindblad, Aa. Sandqvist and S. Jörsäter for a very stimulating meeting and their warm hospitality.

References

Allen, D.A. (1994): in The Nuclei of Normal Galaxies: Lessons from the Galactic Center, eds. R. Genzel, A.I. Harris, Kluwer, Dordrecht, p. 293

Allen, D.A., Hyland, A.R., Hillier, D.J. (1990): MNRAS **244**, 706

Allen, D.A., Sanders, R.H. (1986): Nat **319**, 191

Backer, D. (1994): in The Nuclei of Normal Galaxies: Lessons from the Galactic Center, eds. R. Genzel, A.I. Harris, Kluwer, Dordrecht, p. 403
Balick, B., Brown, R.L. (1974): ApJ **194**, 265
Binney, J.J., Gerhard, O.E., Stark, A.A., Bally, J., Uchida, K.A. (1991): MNRAS **252**, 210
Blitz, L., Spergel, D.N. (1991): ApJ **379**, 631
Blum, R.D., dePoy, D.L., Sellgren, K. (1995b): ApJ **441**, 603
Blum, R.D., Sellgren, K., dePoy, D.L. (1995): ApJ **440**, L17
Chernoff, D.F., Weinberg, M.D. (1990): ApJ **351**, 121
Cotera, A.S., Erickson, E.F., Allen, D.A., Colgan, S.W.J., Simpson, J.P., Burton, M.G. (1994): in The Nuclei of Normal Galaxies: Lessons from the Galactic Center, eds. R. Genzel, A.I. Harris, Kluwer, Dordrecht, p. 217
Davidson, J.A., Werner, M.W., Wu, X., Lester, D.F., Harvey, P.M., Joy, M., Morris, M. (1992): ApJ **387**, 189
Davies, M.B., Benz, W., Hills, J.G. (1991): ApJ **381**, 449
Diehl, R. et al. (1993): A&AS **97**, 181
Downes, D., Martin, A. (1971): Nat **233**, 112
Eckart, A., Genzel, R., Hofmann, R., Sams, B.J., Tacconi-Garman, L.E. (1993): ApJ **407**, L77
Eckart, A., Genzel, R., Hofmann, R., Sams, B.J., Tacconi-Garman, L.E. (1995): ApJ **445**, L26
Eckart, A., Genzel, R., Krabbe, A., Hofmann,R. van der Werf, P.P., Drapatz, S. (1992): Nat **355**, 526
Ekers, R.D., Lynden-Bell, D. (1971): Ap. Lett. **9**, 189
Figer, D. (1995): PhD Thesis, University of California, Los Angeles
Forrest, W.J., Shure, M.A., Pipher, J.L., Woodward, C.A.,(1987): in AIP Conf. Proc. 155 The Galactic Center, ed. D. Backer, p. 153
Geballe, T.R. et al. (1984): ApJ **284**, 118
Genzel, R., Harris, A.I. (1994): The Nuclei of Normal Galaxies: Lessons from the Galactic Center, eds. R. Genzel, A.I. Harris, Kluwer, Dordrecht
Genzel, R., Hollenbach, D., Townes, C.H. (1994): Rep. Progr. Phys. **57**, 417
Genzel, R., Thatte, N., Krabbe, A., Kroker, H., Tacconi-Garman L.E. (1996): ApJ (submitted)
Genzel, R., Townes, C.H. (1987): ARA&A **25**, 377
Genzel, R., Watson, D.M., Crawford, M.K., Townes, C.H. (1985): ApJ **297**, 766
Goldwurm, A. et al. (1994): Nat **371**, 5889
Güsten, R. (1989): in Proc. IAU Symp. 136 The Center of the Galaxy, ed. M. Morris, Kluwer, Dordrecht, p. 89
Güsten, R. et al. (1987): ApJ **318**, 124
Hall, D.N.B., Kleinmann, S.G., Scoville, N.Z. (1982): ApJ **262**, L53
Haller, J.W., Rieke, M.J. (1989): in Proc. IAU Symp. 136 The Center of the Galaxy, ed. M. Morris, Kluwer, Dordrecht, p. 487
Haller, J.W., Rieke, M.J., Rieke, G.H., Tamblyn, P., Close, L., Melia, F. (1996): ApJ **456**, 194
Harris, A.I. et al. (1994): in The Nuclei of Normal Galaxies: Lessons from the Galactic Center, eds. R. Genzel, A.I. Harris, Kluwer, Dordrecht, p. 223
Herbst, T.M., Beckwith, S.V.W., Forrest, W.J., Pipher, J.L. (1993): AJ **105**, 956
Jackson, J. et al. (1993): ApJ **402**, 173
Kormendy, J., Richstone, D. (1995): ARA&A **33**, 581

Koyama, K. et al. (1995): preprint
Krabbe, A., Genzel, R., Drapatz, S., Rotaciuc, V. (1991): ApJ **382**, L19
Krabbe, A. et al. (1995): ApJ **447**, L95
Krichbaum, T.P., Schalinski, C.J., Witzel, A., Standke, K.J., Graham, D.A and Zensus, J.A. (1994): in The Nuclei of Normal Galaxies: Lessons from the Galactic Center, eds. R. Genzel, A.I. Harris, Kluwer, Dordrecht, p. 411
Lacy, J.H., Achtermann, J.M., Serabyn, E. (1991): ApJ **380**, L71
Lacy, J.H., Baas, F., Townes, C.H., Geballe, T.R. (1979): ApJ **227**, L17
Lacy, J.H., Townes, C.H., Geballe, T.R., Hollenbach, D.J. (1980): ApJ **241**, 132
Lacy, J.H., Townes, C.H., Hollenbach, D.J. (1982): ApJ **262**, 120
Lebofsky, M.J., Rieke, G.H. (1987): in AIP Conf. Proc. 155 The Galactic Center, ed. D. Backer, p. 79
Lee, H.M. (1987): ApJ **319**, 801
Lee, H.M. (1994): in The Nuclei of Normal Galaxies: Lessons from the Galactic Center, eds. R. Genzel, A.I. Harris, Kluwer, Dordrecht, p. 335
Lee, H.M. (1995): MNRAS **272**, 605
Libonate, S., Pipher, J.L., Forrest, W.J., Ashby, M.L.N. (1995): ApJ **439**, 202
Lindqvist, M., Habing, H., Winnberg, A. (1992): A&A **259**, 118
Lo, K.Y., Claussen, M, J. (1983): Nat **306**, 647
Lynden-Bell, D., Rees, M. (1971): MNRAS **152**, 461
Melia, F. (1992): ApJ **387**, L25
Mezger, P.G. et al. (1989): A&A **209**, 337
Mezger, P.G. (1994): in The Nuclei of Normal Galaxies: Lessons from the Galactic Center, eds. R. Genzel, A.I. Harris, Kluwer, Dordrecht, p. 415
Miyoshi,M, Moran, J.M., Hernstein, J., Greenhill, L., Nakai, N., Diamond, P. and Inoue, M. (1995): Nat **373**, 127
Moneti, A., Glass, I.,S., Moorwood, A.F.M. (1991): Mem. Soc. Astron. Ital. **62**, 4, 755
Morris, M. (1990): in Galactic and Extragalactic Magnetic Fields, eds. R. Beck, P. Kronberg, R. Wielebinski, Kluwer, Dordrecht, p. 361
Morris, M. (1993): ApJ **408**, 496
Morris, M. (1989): (ed.) Proc. IAU Symp. 136 The Center of the Galaxy, Kluwer, Dordrecht
Najarro, F. et al. (1994): A&A **285**, 573
Narayan, R., Yi, I., and Mahadevan, R. (1995): Nat **374**, 623
Okuda, H. et al. (1990): ApJ **351**, 89
Ozernoy, L., Genzel, R. (1995): in The Galaxy, ed. L. Blitz, Kluwer, Dordrecht, (in press)
Phinney, E.S. (1989): in Proc. IAU Symp. 136 The Center of the Galaxy, ed. M. Morris, Kluwer, Dordrecht, p. 543
Rees, M. (1988): Nat **333**, 523
Rieke, G.H., Lebofsky, M.J. (1982): in AIP Conf. Proc. 83 The Galactic Center, eds. G. Riegler, R.D. Blandford, p. 194
Rieke, G.H., Rieke, M.J. (1988): ApJ **330**, L33
Rieke, G.H., Rieke, M.J. (1994): in The Nuclei of Normal Galaxies: Lessons from the Galactic Center, eds. R. Genzel, A.I. Harris, Kluwer, Dordrecht, p. 283
Sellgren, K., McGinn, M.T., Becklin, E., Hall, D.N.B. (1990): ApJ **359**, 112
Serabyn, E., Güsten, R. (1987): A&A **184**, 133
Serabyn, E., Lacy, J. (1985): ApJ **293**, 445
Serabyn, E., Lacy, J., Townes, C.H., Bharat, R. (1988): ApJ **326**, 171

Skinner, G.K. (1993): A&AS **97**, 149
Simon, M. et al. (1990): ApJ **360**, 95
Simons, D.A., Hodapp, K.W., Becklin, E.E. (1990): ApJ **360**, 106
Sofue, Y. (1994): in The Nuclei of Normal Galaxies: Lessons from the Galactic Center, eds. R. Genzel, A.I. Harris, Kluwer, Dordrecht, p. 43
Sunyaev, R.A., Markevitch, M., Pavlinsky, M. (1993): ApJ **407**, 606
Tamblyn, P., Rieke, G.H., Hanson, M.M., Close, L.M., McCarthy, D.W., Rieke, M.J. (1996): ApJ **456**, 206
Wollman, E. (1976): PhD Thesis, Univ. of California, Berkeley
Yusef-Zadeh, F., Morris, M., and Chance, D. (1984): Nat **310**, 557
Zylka, R., Mezger, P.G., Lesch, J. (1992): A&A **261**, 119

Radio Continuum and Molecular Gas in the Galactic Center

Yoshiaki Sofue

Institute of Astronomy, University of Tokyo, Mitaka, Tokyo 181, Japan; E-mail: sofue@mtk.ioa.s.u-tokyo.ac.jp

Abstract. Nonthermal radio emission in the Galactic Center reveals a number of vertical structures across the Galactic plane, which are attributed to poloidal magnetic field and/or energetic outflow. Thermal radio emission comprises star forming regions distributed in a thin, dense thermal gas disk. The thermal region is associated with a dense molecular gas disk, in which the majority of gas is concentrated in a rotating molecular ring. Outflow structures like the radio lobe is associated with rotating molecular gas at high speed, consistent with a twisted magnetic cylinder driven by accretion of a rotating gas disk.

1 Radio Continuum Emission

1.1 Flat Radio Spectra

The radio emission from the Galactic Center (GC) is a mixture of thermal and nonthermal emissions. The conventional method to investigate the emission mechanism is to study the spectral index, either flat (thermal) or steep (nonthermal). Howerver, the spectral index in the central 3°-region has been found to be almost everywhere flat (Sofue 1985), even in regions where strong linear polarization has been detected. Therefore, a flat spectrum observed near the Galactic Center can no longer be taken as an indicator of thermal emission.

1.2 Infrared-to-Radio Intensity Ratio

Separation of thermal and nonthermal radio emission can be done by comparing far-IR (e.g. 60 μm) and radio intensities (both in Jy str^{-1}): thermal (H II) regions have high IR-to-radio ratio, $R = I_{\rm FIR}/I_{\rm R} \simeq 10^3$, while nonthermal regions have small IR-to-radio ratio, $R < 0 \sim 300$. Using this method, thermal and nonthermal emission regions have been distinguished in a wide area (Reich et al. 1987). The region near the Galactic plane is dominated by thermal emission and many H II regions like Sgr B2. These regions are closely associated with dense molecular clouds related to star formation in the clouds. On the other hand we find that many of the prominent features like the radio Arc, Sgr A and regions high above the Galactic plane, including the Galactic Center Lobe (GCL), are nonthermal.

1.3 Linear Polarization

A direct and more convincing way to distinguish synchrotron radiation is to measure the linear polarization. However, extremely high Faraday rotation toward the Galactic Center causes depolarization due to finite-beam and finite-bandwidth effects. This difficulty has been resolved by the develpoment of a multi-frequency, narrow-band Faraday polarimeter (Inoue et al. 1984) as well as by high-resolution and high-frequency observations using the VLA (Yusef-Zadeh et al. 1986). Very large rotation measure ($RM >\sim 10^3$ rad m^{-2}) and high degree (10 – 50%) polarization have been observed along the radio Arc and in the GCL (Inoue et al. 1984; (Tsuboi et al. 1986; Seiradakis et al. 1985; Sofue et al. 1986; Reich 1988; Haynes et al. 1992).

Linar polarization as high as $p \sim 50$ % has been detected along the Arc at mm wavelengths (Reich et al. 1988). This is nearly equal to the theoretical maximum, $p_{\max} = (\alpha + 1)/(\alpha + 7/3) \simeq 47$ %, for the Arc region, where the spectral index is $\alpha \simeq +0.2$. This implies that the magnetic field is almost perfectly aligned, consistent with the VLA observations showing straight filaments, suggestive of highly ordered magnetic field (Yusef-Zadeh et al. 1984; Morris 1993). From linear polarization it is clear that the radio emission from the radio Arc is nonthermal in spite of its flat radio spectra.

2 Radio Continuum Morphology

2.1 Thermal Disk and Star Formation

The nuclear disk, about 50 pc thick and 200 pc in radius, comprises numerous clumps of H II regions, most of which are active star-forming (SF) regions and are detected in the H recombination lines (Mezger and Pauls 1979). Typical H II regions are named Sgr B, C, D and E. The total H II mass of 2×10^6 $M_\odot$ has been estimated, and the production rate of Ly continuum photons of 3×10^{52} s^{-1} is required to maintain this amount of H II gas (Mezger and Pauls 1979). However, if we take the GC distance of 8.5 kpc and a more accurate thermal/nonthermal separation, we estimate these to be $\sim 10^6$ $M_\odot$ and 1.5×10^{52} s^{-1}, respectively. The SF rate of the central few hundred pc region amounts, therefore, to several % of the total SF rate of the Galaxy.

2.2 Thermal Filaments

Complex thermal filaments connect (bridge) Sgr A with the radio Arc (Yusef-Zadeh et al. 1984). Recombination (Pauls et al. 1976; Yusef-Zadeh et al. 1986) and molecular line observations (Güsten 1989) indicate their thermal characteristics. However, large Faraday rotation is detected in the bridge, indicating co-existence of magnetic fields along the thermal filaments (Sofue et al. 1987). Velocity dispersion of the thermal filaments increases drastically near the Arc, indicative of violent interaction with the Arc (Pauls et al. 1976). Yusef-Zadeh and Morris (1988) also argue that the Arc (straight filaments) and the arched filaments are interacting. A magneto-ionic jet from Sgr A colliding with the ambient poloidal magnetic field would explain this exotic structure (Sofue and Fujimoto 1987).

2.3 Radio Arc and Vertical Magnetic Fields

The radio Arc comprises numerous straight filaments perpendicular to the Galactic plane, and extends for more than $\sim$ 100 pc (Downes et al. 1978; Yusef-Zadeh et al. 1984; Morris 1993). The magnetic field direction is parallel to the filaments and vertical to the Galactic plane (Tsuboi et al. 1986). Field strength as high as $\sim$ 1 mG has been estimated in the Arc and in some filaments (Morris 1993). The life-time of cosmic-ray electrons in the Arc is estimated to be as short as $\sim$ 4 000 years (Sofue et al. 1992), and so the straight filaments may be transient features, temporarily illuminated by recently accelerated high-energy electrons.

The higher latitude extension of the Arc, both toward positive and negative latitudes, is also polarized by 10 to 20% (Tsuboi et al. 1986; Sofue et al. 1987). The rotation measure reverses across the Galactic plane, indicating a reversal of the line-of-sight component of the magnetic field. This is consistent with a large-scale poloidal magnetic field twisted by the disk rotation (Uchida et al. 1985).

2.4 Galactic Center Lobe and Large-Scale Ejection

The Galactic Center Lobe (GCL) is a two-horned vertical structure, probably a cylinder of about 200 pc in diameter (Sofue and Handa 1984; Sofue 1985: Fig. 1). The eastern ridge of the GCL is an extension from the radio Arc, and is strongly polarized. The western ridge emerges from Sgr C. An MHD acceleration model, in which the gas is accelerated by a twist of poloidal magnetic field by an accreting gas disk, has been proposed (Uchida et al. 1985; Uchida and Shibata 1986). High-velocity molecular gas has been found to be associated with the GCL (Sofue 1996: Fig. 1): Molecular gas in the eastern GCL ridge is receding at $V_{\rm lsr} \sim +100$ km s^{-1}, and the western gas is approaching at ~ -150 km s^{-1}, indicating rotation of the GCL. This is consistent with the twisted magnetic cylinder model.

A much larger scale ejection has been found in radio, which emanates toward the halo, reaching as high as $b \sim 25°$ (Sofue et al. 1988). This feature, which is 4 kpc long and some 200 pc in diameter, may be cylindrical in shape and extends roughly perpendicular to the Galactic plane. This structure might be a jet, or it might be magnetic tornado produced by the differential rotation between the halo and the nuclear disk.

2.5 Huge Galactic Bubble by Starburst: North Polar Spur

The radio North Polar Spur (NPS) traces a giant loop on the sky of diameter about 120°, drawing a huge Ω over the Galactic Center (Fig. 2). The Ω-shape

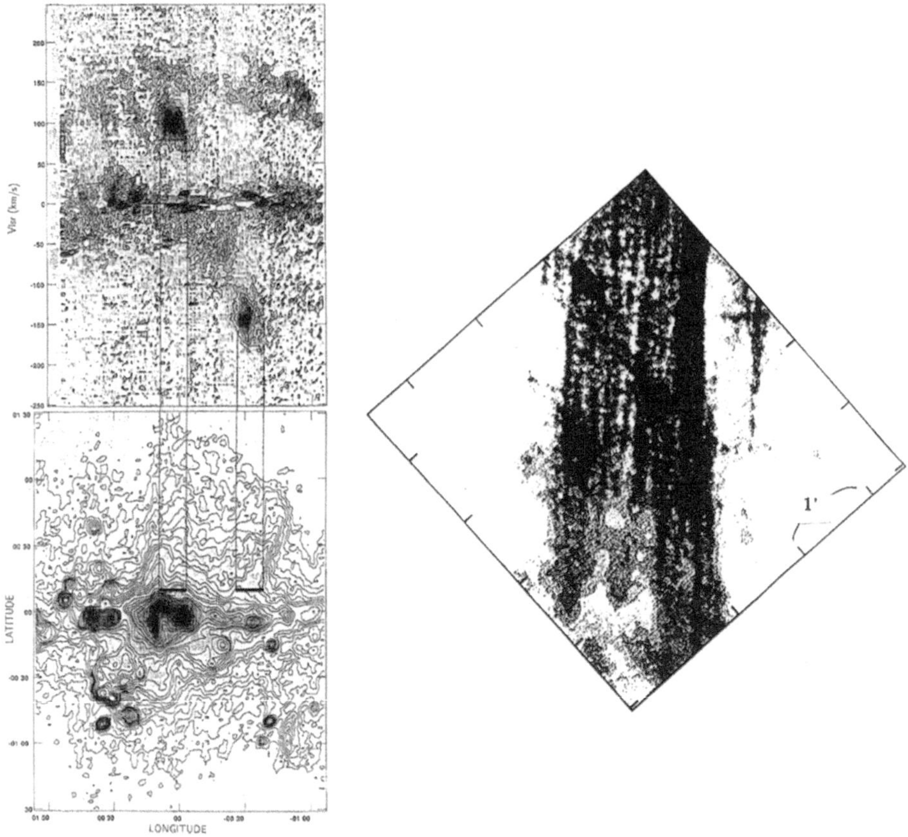

Fig. 1. 10 GHz radio map of the Galactic Center 2.2° ×3° (**left bottom**), in comparison with a $l-V_{lsr}$ plot of the ^{13}CO emission at $b = 8'$ (**top**) showing high-velocity rotating gas in the GCL. A VLA map (**right**) at 5 GHz shows vertical magnetic field filaments in the Arc (tick mark interval is 1′, Yusef-Zadeh 1986)

can be simulated by a shock front due to an explosion (sudden energy input) at the Galactic Center (Sofue 1994). In this model, the distance to NPS is several kpc. The X-ray intensity variation as a function of latitude indicates that the source is more distant than a few kpc, beyond the H I gas disk, consistent with the Galactic Center explosion model, but inconsistent with the local supernova remnant hypothesis.

Hence, the NPS is naturally explained, if the Galaxy experienced an active phase 15 million years ago associated with an explosive energy release of some 10^{56} erg (Sofue 1994). This suggests that a starburst had occurred in our Galactic Center, which involved $\sim 10^5$ supernovae during a relatively short period (e.g., 10^6 yr).

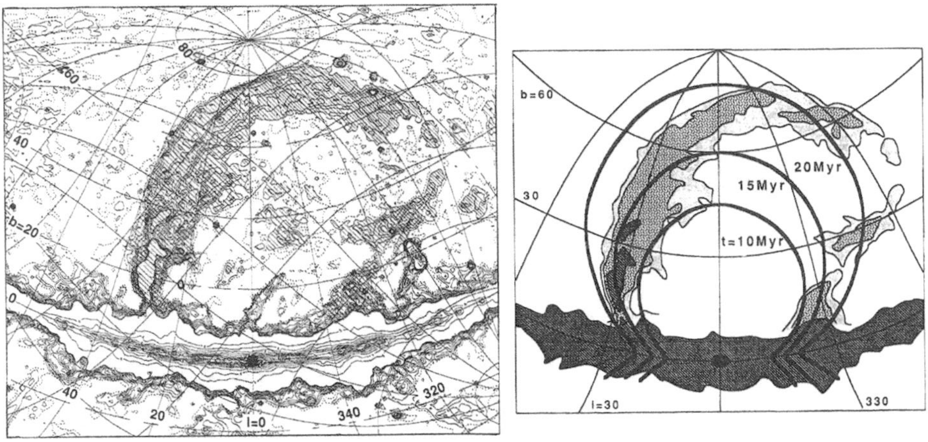

Fig. 2. **Left**: North Polar Spur at 408 MHz after background subtraction (Haslam et al. 1982; Sofue 1994). **Right**: A shock wave associated with a starburst of total energy release of 10^{55} ergs is shown at 10, 15 and 20 million years. The front at 15 million years can fit the radio shell

3 Molecular Arms and the 120-pc Ring

Various molecular features have been recognized in the central $\sim 100 - 200$ pc region: molecular rings and arms of a few hundred pc scale (Scoville et al.1974; Heiligman 1987), shell structures and complexes around H II regions (e.g. Hasegawa et al. 1993), and an expanding molecular ring of 200 pc radius (Scoville 1972; Kaifu et al.1972). Binney et al. (1991) have noticed a "parallelogram" and interpreted it in terms of non-circular kinematics of gas in an oval potential, instead of an expanding ring. However, the gas in this parallelogram shares only a minor fraction of the total gas mass ($\sim$ 10 %).

Figure 3 shows the total intensity map integrated over the full range of the velocity (Bally et al. 1987; Sofue 1995). The total molecular mass in the $|l| < 1°$ region is estimated to be $\sim 4.6 \times 10^7$ $M_\odot$ for a new conversion factor (Arimoto et al. 1995). The molecular mass of the "disk" component is $\sim 3.9 \times 10^7$ $M_\odot$, which is 85% of the total in the observed region. The expanding ring (or the parallelogram) shares the rest, only 7×10^7 $M_\odot$ (15%) in the region.

The H I mass within the central 1 kpc is only of several 10^6 $M_\odot$ (Liszt and Burton 1980). Hence, the central region is dominated by a molecular disk of $\sim$ 150 pc ($\sim 1°$) radius, outside of which the gas density becomes an order of magnitude smaller. The total gas mass within 150 pc is only a few percent of the dynamical mass ($M_{\rm dyn} = RV_{\rm rot}^2/G \sim 8 \times 10^8$ $M_\odot$ for a radius $R \sim 150$ pc and rotation velocity $V_{\rm rot} \sim 150$ km s^{-1}. This implies that the self-gravity of gas is not essential in the Galactic Center.

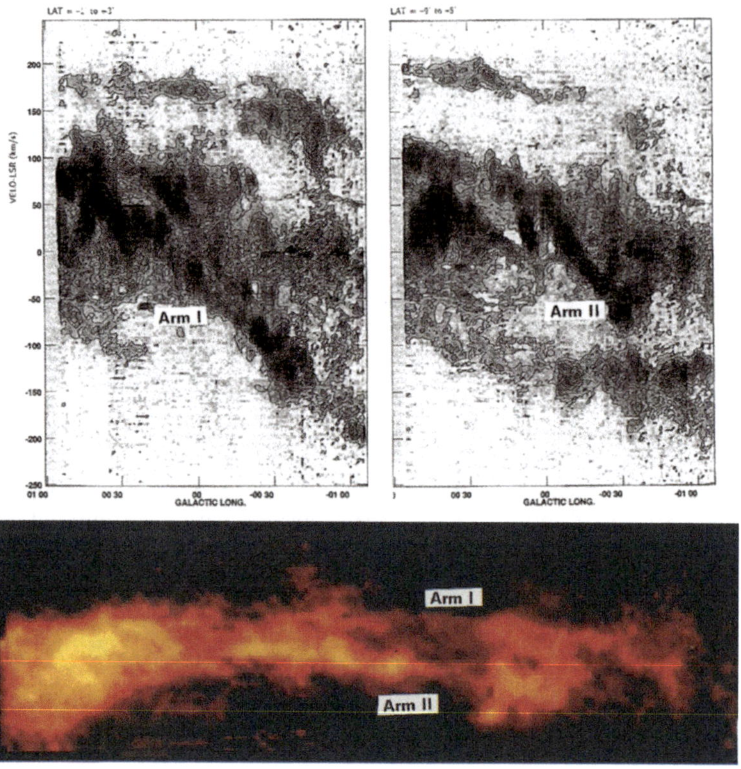

Fig. 3. ^{13}CO intensity map (Bally et al. 1987; Sofue 1995), showing a highly-tilted ring structure of the molecular gas (**bottom**), in comparison with an (l, V) diagram at $b = 2'$ and $-5'$ (**top**). Arms I and II correspond to the upper and lower parts of the tilted ring

Figure 3 shows (l, V) diagrams in the Galactic plane, where the foreground components have been subtracted (Sofue 1995). The major structures of the "disk component" near the Galactic plane are "rigid-rotation" ridges, which we call "arms". The most prominent arm is found as a long and straight ridge, slightly above the Galactic plane at $b \sim 2'$, marked as Arm I in the figure. Its positive longitude part is connected to the dense molecular complex Sgr B. Another prominent arm is seen at negative latitude at $b \sim -6'$, marked as Arm II.

Arms I and II compose a bent ring of radius 120 pc with an inclination 85°. We call this ring the 120-pc Molecular Ring. It is possible to deconvolve the (l, V) diagram into a spatial distribution in the Galactic plane by assuming approximately circular rotation and using the velocity-to-space transformation (VST). Figure 4 shows a thus-obtained possible "face-on" map of the molecular gas for $V_0 = 150$ km s^{-1}. H II regions are also plotted, showing that H II regions lie along the molecular complexes along the arms in the ring.

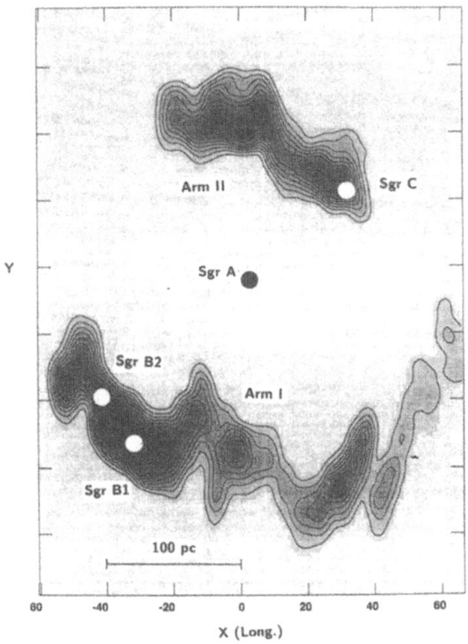

Fig. 4. Possible deconvolution of the CO (l, V) diagrams for Galactic Center into a face-on view

4 Discussion

We have reviewed various features in the central 150 pc region of the Galaxy in radio continuum, both in thermal and nonthermal, and in the CO line. The thermal radio emission and molecular gas are distributed within a thin disk of 150 pc radius and 30 pc thickness. The nonthermal radio emission is distributed in a wider area, often extending far from the Galactic plane, comprising vertical structures associated with poloidal magnetic fields.

We estimate some energetics among the variuos ISM in the central 150 pc region: Molecular gas has turbulent energy density of a few 10^{-8} erg cm^{-3}, when averaged in the central 150 pc radius disk. Energy densities due to H II gas and Ly continuum photon are of the order of $\sim 10^{-10}$ erg cm^{-3}. On the other hand, the magnetic field energy is as high as $\sim 4 \times 10^{-8}$ erg cm^{-3} for ~mG field strength. The molecular gas disk and magnetic field appear to be in an energy balance with each other, with the magnetic energy dominating. The averaged star formation rate compared to the molecular gas amount, or the SF efficiency, is much lower than that observed in the outer disk of the Galaxy. If the Galaxy had experienced a starburst 15 million years ago such as that suggested for the cause of the huge NPS shell, the present Galactic Center may be in a quiet phase, probably in a pumping-up phase for the next burst.

References

Arimoto, N., Sofue, Y., Tsujimoto, T. (1994): submitted to ApJ
Bally, J., Stark, A.A., Wilson, R.W., Henkel, C. (1988): ApJ **324**, 223.
Binney, J.J., Gerhard, O.E., Stark, A.A., Bally, J., Uchida, K.I. (1991): MNRAS **252**, 210.
Downes, D., Goss, W.M., Schwarz, U.J., Wouterloot, J.G.A. (1978): A&AS **35**, 1
Güsten, R. (1989): in The Center of the Galaxy, ed. M. Morris, Kluwer, Dordrecht, p. 89
Hasegawa, T., Sato, F., Whiteoak, J. B., Miyawaki, R. (1993): ApJ **419**, L77
Haslam, C.G.T., Salter, C.J., Stoffel, H., Wilson, W.E. (1982): A&AS **47**, 1
Haynes, R. F. ,, Stewart, R. T., Gray, A. D., Reich, W., Reich, P., Mebold, U. (1992): A&A
Heiligman, G. M. (1987): ApJ **314**, 747
Kaifu, N., Kato, T., Iguchi, T. (1972): Nat **238**, 105
Liszt, H. S., Burton, W. B. (1980): ApJ **236**, 779
Mezger, P.G., Pauls, T. (1979): in Proc. IAU Symp. 84 The Large-Scale Characteristics of the Galaxy, ed. W.B. Burton, Reidel, Drodrecht, p. 357
Morris, M. (1993): in The Nuclei of Normal Galaxies, eds. R. Genzel, A.I. Harris, Kluwer, Dordrecht
Morris, M., Yusef-Zadeh, F. (1985): AJ **90**, 2511
Pauls, T., Downes, D., Mezger, P.G., Churchwell, W. (1976): A&A **4**, 407
Reich, W., Sofue, Y., Fürst, E. (1987): PASJ **39**, 573
Scoville, N.Z. (1972): ApJ **175**, L127
Seiradakis, J.H., Lasenby, A.N., Yusef-Zadeh, F., Wielebinski, R., Klein, U. (1985): Nat **17**, 697
Shibata, K., Uchida, Y. (1987): PASJ **39**, 559
Sofue, Y. (1985): PASJ **37**, 697
Sofue, Y. (1989): in Proc. IAU Symp. 136 The Center of the Galaxy, ed. M. Morris, Reidel, Dordrecht, p. 213
Sofue, Y. (1994): ApJ **431**, L91
Sofue, Y. (1995): PASJ
Sofue, Y. (1996): ApJ Letters, in press
Sofue, Y., Fujimoto, M. (1987): ApJ **319**, L73
Sofue, Y., Handa, T. (1984): Nat **310**, 568
Sofue, Y., Murata, Y., Reich, W. (1992): PASJ **44**, 367
Sofue, Y., Reich, W., Inoue, M., Seiradakis, J.H. (1987): PASJ **39**, 359
Tsuboi, M., Inoue, M., Handa, T., Tabara, H., Kato, T., Sofue, Y., Kaifu, N. (1986): AJ **92**, 818
Uchida, Y., Shibata, K. (1986): PASJ **38**
Uchida, Y., Shibata, K., Sofue, Y. (1985): Nat **317**, 699
Yusef-Zadeh, F., Morris, M. (1988): ApJ **326**, 574
Yusef-Zadeh, F., Morris, M., Chance, D. (1984): Nat **310**, 557

The Galactic Center Dynamics

Alexei M. Fridman[1], Oleg V. Khoruzhii[1], Valentin V. Lyakhovich[1], Leonid Ozernoy[2,3], and Leo Blitz[4]

[1] Institute of Astronomy, Russian Academy of Sciences, 48 Pyatnitskaya St., Moscow, 109017, Russia
[2] Institute for Computational Sciences & Informatics and Department of Physics & Astronomy, George Mason U., Fairfax, VA 22030-4444, USA
[3] Lab. for Astronomy and Solar Physics, NASA/GSFC, Greenbelt, MD 20771, USA
[4] Department of Astronomy, U. of Maryland, College Park, MD 20742, USA

Abstract. The innermost central part of the Galaxy has very complicated morphology and kinematics. We make an emphasis on the dynamical nature of the 'mini-spiral'. Two cases are considered: (i) the putative black hole lies at the center of the dense stellar cluster and (ii) it is off the center. In the first case, the mechanism of the spiral arm generation can be over-reflection instability, and in the second case the axial asymmetry of the gravitational potential serves as a source of excitation of the density wave. In the both cases, the generated spiral density wave is consistent with the observed 'mini-spiral' if the upper limit to the black hole mass is $\simeq 10^5 M_\odot$.

1 Introduction

The region inside the circumnuclear (molecular) ring, the 'cavity', is filled with atomic and ionized gas, whose densest part forms the so-called 'mini-spiral'. Below, after summarizing the basic data on the central few parsecs and concisely reviewing the available concepts on the origin of the 'mini-spiral', we consider two possible hydrodynamical mechanisms able to generate the latter.

2 Basic Observational Data

As two detailed observational reviews on the Galactic Center (by Genzel and Sofue) are presented in these Proceedings we only list here in Tables 1 to 5 the main parameters of gas and stars in the innermost region.

Table 1. Dust

Measured parameter	Numerical value	References
Three components revealed with temperatures	40 K ($Q \sim 30''$) 170 K ($Q \sim 15''$) 400 K ($Q \sim 3''$)	Zylka et al. (1995)
Total mass inside 1 pc	1–6 $M_\odot$	Zylka et al. (1995)

Table 2. The molecular ring ('Circumnuclear Disk')

Outer radius	7 pc	Genzel (1989)
	8 pc	Roberts & Goss (1993)
	12 pc	Zylka et al. (1995)
Inner radius	1.5 – 2 pc	consensus
Thickness of the disk		Güsten et al. (1987)
at the inner edge	0.4 pc	Serabyn et al. (1986)
at the outer edge (7pc)	2 pc	Zylka et al. (1995)
Total mass (7pc)	$\sim 10^4$ $M_\odot$	Genzel et al. (1985)
(10pc)	$\sim 10^5$ $M_\odot$	Blitz et al. (1993)
Temperature	300 K	Harris et al. (1985)
	> 100 K (CO)	Lugsten et al. (1987)
		Sutton et al. (1990)
		Jackson et al. (1993)
	50 – 200 K (HCN)	Güsten et al. (1987)
		Wright et al. (1989)
		Jackson et al. (1993)
Disk consisting of dense clumps of the size	0.15 pc	Jackson et al. (1993)
Density in clumps	$10^6 - 10^8$ cm^{-3}	
Average density	$10^4 - 10^5$ cm^{-3}	
Rotation velocity	110 km s^{-1}	Genzel et al. (1985)

Table 3. Ionized gas

Mass and density inside 2 pc	60 $M_\odot$	Lo & Claussen (1983)
(in the minispiral only)	50 $M_\odot$, 10^4 cm^{-3}	Zylka et al. (1995)
Average density	$2.5 \cdot 10^3$ cm^{-3}	
Temperature (HII region)	5 000 – 7 000 K	Roberts & Goss (1993)
Line-of-sight magnetic field	10 mG	Aitken (1989)
in the Northern arm	15 mG	Roberts et al. (1991)
	$\lesssim$ 8 mG	Roberts & Goss (1993)

Table 4. Atomic gas

Mass and density inside 1.5 pc	300 $M_\odot$	Jackson et al. (1993)
	200 $M_\odot$, 10^4 cm^{-3}	Zylka et al. (1995)
Temperature	200 K	Jackson et al. (1993)

Table 5. The central stellar cluster

Two components revealed		
core radius, r_c, for young stars	0.3 pc	Eckart et al. (1995)
core radius for old stars	0.8 pc	Zylka et al. (1995)
Stellar density		
IR photometry + modelling	$\rho \gtrsim 10^6\ M_\odot pc^{-3}$ at $r < r_c$,	McGinn et al. (1989)
	$\rho \propto r^{-1.8}$ at $r > r_c$	Bailey (1980)
star count + modelling	$\rho \sim 10^7\ M_\odot pc^{-3}$ at $r < r_c$, $\rho \propto r^{-2}$ at $r > r_c$	Eckart et al. (1993)

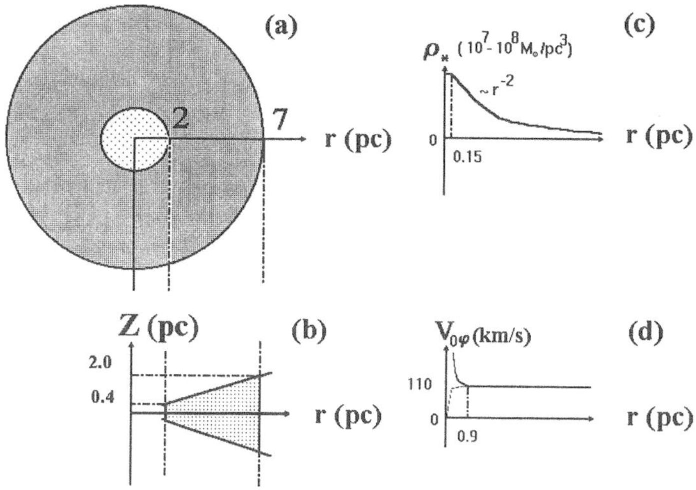

Fig. 1. a Distribution of gas in the innermost central region; **b** the thickness of the molecular ring, $H(r)$ (Güsten et al. 1987); **c** density run of the central stellar cluster; **d** the azimuthal rotation velocity of the disk as a function of r, both in case (i) when the black hole is located at the stellar cluster's center (solid line) and in case (ii) when the black hole is located at the radius R_0 where $M_* \gg M_{BH}$ (dotted line)

From Fig. 1b it follows that $H = \alpha \cdot r$, where $\alpha = \text{const}$, which implies that the molecular ring is isothermal, i.e. the sound speed $c_s = \text{const}$. Indeed, equilibrium of a ring rotating with angular velocity $\Omega_0(r)$ in the field of a spherically-symmetric mass distribution results, according to the Newton theorem, in $H(r) \simeq c_s/\Omega_0(r)$, i.e. $c_s \simeq \alpha r \Omega_0(r) = \alpha V_{0\varphi} = \text{const}$, in accordance with Fig. 1d. Observations indicate that the atomic disk, ionized disk, and dust ring are each isothermal, indeed (e.g. Zylka et al. 1995), although they have different temperatures (see Tables 1 – 4). The innermost part of the dust disk overlaps with the atomic (ionized) disk and its outer part overlaps with the molecular ring. Possibly, all the flat subsystems in the central region form a single, although highly inhomogeneous, disk.

3 Dynamics

Although the overall dynamics of the central region of the Galaxy may be as complicated as its morphology, the corner-stone of many dynamical problems is apparently the origin of the 'mini-spiral'. Available hypotheses on the nature of the 'mini-spiral' can be divided into the three categories:

1) *Inflow* (Lo and Claussen 1983; Ekers et al. 1983; Quinn and Sussman 1985; Serabyn and Lacy 1985; Serabyn et al. 1988; Jackson et al. 1993). It is assumed that the gaseous clouds in the central region lose, due to collisions, their angular momenta and fall into the central parsec forming a 'mini-spiral'. As soon as they arrive there, a strong UV radiation from the center dissociates and ionizes the molecular gas. Accounting for the atomic gas alone, its mass $\approx 200 - 300\ M_\odot$ inside the central 1.5 pc (Zylka et al. 1995; Jackson et al. 1993) and the inflow time $\sim 10^4$ yr yield the inflow rate $\approx (2-3) \cdot 10^{-2}\ M_\odot\ yr^{-1}$ (Jackson et al. 1993).

This inflow cannot reach an immediate vicinity of the central black hole (BH) and feed the latter owing to the ram pressure of the wind from the hot stars in the central parsec (Ozernoy and Genzel 1996). However, the wind itself can serve as a source of feeding the BH due to Bondi-Hoyle accretion, especially if the BH mass, $M_{\rm BH}$, is high enough (Ozernoy 1989). If $M_{\rm BH}$ were as high as $\sim 10^6\ M_\odot$, the X-ray emission of Sgr A* would significantly exceed what is observed from the direction toward the Galactic Center (Mastichiadis and Ozernoy 1994; Goldwurm et al. 1994)[1]. Although Narayan et al. (1995) argue that advection of radiation could appreciably reduce the accretion luminosity, a low efficiency alone seems to be unable to resolve the above contradiction (Ozernoy and Genzel 1996).

2) *Outflow from the center* (Brown 1982; Ekers et al. 1983; Heyvaerts, Norman and Putritz 1988; Schwarz, Bregman and van Gorkom 1989). An assumption that the 'mini-spiral' is a double-side precessing jet (Brown 1982) is in contradiction with the absence of symmetry (e.g. Lo 1986): Sgr A* as a natural source of the outflow is displaced from the center of the 'mini-spiral'.

3) *The 'mini-spiral' as the wave*: The Northern Arm and Western Arc are a single structure rotating in the Circumnuclear Disk plane + some flows (Lacy, Achtermann and Serabyn 1991). Analysis of the line-of-sight velocity field of the 'mini-spiral' has led Lacy et al. (1991) to the conclusion that the Northern Arm and Western Arc are a coherent structure in the form of an Archimedes spiral rotating in the plane of the Circumnuclear Disk. At the same time, the presence

[1] It is worth noting that Lynden-Bell and Rees (1971) suspected the existence of a black hole at the Galactic Center *before* the actual candidate for it (Sgr A*) was revealed.

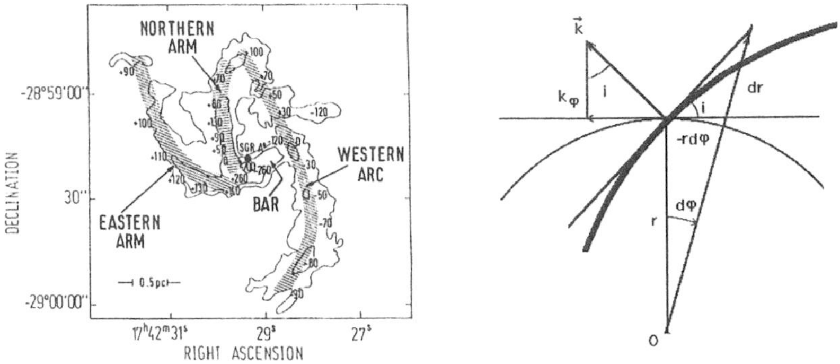

Fig. 2. **a** Line-of-sight velocity field of the 'mini-spiral' (Genzel and Townes 1987). **b** A spiral wave with pitch angle i and wave number k, $\tan i = -\mathrm{d}r/\mathrm{d}\varphi$, $k_\varphi = k \sin i$

of a steep velocity gradient in the region where the Eastern Arm and the Bar join with the Northern Arm (Fig. 2a) indicates, according to Lacy et al., the absence of any dynamical connection between those structures.

We would like to make the next step and consider the formation of that spiral structure. Two cases are worth analyzing: (i) a black hole sits at the center of the stellar cluster and (ii) it is located off the center. The rotation curve of the gaseous disk shown in Fig. 1d has in both cases a kink region.

Suppose, a kink is located at the point $r = R_0$ on the rotation curve. It means that, at this point, there is a jump of the derivative of the rotation velocity: $V'_{0\varphi}(R_0) \propto \Theta(r - R_0)$ and, respectively, $V''_{0\varphi}(R_0) \propto \delta(r - R_0)$, where Θ is the Heaviside step function, δ is the Dirac delta-function. In this case, the hydrodynamical equations that describe the rotating disk dynamics can be reduced to the Schrödinger equation with the potential in the form of a δ-well. The latter has always at least one level, i.e. in the region of $r = R_0$ there is a spiral wave. The generation mechanism for this spiral wave is a corotation resonance. Indeed, the energy of a sound wave in the flow is given by (Landau and Lifshitz 1984) $E = E_0\omega/(\omega - \mathbf{kv}(r)) = E_0\omega/(\omega - m\Omega_0(r))$, $E_0 > 0$, where ω is the eigenfrequency, $\mathbf{k}$ is the wave vector, and we use $\mathbf{kv} = k_\varphi v_{o\varphi} = (m/r) \cdot (r\Omega_0)$, m being the azimuthal wave number. Evidently, the energy E is positive if $\omega/m > \Omega_0(r)$.

In Fig. 3 (Fridman et al. 1993) the form of the potential well corresponds to the rotation velocity profile shown in Fig. 1d, solid line [case (i)]. The wave function has an exponential cutoff at the both edges of the potential well. Therefore, if corotation resonance were absent, the reflection coefficients at both sides of the potential barrier would be equal, with exponential accuracy, to unity. In reality, the wave reflected from the corotation resonance gets from it an additional positive energy. As a result, the reflection coefficient at the left potential barrier exceeds unity[2], i.e. an *over-reflection instability* happens. Orr (1907) was

[2] The potential well lies on the right-hand side of the corotation circle, where the wave energy is positive since $\omega/m > \Omega_0(r)$.

the first who discovered an over-reflection phenomenon in hydrodynamics, and in physics of galaxies an over-reflection was investigated, for the first time, by Goldreich and Lynden-Bell (1965) and by Julian and Toomre (1966).

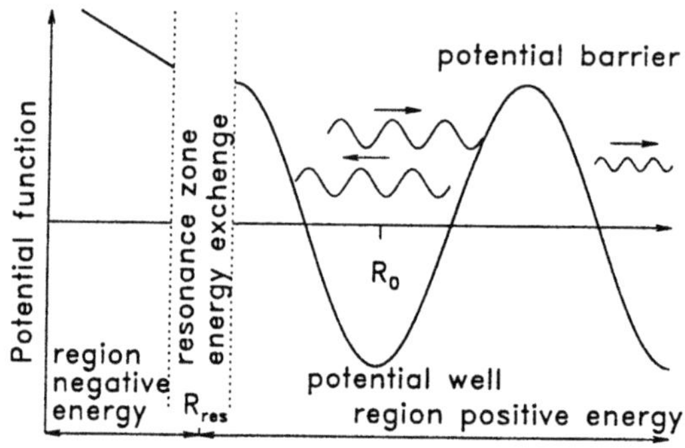

Fig. 3. A schematic sketch of the over-reflection phenomenon (Fridman et al. 1994)

The kink on the rotation curve shown in Fig. 4a causes the generation of the one-arm spiral wave that coincides with the Northern Arm and the Western Arc (Fig. 4b).

Let us turn now to case (ii)[3] when a black hole (BH) is located on the same distance R_0 from the stellar cluster's center and has a mass $M_{\rm BH} \ll M_*(R_0)$. The BH excites a sound wave. Using the azimuthal pattern speed $\Omega_{\rm p} = \omega/m$, $k_\varphi = k \sin i$, where i is the pitch angle (Fig. 2b), one obtains that, outside of the resonance zone, in the 'wave-zone' where $k \approx \omega/c$ and $\tan i \approx \sin i$

$$\Omega_{\rm p} = \frac{\omega}{m} = \frac{\omega}{k_\varphi r} \approx \frac{c_{\rm s}}{r \sin i} \approx -c_{\rm s}\frac{{\rm d}\varphi}{{\rm d}r}. \tag{1}$$

This expression keeps its validity for the case of the over-reflection instability too.

For an isothermal disk ($c_{\rm s}$ = const), (1) describes an Archimedes spiral that fits the observations (Lacy et al. 1991). Let us estimate the value of ${\rm d}\varphi/{\rm d}r$. The wave's phase velocity, $\Omega_{\rm p}$, coincides with the angular velocity at the position of the BH. Let the BH be on the distance of 0.15 pc from the stellar cluster's center, $R_{\rm BH} \simeq 0.15$ pc. According to Tables 2 and 5, $M_*(0.15~{\rm pc}) \sim 10^5~{\rm M}_\odot$, so that one concludes that $M_{\rm BH} < 10^5~{\rm M}_\odot$. From the equilibrium condition $GM/R^2 \simeq V_\varphi^2/R$, it follows that $V_{{\rm BH}\varphi} \approx 50~{\rm km~s^{-1}}$, and we obtain

$$-\frac{{\rm d}r}{{\rm d}\varphi} = \frac{c_{\rm s}}{\Omega_{\rm p}} = \frac{c_{\rm s} \cdot R_{\rm BH}}{V_{{\rm BH}\varphi}} \simeq 0.15~{\rm pc~rad^{-1}}. \tag{2}$$

[3] Considered jointly with D.Bisicalo and O.Kuznetsov

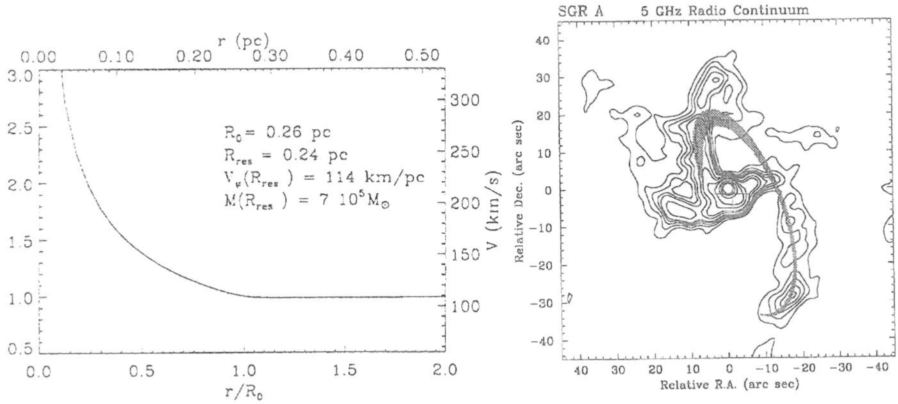

Fig. 4. a The rotation curve used for modelling of the spiral wave. **b** The mini-spiral: the contour map represents the observational data (Blitz et al. 1993), the filled region is the computer simulation (Fridman et al. 1996), the cross is the center of rotation. The offset for the rotational center with respect to Sgr A* is 0.″6 W and 1.″3 S [the rotational center coincides with the center of the stellar cluster, 0.″6 W ± 0.″7, 1.″3 S ± 1.″0 (3σ errors), as given by Eckart et al. (1993)]. The sound speed $c_s = 50$ km s^{-1}. The inclination of the disk is 60°, the position angle of the disk is 20°

This result turns out to be in a fair agreement with observational data for the 'mini-spiral' (Lacy et al. 1991).

Acknowledgements. A.F., O.H. and V.L. thank the ISF (grants M-74000, M-74300) and the RFBR (grant 93-02-17248) for financial support.

References

Aitken, D.K. (1989): in Proc. IAU Symp. 136 The Center of the Galaxy, ed. M. Morris, Kluwer, Dordrecht, p. 457

Bailey, M.E. (1980): MNRAS **190**, 217

Blitz, L., Binney, J., Lo, K.Y., Ho, P.T.P. (1993): Nat **361**, 417

Brown, R.L. (1982): ApJ **262**, 110

Eckart, A., Genzel, R., Hoffmann, R., Sams, B.J., Tacconi-Garman, L.E. (1993): ApJ **407**, L77

Eckart, A., Genzel, R., Hoffmann, R., Sams, B.J., Tacconi-Garman, L.E. (1995): ApJ **445**, L23

Ekers, R.D., van Gorkom, J.H., Schwarz, U.J., Goss, W.M. (1983): A&A **122**, 143

Fridman, A.M., Khoruzhii, O.V., Lyakhovich, V.V., Ozernoy, L., Blitz, L. (1994): in Physics of the Gaseous and Stellar Disks of the Galaxy, ed. I. King, ASP Conference Series **66**, 305

Fridman, A.M., Khoruzhii, O.V., Lyakhovich, V.V., Ozernoy, L., Blitz, L. (1996): in Proc. IAU Symp. 169 Unsolved Problems of the Milky Way, ed. L. Blitz (in press)
Genzel, R. (1989): in Proc. IAU Symp. 136 The Center of the Galaxy, ed. M. Morris, Kluwer, Dordrecht, p. 393
Genzel, R., Townes, C.H. (1987): ARA&A **25**, 377
Genzel, R., Watson, D. M., Crawford, M. K., Townes, C.H. (1985): ApJ **297**, 766
Goldreich, P., Lynden-Bell, D. (1965): MNRAS **130**, 125
Goldwurm, A., Cordier, B., Paul, J., Ballet, J., Bouchet, L., Roques, J.-P., Vedrenne, G., Mandrou, P., Sunyev, R., Churasov E., Gilfanov M., Finogenov A., Vikhlinin A., Dyachkov A., Khavenson, N., Kovtunenko, V. (1994): Nat **371**, 586
Güsten, R., Genzel, R., Wright, M. C. H., Jaffe, D.T., Stutzki, J., Harris, A. I. (1987): ApJ **318**, 124
Heyvaerts, J., Norman C., Pudritz, R. (1988): ApJ **330**, 718
Jackson, J.M., Geis, N., Genzel, R., Harris, A.I., Madden, S., Poglitsch, A., Stacey, G., Townes, C.H. (1993): ApJ **402**, 173
Julian, W.H., Toomre, A. (1966): ApJ **146**, 810
Krabbe, A., Genzel, R., Drapatz, S., Rotaciuc, V. (1991): ApJ **382**, L 19
Lacy, J. H., Achtermann, J. M., Serabyn, E. (1991): ApJ **380**, L 71
Landau, L.D., Lifshitz, E.M. (1984): Fluid Mechanics, Pergamon Press, Oxford
Lo, K.Y. (1986): Science **233**, 1394
Lo, K.Y., Claussen, M.J. (1983): Nat **306**, 647
Lugten, J.B., Stacey, G.J., Harris, A.I., Genzel, R., Townes, C.H. (1987): in AIP Proc. 155 The Galactic Center, ed D.C. Becker, AIP, New York, p. 118
Lynden-Bell, D., Rees, M.J. (1971): MNRAS **152**, 461
Mastichiadis A., Ozernoy, L.M. (1994): ApJ **426**, 599
McGinn, M.T., Sellgren, K., Becklin, E.E., Hall, D.N.B. (1989): ApJ **338**, 824
Narayan, R., Yi, I., Mahadevan, R. (1995): Nat **374**, 623
Orr, W.McF. (1907): Proc. Roy. Irish Acad., Sect A. **27**, p. 9 and p. 69
Ozernoy, L.M. (1989): in Proc. IAU Symp. 136 The Center of the Galaxy, ed. M. Morris, Kluwer, Dordrecht, p. 555
Ozernoy, L., Genzel, R. (1996): in Proc. IAU Symp. 169 Unsolved Problems of the Milky Way, ed. L. Blitz (in press)
Quinn, P.J., Sussman, G.J. (1985): ApJ **288**, 377
Roberts, D.A., Goss, W.M. (1993): ApJS **86**, 133
Schwarz, U. L., Bregman J.D., van Gorkom J.H. (1989): A&A **215**, 33
Serabyn, E., Güsten, R., Walmsley, C.M., Wink, J.E., Zylka, R. (1986): A&A **169**, 85
Serabyn, E., Lacy, J.H. (1985): ApJ **293**, 445
Serabyn, E., Lacy, J.H., Townes, C.H., Bharat, R. (1988): ApJ **326**, 171
Sutton, E.C., Danchi, W.C., Jaminet, P.A., Masson, C.R. (1990): ApJ **384**, 503
Wright, M.C.H., Marr, J., Backer, D.C. (1989): in Proc. IAU Symp. 136 The Center of the Galaxy, ed. M. Morris, Kluwer, Dordrecht, p. 407
Zylka, R., Mezger, P.G., Ward-Thompson, D., Duschl, W.J., Lesch, H. (1995): A&A **297**, 83

Observational Evidence for the AGN Paradigm

Andrew S. Wilson[1,2]

[1] Astronomy Department, University of Maryland, College Park, MD 20742, U.S.A.
[2] Space Telescope Science Institute, 3700 San Martin Drive, Baltimore, MD 21218, U.S.A.

Abstract. The paradigm for active galactic nuclei (AGN) comprises a massive black hole accreting material through a disk. A competing model, in which no black hole is present and the luminosity is supplied by stellar processes ("starburst" model), has been developed for radio-quiet active galaxies (i.e. Seyfert galaxies, LINERs and radio-quiet quasars). In this article, I discuss recent observational work on the nuclear structures of Seyfert galaxies and LINERs with a view to discriminating between these two models. Seyfert galaxies are, to first order, cylindrically symmetric objects. Strong evidence for compact ($<$ pc) nuclear disks is now available through observations of broad FeKα emission, broad, double-peaked optical recombination lines and H_2O megamasers. Near infrared imaging reveals the outer parts of the disks. Bi-polar structures, collimated by the nuclear disks, include radio jets and lobes, high velocity ionized gas (the narrow line region) and ambient gas photoionized by a collimated nuclear radiation source. In the best studied Seyferts, the collimation of the radio jets is better than a few degrees. Compact accretion disks and highly collimated radio ejecta are consistent with the black hole model, but are neither observed in galaxies with (only) nuclear starbursts nor expected in starburst models. It remains to be shown, however, that accretion disks and highly collimated jets are prevalent in the population of radio-quiet AGN.

1 Introduction

The paradigm for active galactic nuclei (AGN) is a massive black hole accreting material through a disk (Lynden-Bell 1969). If we leave aside models outside the realm of conventional physics (such as the "white holes" of Ambartsumian and Hoyle), this paradigm has never been seriously challenged for radio-loud AGN i.e. BL Lac objects, radio galaxies and quasi-stellar radio sources. On the other hand, a competing picture has developed for radio-quiet objects – LINERs, Seyfert galaxies and quasi-stellar objects – in which no black hole is present and the objects are instead powered by hot stars and entities resulting from them, such as supernovae and supernova remnants. This "starburst" model for AGN has been investigated over the past two or three decades by a number of workers (e.g. Harwit and Pacini 1975; Weedman 1983), and most recently by Terlevich and his colleagues (e.g. Terlevich 1992; Terlevich et al. 1992). In order to distinguish between these two completely different scenarios, it is necessary to obtain direct evidence for or against accreting massive black holes in active galactic nuclei. Over the last year, some spectacular observations favoring the accretion disk – black hole model have been obtained, and I should like to review these, and other, pieces of evidence in this paper, concentrating on radio-quiet objects.

I divide the observational evidence into two kinds – that which shows an accretion disk directly (or is, at least, best interpreted in terms of emission from a disk) and that for which the evidence is indirect, involving bi-polar phenomena which are presumably collimated by an accretion disk. In the former category (discussed in Sect. 2) are broad, redshifted FeKα lines from Seyfert galaxies (which originate from disk scales of order 6 – 20 R_G, where $R_G = GM/c^2$), broad, double-peaked optical hydrogen recombination lines from radio galaxies and LINERs (which probe disks scales of 100 – $10^4 R_G$), H_2O megamasers (from scales of 10^4 – $10^6 R_G$, or 0.1 – 10 pc) and thermal dust emission from Seyferts (scales of order 10^7 – $10^8 R_G$, or 100 pc – 1 kpc). In the first two of these observations, the disk is not spatially resolved and its existence is inferred from the line profiles. In the last two, the disk is directly mapped out by the observations. Bi-polar phenomena in Seyfert galaxies include radio synchrotron and emission-line jets and lobes and highly ionized, V-shaped, gaseous structures apparently photoionized by a collimated nuclear source; these phenomena are reviewed in Sect. 3. Concluding remarks are given in Sect. 4.

2 Direct Evidence for Accretion Disks

2.1 FeKα Lines from Seyfert Galaxies

Tanaka et al. (1995) have recently found a broad emission feature, extending from 5 to 7 keV, in the Seyfert 1 galaxy MCG–6–30–15 (Fig. 1). The only reasonable identification is broad (full width at zero intensity $\approx$ 100 000 km s^{-1}), redshifted FeKα emission. The line is also asymmetric, with a relatively narrow core around 6.4 keV and a broad red wing. Tanaka et al. show that the profile can be modelled in terms of emission from an accretion disk between 3 and 10 Schwarzschild radii from a Schwarzschild or Kerr black hole. The physical effects governing the shape of the profile from such a disk include the gravitational redshift and transverse Doppler effects, which produce a profile skewed to the red, plus aberration, time dilation and blueshift, which make the blue horn stronger than the red. Expected profiles, taking into account these relativistic effects, have been calculated by Fabian et al. (1989). It is believed that the FeKα line arises through fluorescence and back-scattering from optically thick material in the disk illuminated by a primary hard X-ray source above it.

Broad FeKα lines have now been found in some half a dozen Seyfert galaxies (e.g. Mushotzky et al. 1995; Fabian et al. 1995). The full width at half maximum (FWHM) ranges from 10 000 to 50 000 km s^{-1}, but only in the case of MCG–6–30–15 is the signal to noise good enough to define the line profile reasonably well. Fabian et al. (1995) have recently presented a critical assessment of the kinematic disk interpretation of these lines. Non-kinematic line broadening mechanisms, such as Comptonization, are shown to be implausible. Perhaps the most serious competitor to the disk model is outflow, possibly in the form of a jet, either

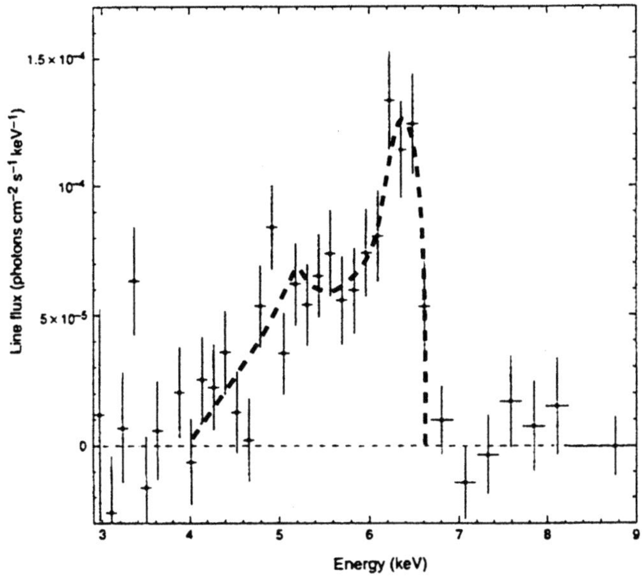

Fig. 1. From Tanaka et al. (1995). The line profile of FeKα in the X-ray emission from MCG–6–30–15. The dashed line shows the best fit line profile of an externally-illuminated accretion disk orbiting a Schwarzschild black hole

perpendicular to the line of sight or within $\approx$ 10 Schwarzschild radii of the black hole, with the observed redshift resulting from the transverse Doppler and gravitational redshifts, respectively. Such a model cannot be ruled out, but the absence of any strongly blueshifted lines argues against a jet model. Further, the Seyfert galaxies in which broad FeKα lines have been seen do not produce high power radio jets. While higher signal to noise and monitoring of the line profiles are certainly desirable, the relativistic disk interpretation is favored at the moment.

2.2 Double-Peaked, Broad, Optical Recombination Lines

A phenomenon closely related to the FeKα line profiles is found in the optical spectra of some AGN. Eracleous and Halpern (1994, hereafter EH) have completed a comprehensive study of the broad Hα emission lines of radio galaxies and radio-loud quasars. They find that some 10% of radio-loud objects exhibit broad, double-peaked emission (e.g. Fig. 2).

These double-peaked emitters have several distinctive properties in comparison with the more common single-peaked broad line emitters: (1) the average FWHM of the double-peaked lines is roughly twice as large as that of the rest of the sample (12 500 km s^{-1} versus 5 700 km s^{-1}), (2) the optical continuum near Hα comprises 20%–100% starlight, versus less than 20% for the other objects, (3)

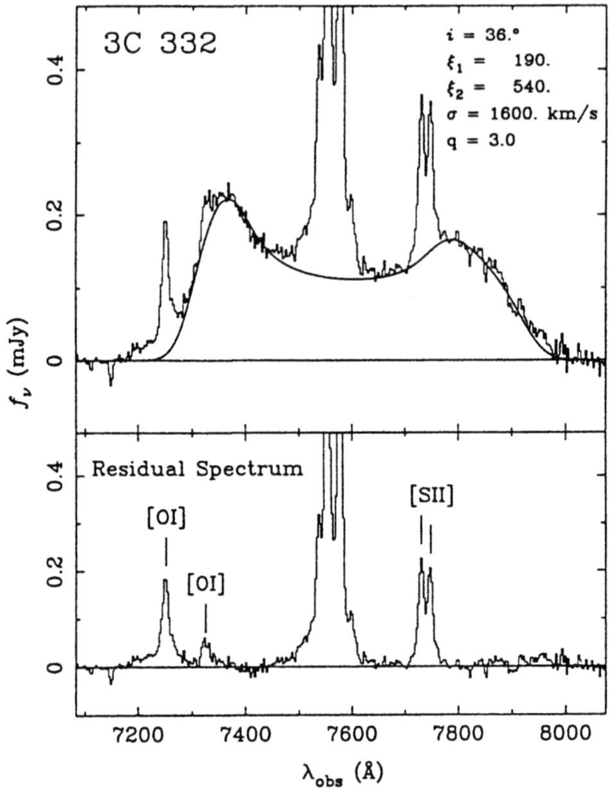

Fig. 2. From Eracleous and Halpern (1994). The broad Hα profile (after continuum subtraction) of 3C 332 and its best fit circular, relativistic Keplerian disk model. The fitting parameters are listed in the upper right: i is the disk inclination, ξ_1 and ξ_2 are the inner and outer radii (in units of GM/c^2), σ is a local broadening parameter and q is the power-law exponent of the variation of line emissivity with radius (emissivity $\propto \xi^{-q}$). The lower panel shows the residual spectrum resulting from the subtraction of the model profile from the observed profile

the double-peaked emitters have very strong lines from low ionization species, such as [O I]λ6300 and [S II]λ6724, and high [O I]λ6300/[O III]λ5007 ratios.

In the entire sample of radio-loud objects, the broad Hα lines are preferentially redshifted with respect the narrow lines. The mean shift at half maximum is $< \Delta\lambda/\lambda >= (6 \pm 2) \times 10^{-4}$. This result contrasts with studies of optically selected samples, which find equal numbers of blueshifts and redshifts (e.g. Sulentic et al. 1990; Boroson and Green 1992). For the double-peaked objects, the blue peak is stronger than the red in 13 out of the 18 cases. It is notable that most double-peaked emitters are powerful Fanaroff-Riley class II (Fanaroff and Riley 1974) radio sources.

EH discuss their results in terms of a model in which an inner, ion-supported torus produces a hard ionizing continuum, which illuminates a circular outer disk (Chen and Halpern 1989). The Hα emission is envisaged to come from the outer disk. Such external illumination of the outer disk appears necessary for energetic reasons (EH). The model makes specific predictions about the line profile shapes: the rotation of the disk causes the double-peaked structure, relativistic effects make the blue peak stronger than the red, and there is a net redshift of the entire line due to the gravitational redshift and transverse Doppler effect. Interpreted in this way, the observed redshift corresponds to Keplerian motion at $\simeq 2500\ R_G$. EH show that many, but not all, of the double-peaked line profiles can be fitted with such a circular, Keplerian disk model. The primary source of radiation from the ion torus is electron synchrotron radiation in the near infrared (Rees et al. 1982), which is boosted to higher energies through inverse Compton scattering. The synchrotron self-absorption frequency is expected to be about 10^{13} Hz and the inverse Compton X-ray spectrum should show a knee near 100 keV. Indeed, several of the double-peaked emitters have flat or inverted spectra between 25 μm and 60 μm (Halpern and Eracleous 1994). As emphasised by Begelman (1985), the important feature of an ion torus is not so much the spectral shape, but the low radiative efficiency. This ties in nicely with EH's observations. The large contribution of starlight to the optical continuum spectrum of the double-peaked emitters is related to the absence of the usual blue/UV component seen in AGN; this "big blue bump" is sometimes interpreted as emission from the inner parts of an optically thick accretion disk. Thus, if this inner disk is replaced by an ion torus, the blue/UV emission should be much weaker. Further, the ionizing continuum of the ion torus is harder and more dilute than the conventional thin disk. Gas ionized by this spectrum will have a lower ionization parameter and thus strong, low-ionization narrow lines (LINER-like – Halpern and Steiner 1983; Ferland and Netzer 1983).

Alternative explanations of the double-peaked profiles include binary black holes and bi-conical radial flows. Binary black holes may result from galaxy mergers (Begelman, Blandford and Rees 1980), and Gaskell (1983) has suggested that each member of the binary may have its own broad line region, so the orbital motion of the binary would produce displaced peaks. This hypothesis may be tested observationally by searching for the change in the velocities of the peaks associated with the orbital motion. The absence of radial velocity variations in Arp 102B and 3C 332 places lower limits on the masses of the hypothesised binaries in these galaxies of $4 \times 10^9\ M_\odot$ and $2 \times 10^{10}\ M_\odot$, respectively, which renders the binary black hole model an unlikely explanation for the double peaks in these objects (Halpern and Filippenko 1991). Additionally, high velocity gas would be expected to orbit close to each of the two black holes and give rise to double peaks in the wings of the observed line. Low velocity gas orbits the center of mass of the binary and produces a single core. Thus widely spaced peaks separated by the orbital velocity are *not* produced by gas gravitationally bound to a binary black hole (Chen, Halpern and Filippenko 1989). Double stream models for double-peaked profiles have been published by Zheng. Binette and Sulentic (1990) and Zheng, Veilleux and Grandi (1991). The broad line emission originates from material in a bi-cone moving outwards from the central object. Such models can produce a wide variety of profile shapes but, given our ignorance the dynamics of such outflows, the models are of necessity purely kinematic and the parameters ad hoc. Studies of the variability of double-peaked profiles and, in particular, their response to variations in the ionizing continuum may be the best way to distinguish between disk and outflow models.

A recent discovery is that double-peaked, broad Hα emission may appear in AGN which previously had no such emission. Such "new" double-peaked lines have been found in NGC 1097 (Storchi-Bergmann, Baldwin and Wilson 1993), Pic A (Halpern and Eracleous 1994) and M 81 (Bower et al. 1996). In NGC 1097 and Pic A, the wings of the new lines extend to $\simeq \pm$ 11 000 km s^{-1}. While Pic A is a radio galaxy, NGC 1097 and M 81 are spirals with low luminosity active nuclei. Nevertheless, both NGC 1097 and M 81 show jet-like structures: NGC 1097 exhibits four faint, large-scale optical jets (Wolstencroft and Zealey 1975), while M 81 has an elongated radio core with linear dimensions 1 000–4 000 AU, depending on frequency (Bartel et al. 1982; Bietenholz et al. 1996). Thus double-peaked optical lines show a very strong association with nuclei that generate jets *whether or not the jet is a powerful radio source.* All three of the galaxies with newly-appeared, double-peaked lines have strong, low ionization forbidden-lines (LINER class or transition between LINER and Seyfert type line ratios), reinforcing EH's conclusion that this type of narrow line spectrum is associated with the double-peaked broad lines.

When discovered in Nov. 1991, the blue wing of the broad Hα profile of NGC 1097 extended further from systemic than did the red wing, and the red peak was brighter than the blue (the same was true of the double-peaked broad Hα profile of M 81 when discovered – see Bower et al. 1996). These two characteristics conflict with the predictions of the circular relativistic disk model (Storchi-Bergmann, Baldwin and Wilson 1993). Over the subsequent 2.3 years, the flux of the entire Hα line in NGC 1097 has decreased by a factor of ≈ 2 and the red peak has declined more than the blue, so that in 1994 the profile appeared to be symmetric (see Fig. 3 and Storchi-Bergmann et al. 1995). The decline in broad Hα flux was accompanied by a steepening of the broad line Balmer decrement. If the increase in the Balmer decrement is the result of increased reddening, the decline in the Hα flux is roughly accounted for. Alternatively, the decline in the line flux could reflect a decline in the ionizing continuum.

The binary black hole model is unlikely to apply to NGC 1097, because the velocities of the two peaks did not change during the 2.3 years of monitoring. The corresponding lower limit to the total mass of the binary is $\approx 10^9$ $M_\odot$, which is probably too large for a mildly active galaxy like NGC 1097 (Storchi-Bergmann et al. 1995). Bi-polar outflows remain a viable and attractive model. A modification of the disk model has been made by Storchi-Bergmann et al. (1995) and Eracleous et al. (1995), who propose that the lines originate in an elliptical disk (one of Alar Toomre's "last refuge of a scoundrel"). In such a model, the red peak can be stronger than the blue, as is seen in about 1/4 of the double-peaked emitters. Eracleous et al. (1995) calculate the line profiles from

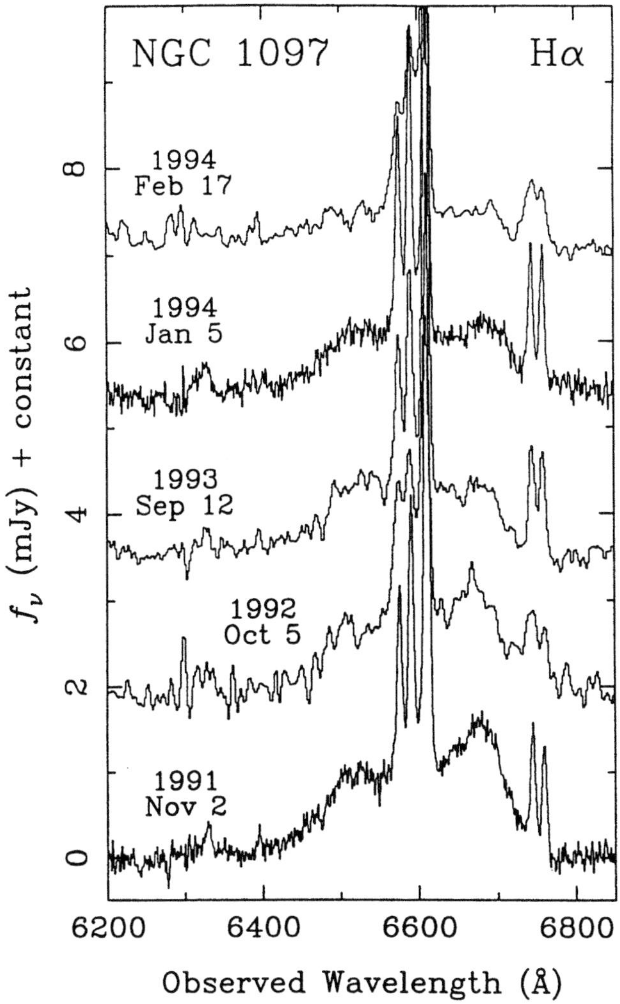

Fig. 3. From Storchi-Bergmann et al. (1995). The observed broad Hα profiles of the nucleus of NGC 1097 from 1991 Nov 2 to 1994 Feb 17. The underlying stellar population spectrum has been subtracted and successive spectra have been shifted vertically (by 1.8 mJy) for clarity

an elliptical, relativistic disk, and show that this model can fit a good number of the profiles that cannot be described by a circular disk model. In particular, NGC 1097 is best described by an eccentric ring, rather than a disk, with a width smaller than its radius.

Eracleous et al. (1995) offer two plausible scenarios for the formation of an elliptical disk in a galactic nucleus. In the first, a binary black hole is present. The tidal field of the secondary is responsible for the eccentricity of the accretion disk around the primary. This effect is analogous to the similar tidal distortion of the accretion disk around the white dwarf by the secondary star in a cataclysmic variable. It is, of course, unclear how a black hole binary would form in a spiral galaxy like NGC 1097. The second scenario discussed by Eracleous et al. (1995) is inspired by the appearance of the double-peaked lines in NGC 1097. The model involves the formation of a transient accretion disk through capture and disruption of a passing star by the black hole. The disruption of a 1 $M_\odot$ star is expected to produce a well-defined elliptical disk within a viscous timescale (Syer and Clarke 1992, 1993). The eccentric outer disk (whence the Hα is emitted) is not strongly affected by relativistic differential precession, which tends to circularise the disk within $\approx 200\ R_G$. This model is particularly attractive for NGC 1097 because (1) the mild level of activity suggests a low mass hole (perhaps $\approx 10^6\ M_\odot$), which can readily disrupt stars before accreting them, (2) the line profile is best fitted by a narrow annulus, which could represent the postdisruption debris, and (3) the structure and properties of the elliptical ring are expected to change on a timescale comparable to the viscous time, and hence the broad recombination profiles should vary.

In conclusion, it is at present not possible to choose between disk and outflow models for broad, double-peaked line profiles. Simple models of both types can describe much of the present data. Unfortunately (or fortunately), both models can be made much more complex in an attempt to fit difficult line profiles. For example, disks can have arbitrarily chosen inner and outer radii, can be eccentric or warped, and may precess or contain spiral arms or hot spots, all of which will affect the line profiles. Jets and outflows can be asymmetric, decelerate and suffer differential obscuration. As Eracleous et al. (1995) emphasise, the best method for discriminating between the models may be studies of profile variability, and especially the response to changes in the ionizing radiation. Another test involves the line polarization, since light emitted by a thin disk with electron scattering opacity emerges polarized parallel to the plane of the disk (e.g. Antonucci 1988). The polarization of the line should thus be perpendicular to the radio axis. A recent study of Arp 102B finds that the polarization of the double-peaked broad Hα line is not in accord with the relativistic disk model (Antonucci, Hurt and Agol 1996).

2.3 H_2O Megamasers

NGC 4258. Recent VLBI observations (Greenhill et al. 1995a; Miyoshi et al. 1995) of the LINER galaxy NGC 4258 have shown that the water vapor megamasers in this galaxy arise in a thin, edge-on gaseous annulus at a galactocentric radius of 0.13 – 0.26 pc. Maser emission is observed both near systemic velocity, arising from clouds at the near side of the disk and along our line of sight to a central opaque core (hypothesised to be a source of continuum radiation at 22 GHz), and from "satellite lines" with velocities of $\pm$ 900 km s^{-1} w.r.t. systemic (Nakai, Inoue and Miyoshi 1993), arising from gas at the tangent points with rotational velocities directed towards and away from Earth. The satellite lines show an accurately Keplerian rotation curve (Fig. 4). The recessional velocities of the near-systemic features are observed to be increasing at a rate of about 9 km s^{-1} yr^{-1} (Haschick, Baan and Peng 1994; Greenhill et al. 1995b). This increase of velocity is believed to result from the centripetal acceleration of clumps of gas in the annulus as they move across our line of sight to the central core (Watson and Wallin 1994; Haschick, Baan and Peng 1994; Greenhill et al. 1995b). Figure 5 shows a cartoon of the geometry for a nearly edge-on disk.

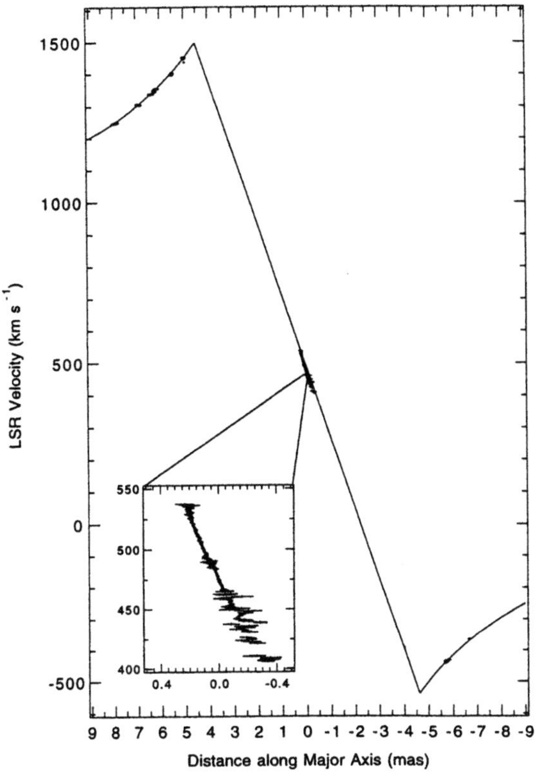

Fig. 4. From Miyoshi et al. (1995). Line of sight velocity versus distance along the major axis (position angle = 86°) of the masering disk in NGC 4258. Inset: the data near the systemic velocity of the galaxy. The position errors are only visible on the scale of the inset. Note the excellent fit of the plotted Keplerian curve to the satellite lines. The linear dependence of the systemic emission is a consequence of the change in projection of the rotation velocity

The most important information obtainable from these observations can be summarised in the following simplified way. Suppose that the masers arise in a thin, circular annulus viewed edge-on. Let θ be the observed angular dimension along the projected edge-on disk, V the observed recession velocity of the masers

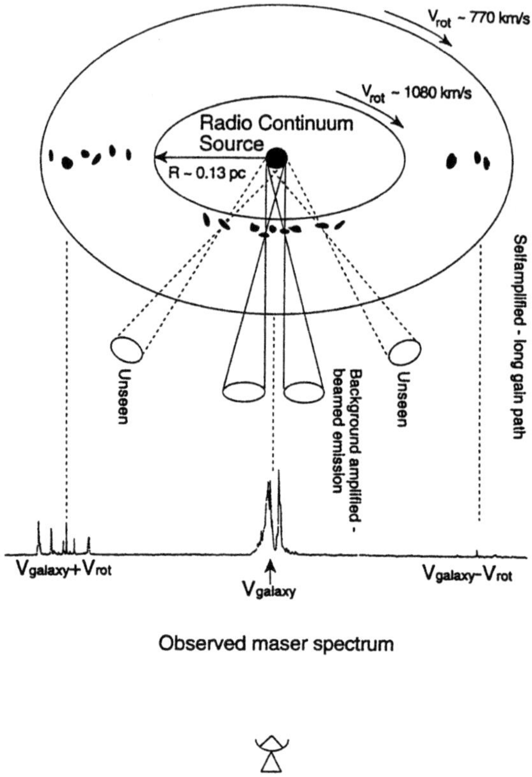

Fig. 5. From Moran et al. (1995). Cartoon of the maser geometry for the nearly edge-on disk of NGC 4258. The view is from slightly above the plane containing the disk and Earth. The masing region occupies an annulus, with a fractional radial thickness of 0.5, that probably lies within a more extensive dusty torus. Masers in the systemic velocity group become visible to us when they pass in front of the continuum source, which they amplify, and whose diameter may be inferred from the angular extent of the maser features. However, the high-velocity features do not amplify a background source, relying instead on long velocity-coherent gain paths through the disk. The emission that we see is beamed anisotropically in and near the plane of the disk

and V_{gal} the systemic velocity of NGC 4258. Intrinsic properties of interest include the distance to the galaxy, D, the rotational velocity of the annulus, U, and the radius of the annulus, r. We can relate the information obtainable from the various observations to the intrinsic properties as follows.
a) VLBI mapping of the systemic features: $(\mathrm{d}V/\mathrm{d}\theta)_{\rm syst} = UD/r$.
b) Monitoring of the time dependence of the velocity of individual clumps of gas in the systemic features: $(\mathrm{d}V/\mathrm{d}t)_{\rm syst} = U^2/r$.
c) VLBI measurement of the angular radius of the satellite lines: $\theta_{\rm sat} = r/D$.
d) Measurement of the recessional velocity of the satellite lines: $V_{\rm sat} - V_{\rm gal} = U$.
e) VLBI mapping of the systemic features at more than one epoch: $(\mathrm{d}\theta/\mathrm{d}t)_{\rm syst} = U/D$ (this measurement has not yet been made).

Combination of measurements a) through d) gives $D = 6.4$ Mpc, $r = 0.13 - 0.25$ pc, $U = 800 - 1\,100$ km s^{-1}, and a central mass $M = rU^2/\mathrm{G} = 3.6 \times 10^7$ M$_\odot$ (Greenhill et al. 1995a, b; Miyoshi et al. 1995). What is impressive here is the high degree of internal self-consistency in the model. If instead of obtaining D from the maser observations we assume it is known, measurements a) through d) give four equations for two unknowns. For example, measurement of the angular radius (c) and velocity (d) of the satellite lines allows the angular change in velocity (a) and the acceleration (b) of the systemic components to be correctly predicted. This self-consistency gives confidence in the model and argues strongly against alternative theories in which the masering gas is moving radially.

The discovery of the masing disk in NGC 4258 has stimulated important theoretical work on both the disk itself and the binding mass, as I now discuss. The disk is unresolved in the vertical direction ($h/r \leq 0.0025$, where h is the vertical thickness of the disk; Moran et al. 1995). For hydrostatic equilibrium, $h/r = c_s/U$, where c_s is the sound speed. The observed limit on h/r translates to an upper limit on the temperature $T < 1\,000$K (Moran et al. 1995). This temperature is compatible with those needed in models of the H_2O maser emission. If, instead, the vertical pressure is magnetic, then c_s should be replaced by the Alfvén speed $V_A = B/(4\pi\rho)^{1/2}$, where B is the magnetic field and ρ the gas density. For a molecular density of 10^{10} cm^{-3} (the maximum allowed to avoid quenching of the water maser), then $B < 0.2$ Gauss (Moran et al. 1995).

Application of the Toomre stability criterion for rotating, self-gravitating disks (Toomre 1964; Binney and Tremaine 1987) shows that the disk is near the borderline of stability (Moran et al. 1995; Maoz 1995b). Maoz (1995b) suggests that the separation of the satellite emission sources into several distinct clumps and the regularity of the distance intervals between them is indicative of spiral structure. The increased density at the wave crests of the spiral pattern is hypothesised to provide increased amplification of the maser.

As the disk is slightly warped (Miyoshi et al. 1995), it can be illuminated by a central X-ray source. These X-rays are expected to heat the molecular gas in the midplane at the observed radius of the water masers to 200 – 1 000 K (Neufeld, Maloney and Conger 1994). At temperatures above 300 K, the water abundance is enhanced by chemical reactions of O and OH with H_2 (Neufeld, Maloney and Conger 1994). The maser emission in the 22 GHz line probably results from collisional excitation in this gas (Neufeld, Maloney and Conger 1994; Neufeld and Maloney 1995). Given the central mass and the estimated luminosity of the nucleus, the inferred Eddington ratio is small: $L/L_E \approx 10^{-4}$. Modelling of the disk by Neufeld and Maloney (1995) suggests that material accretes through the disk at a rate of $7 \times 10^{-5}\alpha$ M$_\odot$ yr^{-1}, where α is the usual dimensionless number parameterising the viscosity. This value indicates that the efficiency for the conversion of rest mass energy into radiation is $\approx 0.1\alpha^{-1}$, in agreement with theoretical expectations for a standard accretion disk if $\alpha \approx 1$ (see Lasota et al. 1995 for a different picture in which the accretion flow is advection dominated).

Maoz (1995a) has evaluated alternatives to a supermassive black hole as the binding mass for the disk. If the entire mass within the inner radius of the masering disk is in a dense star cluster with a Plummer mass density profile, the upper limit to the velocity deviations from a Keplerian curve ($\Delta U/U < 4 \times 10^{-3}$) implies a central mass density of $\rho_0 \geq 4.5 \times 10^{12}$ $M_\odot$ pc^{-3}. At these densities, a cluster which consists of 1.4 $M_\odot$ neutron stars would evaporate in $\leq 1.5 \times 10^8$ yr. In fact, a cluster which consists of any objects with mass ≥ 0.03 $M_\odot$ would evaporate within ≤ 6 Gyr. Further, a cluster which consists of objects with typical mass ≤ 0.03 $M_\odot$ is also ruled out by constraints on the physical collision timescale, unless the objects are extremely dense. Maoz (1995a) also considers the hypothesis that the mass within the inner edge of the disk is in the form of a relatively light black hole surrounded by a massive density cusp in the background mass distribution. Maoz shows that this also is not a viable possibility, and concludes that the black hole mass must be at least 98% of that inferred from the accurately Keplerian fit to the rotation curve of the disk (i.e. 3.6×10^7 $M_\odot$).

Other H_2O Megamasers. Gallimore et al. (1995) have recently found that the kinematics of the brightest megamasers in NGC 1068 may be described by an edge-on disk around a central mass concentration of $(4.4 \pm 0.8) \times 10^7$ $M_\odot$. Higher resolution observations with the VLBA are desirable to confirm these results and check whether the rotation curve is Keplerian.

Wilson, Braatz and Henkel (1995) report monitoring of the brightest maser spike in the LINER nucleus of NGC 2639 and find a redward velocity drift of 6.6 ± 0.4 km s^{-1} yr^{-1} over a period of 1.4 yr. If this acceleration represents the centripetal acceleration of the near side of an edge-on Keplerian disk, as is the case in NGC 4258, the mass of the central object is 1.5×10^7 $(r/0.1\ pc)^2$ $M_\odot$ (cf. observation b above for NGC 4258). Further observations (i.e. a, c, or d in the NGC 4258 list) are needed to determine r or v, and hence M, uniquely. Unfortunately, no satellite lines have been detected so far and the maser is too weak for VLBI observations.

2.4 Direct Imaging of Dusty Disks

Recently, high-resolution near-infrared imaging has revealed red structures straddling the nucleus perpendicular to the radio and emission-line axes in a handful of Seyfert 2 galaxies (Simpson et al. 1995). An example is shown in Fig. 6. The grey scale represents the ratio of [O III]λ5007 to Hα+[N II]λλ6548, 6584 in the Seyfert 2 galaxy Mkn 348. Regions of high excitation gas are seen to the north and south of the nucleus. The axis of the high excitation nebulosity aligns well with the radio axis and is perpendicular to the optical polarization, as expected if the nucleus is blocked from view by a dusty torus and the optical light is scattered by regions above and below the torus (Miller and Antonucci 1983). The contours represent the $J-K$ color; a red band, extending about 1 kpc with $J-K = 2.0$ at the center, is seen aligned perpendicular to the axis of the

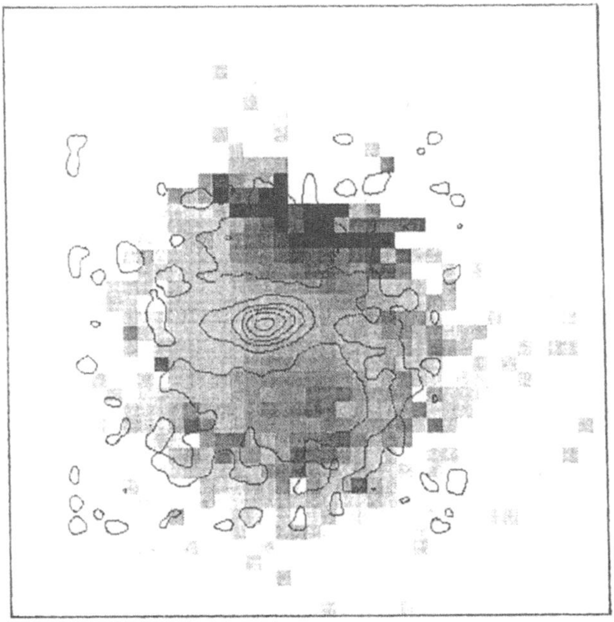

Fig. 6. From Simpson et al. (1996). *Greyscale:* ratio of [O III]λ5007 to Hα+[N II]λλ6548, 6584 flux in the Seyfert 2 galaxy Mkn 348. The greyscale is linearly scaled between 0.3 (white) and 3 (black). Areas of low signal-to-noise and with a ratio below 0.3 have been blanked out. Regions of high excitation gas can be seen to north and south of the nucleus. *Contours:* $J - K$ color, with contour interval 0.2 magnitudes; the centermost (reddest) contour represents $J - K = 2.0$. The figure covers 15″ × 15″ (4.0 kpc × 4.0 kpc)

high excitation gas and radio emission (Simpson et al. 1996). This red band is associated with excess emission at K band and may represent thermal emission from hot ($T \approx 700\,\mathrm{K}$) dust in a torus or disk viewed edge-on. It is tempting to speculate (Simpson et al. 1996) that this probable dusty torus represents the outer part of the toroidal structure invoked to hide the broad line region in this object (Miller and Goodrich 1990).

3 Bi-Polar Phenomena – Indirect Evidence for Accretion Disks

3.1 The Radio Emission of Seyfert Galaxies

The first evidence that the nuclei of Seyfert galaxies eject material in opposite directions came from radio mapping of their nuclei. Sub-arc second radio maps have been made of three samples of Seyferts with the VLA – a radio flux-limited sam-

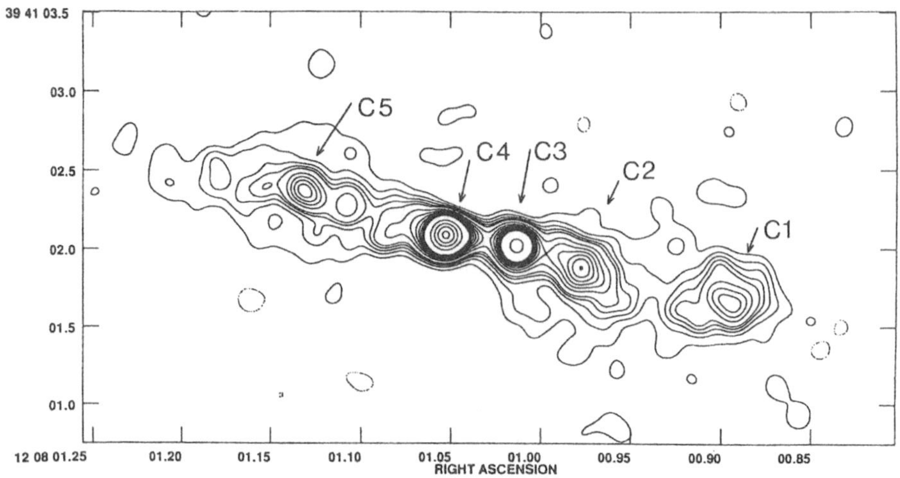

Fig. 7. From Pedlar et al. (1993). Contour map of the radio emission of NGC 4151 at 5 GHz. The length of the jet is ≈ 270 pc

ple of Markarian Seyferts (Ulvestad and Wilson 1984a), a distance-limited sample containing all known Seyferts with recessional velocity below 4 600 km s^{-1} (Ulvestad and Wilson 1989), and an optical magnitude-limited sample complete to a (total) galaxy Zwicky magnitude of $m_{\rm Zw} \leq 14.5$ (Kukula et al. 1995). Analysis of the last dataset is in preparation, so the following discussion is based on the first two surveys. A significant number of the radio nuclei are too compact to map at the $0\farcs4$ (120 pc at recessional velocity 4 600 km s^{-1}) resolution of the VLA at 6 cm: in the Markarian Seyferts (distance-limited) sample, 41% (26%) of the nuclei were slightly resolved and 21% (28%) were unresolved. The majority of the remainder have "linear" radio structures – double, triple or jet-like, usually straddling the optical continuum nucleus. Such sources comprise 28% (26%) of these two samples. These sources represent bi-polar collimated ejection from the Seyfert nucleus (e.g. Wilson 1982). A much smaller percentage – 3% (5%) – exhibit "diffuse" ("blob-like" or non-collimated) morphology, which radio emission is believed to be powered by circumnuclear star formation (Wilson 1988). A similarly small percentage – 0% (7%) – show both "linear" and "diffuse" radio structures, with the remainder – 7% (2%) – having morphologies which defy this simple classification.

The collimation of the "linear" radio sources in Seyfert galaxies can be extremely tight. For example, the length-to-width ratios of the radio jets in NGC 4151 and Mkn 3 are ≥15:1 (Pedlar et al. 1993) and ≥25:1 (Kukula et al. 1993), respectively (see Figs. 7 and 9). Most other "linear" radio sources are unresolved

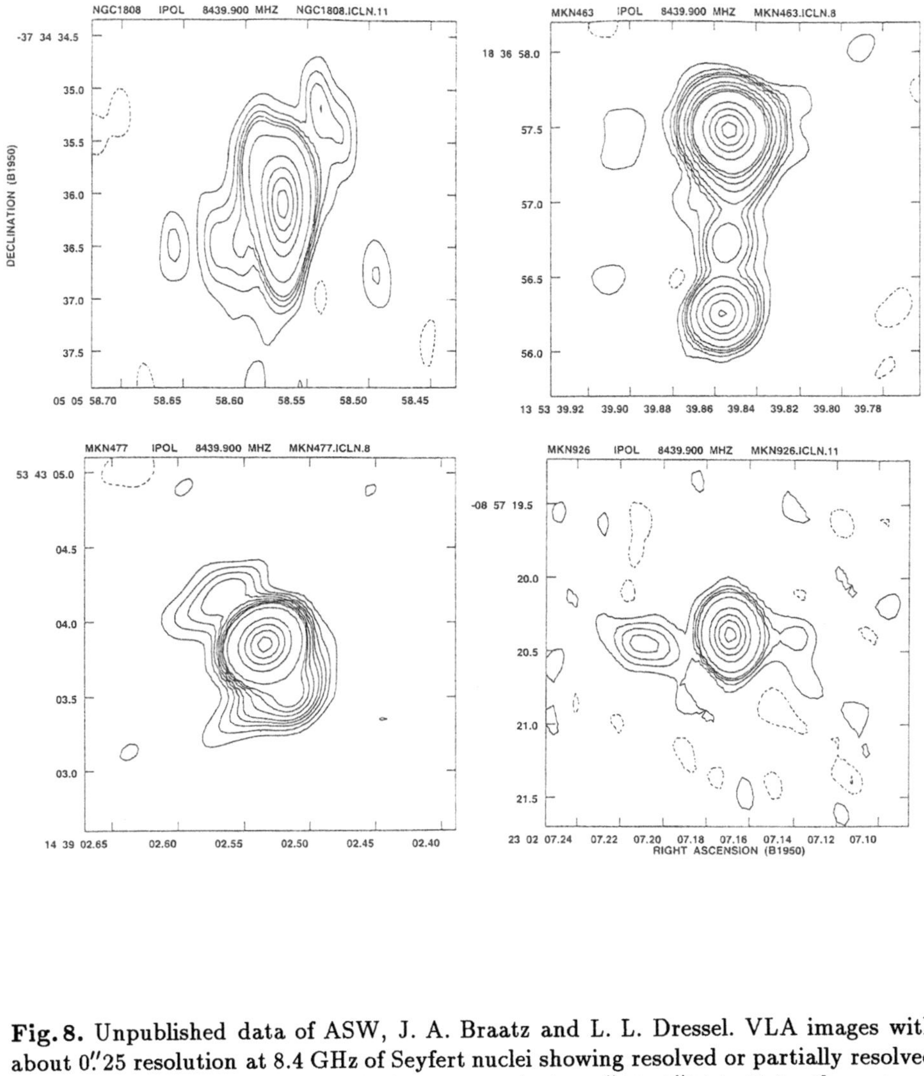

Fig. 8. Unpublished data of ASW, J. A. Braatz and L. L. Dressel. VLA images with about $0''.25$ resolution at 8.4 GHz of Seyfert nuclei showing resolved or partially resolved triple radio sources. **Top left**: NGC 1808, resolution $0''.64 \times 0''.24$ in P.A. 4°, contours at −1 (dotted), 1, 2, 3, 4, 5, 10, 30, 50, 70 and 90% of the peak brightness of 7.8 mJy (beam area)$^{-1}$. **Top right**: Mkn 463E, resolution $0''.26 \times 0''.26$, contours at −0.25 (dotted), 0.25, 0.5, 0.75, 1, 1.5, 3, 5, 7.5, 10, 30, 50, 70, and 90% of the peak brightness of 38.5 mJy (beam area)$^{-1}$. **Bottom left**: Mkn 477, resolution $0''.28 \times 0''.24$ in P.A. −52°, contours at −0.6 (dotted), 0.6, 1.2, 1.8, 2.4, 3, 4, 5, 6, 10, 30, 50, 70, and 90% of the peak brightness of 12.6 mJy (beam area)$^{-1}$. **Bottom right**: Mkn 926 (=MCG–2–58–22), resolution $0''.27 \times 0''.21$ in P.A. 3°, contours at −2.5, −1.25 (dotted), 1.25, 2.5, 3.75, 5, 10, 30, 50, 70, and 90% of the peak brightness of 7.5 mJy (beam area)$^{-1}$

transverse to their axes so, as for NGC 4151 and Mkn 3, only lower limits to their length-to-width ratios can be established. When the radio lobes *are* resolved in the transverse direction, this finite width probably results from interaction with the interstellar medium (e.g. NGC 1068 – Wilson and Ulvestad 1987); the collimation by the nucleus itself is better.

The great majority of the slightly resolved and unresolved sources have steep radio spectra, like the "linear" sources. As the resolution of the VLA at 6 cm is unlikely to correspond to any critical physical scale in the nuclei, most of the slightly resolved and unresolved sources are presumably "linear" sources with angular sizes below the resolution. We (ASW, J. A. Braatz and L. L. Dressel) have recently obtained new VLA observations with somewhat better resolution ($0\farcs25$) and sensitivity than our original survey for about 15 Seyferts. Figure 8 shows maps of four of these galaxies, which reveal resolved or partially resolved triple sources. Two of these galaxies (Mkn 477 and Mkn 926) are in the radio-flux limited sample of Markarian Seyferts and were classified as "slightly resolved" in the original 4.9 GHz observations (Ulvestad and Wilson 1984a). These new observations thus support the notion that most, if not all, Seyfert galaxies exhibit double, triple or jet-like radio structures.

Parenthetically, it is worth recalling that the axes of the "linear" radio sources are misaligned, and may be randomly oriented, with respect to the axes of the disks of the host galaxy (Ulvestad and Wilson 1984b). The same is true of the narrow line and extended narrow line regions (see Sect. 3.2). Thus the disk which collimates the radio outflow is misaligned with respect to the galaxy disk. For example, the masering disk in NGC 4258 is almost perpendicular to the galaxy disk (Miyoshi et al. 1995). This apparently random orientation of the nuclear disks may not be too surprising when we recall that their sizes are much smaller than the $\approx$ 100 pc scale height of the gas layer of the galaxy disk. The nuclear activity presumably results from accretion of material in the inner part of the galaxy onto the black hole, and the resulting disk axis may simply reflect the orbit of the last molecular cloud complex to be captured and accreted by the hole. The axes of prior and future accretion events could then be quite different.

It is instructive to compare these results with similar VLA surveys of the nuclear regions of normal spiral galaxies and galaxies with nuclear starbursts (e.g. van der Hulst, Crane and Keel 1981; Condon et al. 1982; Hummel, van der Hulst and Dickey 1984; review in van der Hulst 1991). Double, triple or jet-like structures are hardly ever found, and when they are, the nucleus almost always shows some Seyfert or LINER characteristics in its optical spectrum. Another distinction between the radio emissions of Seyfert and starburst galaxies has been noted by Norris et al. (1990). Radio interferometric observations using a single baseline interferometer with resolution $\leq$ 40 pc (for the typical distances of the galaxies studied) detected compact cores in some 30% of the Seyfert galaxies but in none of the starbursts.

My conclusion is that the *characteristic* radio emission of Seyferts (i.e. the radio property that distinguishes Seyferts from inactive spirals) results from *bi-polar, well collimated ejection from the inner (< a few pc) nucleus.* A key point is that such radio structures are not found in galaxies with only nuclear starbursts. This is exactly as expected, for the radio emission of starbursts is dominated by supernovae, supernova remnants and relativistic electrons which have leaked out of the remnants. Although starbursts do drive winds out along the minor axes of the galaxy disks, these winds cover a wide angle (Heckman, Armus and Miley 1987, 1990; Chevalier and Clegg 1985); there is no scope here for highly collimated radio jets and lobes.

3.2 The Narrow and Extended Narrow Line Region of Seyfert Galaxies

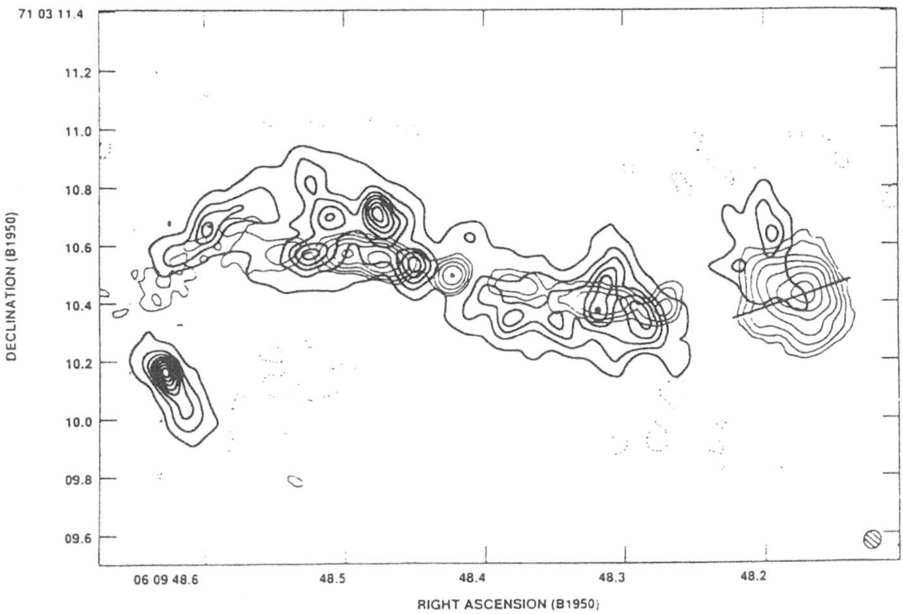

Fig. 9. From Capetti et al. (1995a). Contour map of the radio emission of Mkn 3 (thin lines) superimposed on a contour map of the [O III]$\lambda\lambda$4959, 5007 line emission (thick lines). The high degree of collimation of both the radio emission and the ionized gas is noteworthy. The length of the jet is $\approx$ 620 pc

It has long been known that the narrow line regions of Seyfert galaxies are closely associated with their radio emission. This association includes a similarity of spatial scale, morphological associations based on ground-based imaging in strong optical emission lines, correlations between radio power and both emission-line luminosity and emission-line width and broadening of the lines in the vicinity of the radio clouds (review in Wilson 1991). Only with the launch of the Hubble Space Telescope (HST), however, has it been possible to obtain images of the ionized gas with spatial resolution similar to or better than the radio maps. Such images have been published for about a dozen Seyferts, mostly of type 2. In general, these results confirm the association of the ionized gas with the radio jets and lobes, or at least show that the gas aligns with the radio axis. In detail, a variety of morphological structures are seen. The emission-line images show well collimated jet-like morphologies in Mkn 3 (Capetti et al. 1995a – see Fig. 9), Mkn 463E (Uomoto et al. 1993), Mkn 1066 (Bower et al. 1995), and probably Mkn 6 (Capetti et al. 1995b) and NGC 2110 (Mulchaey et al. 1994). The ionized gas comprises clouds associated with the radio lobes in NGC 5929 (Bower et al. 1994) and Mkn 78 (Capetti et al. 1994); some clouds in the complex emission-line region of NGC 1068 are probably associated with the radio components (Evans et al. 1991; Macchetto et al. 1994). The morphology of the ionized gas (i.e. "jet-like" or "lobe-like") seems to mimic the morphology of the radio emission. V-shaped regions, with apices near the nucleus, of ionized gas are found in NGC 1068 (Evans et al. 1991; Macchetto et al. 1994), NGC 4151 (Evans et al. 1993; Boksenberg et al. 1995) and NGC 5728 (Wilson et al. 1993). These V-shaped regions are usually interpreted in terms of ionization by a collimated nuclear radiation field (Pogge 1989). The axis of collimation aligns tightly with the radio axis (Unger et al. 1987; Wilson and Tsvetanov 1994).

The general interpretation of associations of ionized gas with synchrotron radio components involves compression and acceleration of ambient gas by outwardly moving radio jets and lobes. The compression arises through shock waves driven into the clouds by the jets and this "stirring" of gas by the jets produces the high velocities seen in the gas (e.g. Whittle et al. 1988). Models of this type have been discussed by Pedlar, Dyson and Unger (1985), Wilson and Ulvestad (1987) and Taylor, Dyson and Axon (1992). The gas is known to be photoionized, but the source of ionizing radiation is controversial. This source may be the compact nucleus seen directly in Seyfert 1's and inferred (through unified schemes - see Antonucci 1993) to be present in Seyfert 2's. Alternatively, the shocks associated with the jet - interstellar gas interaction may produce the ionizing photons – so called "photoionizing shocks" (Sutherland, Bicknell and Dopita 1993; Dopita and Sutherland 1995). A critique of these two competing pictures is provided by Morse et al. (1996).

4 Concluding Remarks

Excellent discussions of the starburst model for AGNs have been given by Heckman (1991), Terlevich (1992) and Filippenko (1992). Difficulties for the starburst model raised by Filippenko and Heckman include: 1) the similarity in properties (except at radio wavelengths) between radio-loud and radio-quiet AGN. Everyone agrees that the starburst model cannot apply to radio-loud objects, so the similarities are surprising if the radio-louds are powered by black holes and the radio-quiets by starbursts; 2) the existence of rapid X-ray variability; 3) the nature of the progenitors of the active galaxies. In the starburst hypothesis, most of the luminosity of the AGN comes ultimately from the kinetic energy of supernovae (Terlevich 1992). However, photospheric emission from the hot stars in the pre-supernova phase is 10 – 100 times more luminous than the supernova ejecta. Thus there must be objects in the initial starburst phase with 10 – 100 times the luminosity of AGNs; the existence of these objects is uncertain. Heckman (1991) and Terlevich (1992) suggest the luminous progenitor phase could be shrouded in dust and that luminous IRAS galaxies might be the progenitors of quasars. Closer to home, one might wonder which objects are the starburst progenitors of nearby luminous Seyferts. Such objects would be *nuclear* starbursts with luminosities of $10^{11} - 10^{13}$ $L_\odot$ in *early-type* spirals (where Seyfert nuclei are preferentially located). 4) A SNII produces only 10^{-3} $M_\odot c^2$ of kinetic energy (Filippenko 1992) and leaves a neutron star or black hole of mass ≥ 1 $M_\odot$. This efficiency of production of energy (which could be converted into radiation) per unit "mass left behind" is $\approx$ 100 times lower than gravitational accretion onto a black hole if this proceeds with efficiency 0.1. In the latter picture, the observed amount of quasar radiant energy implies that each luminous galaxy should harbor a black hole with mass $\approx 10^7$ $M_\odot$ on average (Soltan 1982). Thus in the starburst model, masses 100 times larger – $\approx 10^9$ $M_\odot$ per bright galaxy – should be left in the form of neutron stars and black holes. Adding in the low mass stars that formed along with the OB stars, Filippenko (1992) estimates central masses of $10^{10} - 10^{11}$ $M_\odot$ per galaxy left over from the quasar phase. If confined within 100 pc diameter, stellar velocity dispersions of $\sigma \simeq 500$ km s^{-1} are implied, larger than observed (but see Terlevich 1992 for uncertainties in this argument). 5) To avoid problem 4), the starburst must be spatially extended, leading to predictions that quasars should be resolvable by HST.

The evidence for compact accretion disks and highly collimated jets reviewed in this paper applies to a limited number of AGNs. Very broad FeKα lines have been found in some half a dozen Seyferts. Double-peaked broad optical recombination lines are known in about 20 galaxies. A Keplerian accretion disk has been mapped out through H_2O megamaser emission in only NGC 4258. Because of instrumental limitations of angular resolution and sensitivity, truly jet-like (length:width ratios $\geq$10:1) radio sources are known in only a small, but increasing, minority of Seyfert galaxies. Further observations will reveal whether or not these characteristics favoring the accreting black hole model are typical of radio-quiet AGN.

Acknowledgements. I am grateful to the Nobel Foundation for financial support and to the organisers for arranging such a wonderful meeting.

References

Antonucci, R. R. J. (1988): in Supermassive Black Holes, ed. M. Kafatos, Cambridge University Press, p. 26

Antonucci, R. R. J. (1993): ARA&A **31**, 473

Antonucci, R. R. J., Hurt, T., Agol, E. (1996): ApJ Letters, in press

Bartel, N. et al. (1982): ApJ **262**, 556
Begelman, M. C. (1985): in Astrophysics of Active Galaxies and Quasi-Stellar Objects, ed. J. S. Miller, University Science Books, p. 411
Begelman, M. C., Blandford, R. D., Rees, M. J. (1980): Nat **287**, 307
Bietenholz, M. F. et al. (1996): ApJ **457**, 604
Binney, J., Tremaine, S. (1987): Galactic Dynamics, Princeton University Press, p. 363
Boksenberg, A. et al. (1995): ApJ **440**, 151
Boroson, T. A., Green, R. F. (1992): ApJS **80**, 109
Bower, G. A., Wilson, A. S., Heckman, T. M., Richstone, D. O. (1996): AJ (submitted)
Bower, G. A., Wilson, A. S., Morse, J. A., Gelderman, R., Whittle, M., Mulchaey, J. S. (1995): ApJ **454**, 106
Bower, G. A., Wilson, A. S., Mulchaey, J. S., Miley, G. K., Heckman, T. M., Krolik, J. H. (1994): AJ **107**, 1686
Capetti, A., Axon, D. J., Kukula, M., Macchetto, F., Pedlar, A., Sparks, W. B., Boksenberg, A. (1995b): ApJ **454**, L85
Capetti, A., Macchetto, F., Axon, D. J., Sparks, W. B., Boksenberg, A. (1995a): ApJ **448**, 600
Capetti, A., Macchetto, F., Sparks, W. B., Boksenberg, A. (1994): ApJ **421**, 87
Chen, K., Halpern, J. P. (1989): ApJ **344**, 115
Chen, K., Halpern, J. P., Filippenko, A. V. (1989): ApJ **339**, 742
Chevalier, R., Clegg, A. (1985): Nat **317**, 44
Condon, J. J., Condon, M. A., Gisler, G., Puschell, J. J. (1982): ApJ **252**, 102
Dopita, M. A., Sutherland, R. S. (1995): ApJ **455**, 468
Eracleous, M., Halpern, J. P. (1994): ApJS **90**, 1 (EH)
Eracleous, M., Livio, M., Halpern, J. P., Storchi-Bergmann, T. (1995): ApJ **438**, 610
Evans, I. N., Ford, H. C., Kinney, A. L., Antonucci, R. R. J., Armus, L., Caganoff, S. (1991): ApJ **369**, L27
Evans, I. N., Tsvetanov, Z. I., Kriss, G. A., Ford, H. C., Caganoff, S., Koratkar, A. P. (1993): ApJ **417**, 82
Fabian, A. C., Nandra, K., Reynolds, C. S., Brandt, W. N., Otani, C., Tanaka, Y., Inoue, H., Iwasawa, K. (1995): MNRAS **277**, L11
Fabian, A. C., Rees, M. J., Stella, L., White, N. E. (1989): MNRAS **238**, 729
Fanaroff, B. L., Riley, J. M. (1974): MNRAS **167**, 31P
Ferland, G. J., Netzer, H. (1983): ApJ **264**, 105
Filippenko, A. V. (1992): in Relationships between Active Galactic Nuclei and Starburst Galaxies, ed. A. V. Filippenko, PASPC **31**, p. 253
Gallimore, J. F., Baum, S. A., O'Dea, C. P., Brinks, E., Pedlar, A. (1995): preprint
Gaskell, C. M. (1983): in Proc. 24th Liège Astrophysical Colloquium, Quasars and Gravitational Lenses, Institut d'Astrophysique, Université de Liège, p. 473
Greenhill, L. J., Henkel, C., Becker, R., Wilson, T. L., Wouterloot, J. G. A. (1995b): A&A, **304**, 21
Greenhill, L. J., Jiang, D. R., Moran, J. M., Reid, M. J., Lo, K. Y., Claussen, M. J. (1995a): ApJ **440**, 619
Halpern, J. P., Eracleous, M. (1994): ApJ **433**, L17
Halpern, J. P., Filippenko, A. V. (1991): in Testing the AGN Paradigm, eds. S. S. Holt, S. G. Neff, C. M. Urry, AIP Conference Proceedings **254**, American Institute of Physics, New York, p. 57
Halpern, J. P., Steiner, J. E. (1983): ApJ **269**, L37
Harwit, M., Pacini, F. (1975): ApJ **200**, L127

Haschick, A. D., Baan, W. D., Peng, E. W. (1994): ApJ **437**, L35
Heckman, T. M. (1991): in Massive Stars in Starbursts, eds. C. Leitherer, N. R. Walborn, T. M. Heckman, C. A. Norman, STScI Symposium Series Nr. 5, Cambridge University Press, p. 289
Heckman, T. M., Armus, L., Miley, G. K. (1987): AJ **93**, 276
Heckman, T. M., Armus, L., Miley, G. K. (1990): ApJS **74**, 833
Hummel, E., van der Hulst, J. M., Dickey, J. M. (1984): A&A **134**, 207
Kukula, M. J., Ghosh, T., Pedlar, A., Scilizzi, R. T., Miley, G. K., de Bruyn, A. G., Saikia, D. J. (1993): MNRAS **264**, 893
Kukula, M. J., Pedlar, A., Baum, S. A., O'Dea, C. P. (1995): MNRAS **276**, 1262
Lasota, J.-P., Abramowicz, M. A., Chen, X., Krolik, J. H., Narayan, R., Yi, I. (1995): preprint
Lynden-Bell, D. (1969): Nat **223**, 690
Macchetto, F., Capetti, A., Sparks, W. B., Axon, D. J., Boksenberg, A. (1994): ApJ **435**, L15
Maoz, E. (1995a): ApJ **447**, L91
Maoz, E. (1995b): ApJ **455**, L131
Miller, J. S., Antonucci, R. R. J. (1983): ApJ **271**, L7
Miller, J. S., Goodrich, R. W. (1990): ApJ **355**, 456
Miyoshi, M., Moran, J. M., Herrnstein, J., Greenhill, L. J., Nakai, N., Diamond, P., Inoue, M. (1995): Nat **373**, 127
Moran, J. M., Greenhill, L. J., Herrnstein, J., Diamond, P., Miyoshi, M., Nakai, N., Inoue, M. (1995): in Quasars and AGN: High Resolution Imaging, Proc. National Academy of Sciences, in press
Morse, J. A., Raymond, J. C., Wilson, A. S. (1996): PASP (submitted)
Mulchaey, J. S., Wilson, A. S., Bower, G. A., Heckman, T. M., Krolik, J. H., Miley, G. K. (1994): ApJ **433**, 625
Mushotzky, R. F., Fabian, A. C., Iwasawa, K., Kunieda, H., Matsuoka, M., Nandra, K., Tanaka, Y. (1995): MNRAS **272**, L92
Nakai, N., Inoue, M., Miyoshi, M. (1993): Nat **361**, 45
Neufeld, D. A., Maloney, P. R. (1995): ApJ **447**, L17
Neufeld, D. A., Maloney, P. R., Conger, S. (1994): ApJ **436**, L127
Norris, R. P., Allen, D. A., Sramek, R. A., Kesteven, M. J., Troup, E. R. (1990): ApJ **359**, 291
Pedlar, A., Dyson, J. E., Unger, S. W. (1985): MNRAS **214**, 463
Pedlar, A., Kukula, M. J., Longley, D. P. T., Muxlow, T. W. B., Axon, D. J., Baum, S., O'Dea, C., Unger, S. W. (1993): MNRAS **263**, 471
Pogge, R. W. (1989): ApJ **345**, 730
Rees, M. J., Begelman, M. C., Blandford, R. D., Phinney, E. S. (1982): Nat **295**, 17
Simpson, C., Mulchaey, J. S., Wilson, A. S., Ward, M. J., Alonso-Herrero, A. (1996): ApJ Letters, in press
Simpson, C., Wilson, A. S., Ward, M. J., Alonso-Herrero, A., Mulchaey, J. S. (1995): BAAS **27**, 1321
Soltan, A. (1982): MNRAS **200**, 115
Storchi-Bergmann, T., Baldwin, J. A., Wilson, A. S. (1993): ApJ **410**, L11
Storchi-Bergmann, T., Eracleous, M., Livio, M., Wilson, A. S., Filippenko, A. V., Halpern, J. P. (1995): ApJ **443**, 617
Sulentic, J. W., Calvani, M., Marziani, P., Zheng, W. (1990): ApJ **355**, L15
Sutherland, R. S., Bicknell, G. V., Dopita, M. A. (1993): ApJ **414**, 510

Syer, D., Clarke, C. J. (1992): MNRAS **255**, 92
Syer, D., Clarke, C. J. (1993): MNRAS **260**, 463
Tanaka, Y. et al. (1995): Nat **375**, 659
Taylor, D., Dyson, J. E., Axon, D. J. (1992): MNRAS **255**, 351
Terlevich, R. J. (1992): in Relationships between Active Galactic Nuclei and Starburst Galaxies, ed. A. V. Filippenko, PASPC **31**, p. 133
Terlevich, R. J., Tenorio-Tagle, G., Franco, J., Melnick, J. (1992): MNRAS **255**, 713
Toomre, A. (1964): ApJ **139**, 1217
Ulvestad, J. S., Wilson, A. S. (1984a): ApJ **278**, 544
Ulvestad, J. S., Wilson, A. S. (1984b): ApJ **285**, 439
Ulvestad, J. S., Wilson, A. S. (1989): ApJ **343**, 659
Unger, S. W., Pedlar, A., Axon, D. J., Whittle, M., Meurs, E. J. A., Ward, M. J. (1987): MNRAS **228**, 671
Uomoto, A., Caganoff, S., Ford, H. C., Rosenblatt, E. I., Antonucci, R. R. J., Evans, I. N., Cohen, R. D. (1993), AJ **105**, 1308
van der Hulst, J. M. (1991): in The Interpretation of Modern Synthesis Observations of Spiral Galaxies, eds. N. Duric, P. C. Crane, PASPC **18**, p. 215
van der Hulst, J. M., Crane, P. C., Keel, W. C. (1981): AJ **86**, 1175
Watson, W. D., Wallin, B. K. (1994): ApJ **432**, L35
Weedman, D. W. (1983): ApJ **266**, 479
Whittle, M., Pedlar, A., Meurs, E. J. A., Unger, S. W., Axon, D. J., Ward, M. J. (1988): ApJ **326**, 125
Wilson, A. S. (1982): in Proc. IAU Symp. 97 Extragalactic Radio Sources, eds. D. Heeschen, C. Wade, Reidel, Dordrecht, p. 179
Wilson, A. S. (1988): A&A **206**, 41
Wilson, A. S. (1991): in The Interpretation of Modern Synthesis Observations of Spiral Galaxies, eds. N. Duric, P. C. Crane, PASPC **18**, p. 227
Wilson, A. S., Braatz, J. A., Heckman, T. M., Krolik, J. H., Miley, G. K. (1993): ApJ **419**, L61
Wilson, A. S., Braatz, J. A., Henkel, C. (1995): ApJ **455**, L127
Wilson, A. S., Tsvetanov, Z. I. (1994): AJ **107**, 1227
Wilson, A. S., Ulvestad, J. S. (1987): ApJ **319**, 105
Wolstencroft, R. D., Zealey, W. J. (1975): MNRAS **173**, 51P
Zheng, W., Binette, L., Sulentic, J. W. (1990): ApJ **365**, 115
Zheng, W., Veilleux, S., Grandi, S. A. (1991): ApJ **381**, 418

Bar Triggered Nuclear Activity and the Anisotropic Radiation Fields of Active Galactic Nuclei

David J. Axon[1,2] and A. Robinson[3]

[1] Affiliated to the Astrophysics Division of ESA, Space Telescope Science Institute, 3700 San Martin Drive, Baltimore, MD 21218, USA
[2] Nuffield Radio Astronomy Laboratory, Jodrell Bank, Macclesfield, Cheshire, UK.
[3] Division of Physical Sciences, University of Hertfordshire, College Lane, Hatfield, Herts, UK

Abstract. We show observational evidence that double bars are present in Seyfert galaxies, and describe how they fit in with the idea that self-gravitating gas flows in a bar potential play a key role in fueling activity. In addition our results show that in each case the inner bar is closely aligned with the axis of the radiation cones emerging from the nucleus. We propose that the collimating torus of molecular gas responsible for the anisotropy of the radiation field is created at an Inner Inner Lindblad Resonance, and is fed by gas streaming in along the inner nuclear bar, and is thus regulated by the overall dynamical behaviour of the parent galaxy.

1 Introduction

In order to sustain activity in the nucleus of a galaxy, either in a starburst, or an Active Galactic Nucleus (AGN) associated with a supermassive black hole, there must be an adequate supply of fuel. It is generally accepted that the interstellar medium of the host galaxy is the source of this fuel (Phinney 1994). How, and at what rate the gas is delivered to the nucleus is fundamental to the triggering and subsequent evolution of activity in galaxies and thus intimately coupled to the form of their gravitational potential. Numerical simulations (e.g. Combes and Gerin 1985; Sellwood and Wilkinson 1993; Athanassoula 1992) have demonstrated that the torques set up by the non-axisymmetric gravitational potential of a bar are the most likely method of driving gaseous material into the centre of a galaxy. On kiloparsec scales the inflowing gas accumulates in circum-nuclear rings, bar-like features or spiral arms (Athanassoula 1992; Sellwood and Wilkinson 1993), which form provided the bar has Inner Lindblad Resonances (ILR's). Most existing simulations of gas flows in galactic bars lack sufficient spatial resolution to follow the gas flow down to scales comparable with the source of the nuclear activity (e.g., the accretion radius of the black hole $\sim$ 10 pc). Consequently, how the gas accumulating in the reservoir at 1 kpc loses its angular momentum and eventually falls towards the centre is a matter of debate, though a number of theoretical schemes for accomplishing this which invoke the self-gravity of the gas have been proposed, most notably that of

(Shlosman, Frank and Begelman 1989). In their picture, gas swept up into the inner kiloparsec by the stellar bar forms an unstable configuration resulting in a gaseous *bar within a bar* and triggering gas flow along the bar towards the centre.

Here we present direct evidence, based on a large ($\sim$ 350 galaxies) near IR survey of SAB and SB galaxies and AGN that inner stellar and gaseous bars are present in Seyfert galaxies and may be responsible for collimating their nuclear radiation fields.

2 AGN with Inner Bars

A significant fraction (around 20%) of galaxies in the near–IR survey show spectacular double bar structures, in which there is an inner stellar bar in the central $\sim$ kpc which is misaligned at large angles $\sim 60 - 90°$ to the primary bar, two examples of which are shown in Fig. 1. All of these are AGN, either Liners (e.g. NGC 1097, Shaw et al. 1993) or Seyferts (e.g. NGC 5728, Shaw et al. 1993). Many more of the sample show evidence of dramatic isophotal twists, which might be due to the presence of a triaxial bulge, but some of which are likely also to be inner bars viewed at unfavourable projections (Shaw et al. 1995). Whilst these observed isophote twists are not faint, we believe they have not been detected previously because of the dominance on the optical luminosity of OB stars within sites of active star formation. In the near–IR, the influence of these stars and also of dust obscuration are reduced. The observed light distribution more accurately represents the gravitational potential and we are able to probe deep into the potential in the inner regions of the galaxies. Similarly inner bars and isophotal twists have also been found in a sample of Seyferts by Woznaik et al. (1995) using I band CCD imaging, implying that they are a common feature of their inner structure.

The spatial distribution of the gas and dust around the nuclei of these AGN bears a striking resemblance to the predictions of numerical simulations of gas flows induced by barred potentials in which the x_2 family of orbits are well developed (cf. Athanassoula 1992; Shaw et al. 1993; Friedli and Martinet 1993). However, high resolution stellar and gas N–body simulations have shown that such orthogonal inner bar structures are *not stable in pure stellar systems* due to dynamical heating (Shaw et al. 1993). The simulations of Shaw et al. (1993) and Wada and Habe (1995) show that it is possible to create a stable inner bar if the self-grativity of the gas makes a significant contribution to the potential. In this situation, the gas accumulates in a gaseous bar between the two inferred ILR which leads the stellar bar and has sufficient mass to perturb the central stellar component. Since the gas follows the x_2 orbits, the stars also get trapped on the x_2 family of orbits.

Using self-consistent 3D N-body simulations Friedli and Martinet (1993) revealed an alternative situation for maintaining a secondary bar when sufficient mass of gas has accumulated in the x_2 orbit families to decouple it from the main

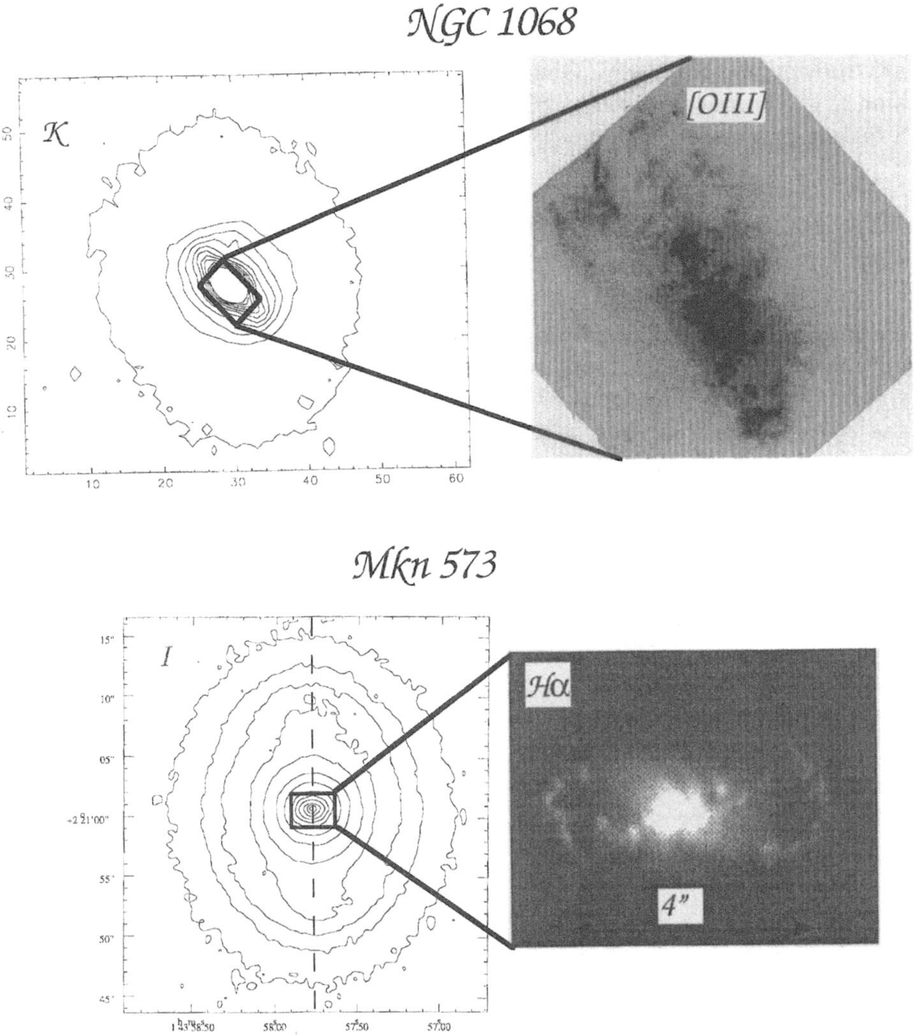

Fig. 1. A comparison between the relative orientations of the inner bars and the direction of the ionization cones in two Seyfert 2 galaxies. Shown at the **top** is NGC 1068, with on the **left** a K band image showing the inner bar, and on the **right** the HST [O III] image taken from Macchetto et al. (1995), showing the inner 7″ of the large scale ionization cone (Unger et al. 1992). The **lower** two panels show Mkn 573 using data taken from the HST-results of Capetti et al. (1996). The I band image is shown on the **left**. Note the double bar structure, very reminiscent of that seen in NGC 5728. The **right** hand panel shows the emission line structure

bar. The two bars then have different pattern speeds, with the inner bar rotating the fastest. Thus not only is an ILR an essential condition for the formation of such secondary bars, but as these N-body studies show they require a significant self-gravitating gas component. Our results therefore provide direct evidence for the kind of structure envisaged by Shlosman, Begelman and Frank (1990) in the AGN fueling process.

3 The Relationship Between Inner Bars and Radiation Cones

The ionizing radiation escaping from the nuclei of many Seyfert galaxies takes the form of a broad cone preferentially orientated along the axis of their kiloparsec-scale radio structure. The locus of intersection of these *radiation cones* with the gas in the disk of the galaxy creates collimated regions of high excitation gas (Unger et al. 1987; Wilson et al. 1991). It is now widely accepted that the radiation cones are produced by *rings or tori* of optically thick material which enshroud the continuum source (e.g. Antonucci and Miller 1985). Most theoretical studies of these tori (e.g. Krolik 1992) have concentrated on calculating their structure and physical conditions and little attention has been paid to the question of how they are formed.

In all the Seyferts in which we have found the inner bars , the direction of the inner bars and the radiation cones are closely aligned. Furthermore, at least two other Seyferts (NGC 3227, Mundell et al. 1995; NGC 4151, Vila-Vilaro et al. 1995) in which inner stellar bars have not as yet been detected show similar morphological connections between their radiation fields and bars.

In NGC 4151, the presence of a circumnuclear gas ring oriented roughly orthogonal to the galactic bar, and the evidence of the galactic rotation curve, point to the presence of at least one, and possibly two, ILRs. The gas ring is clearly associated with a significant increase in the velocity dispersion (cf. Robinson et al. 1994, Fig.17) and the site of enhanced dynamical activity, even though vigorous star formation is not taking place. The numerical simulations suggest that the dust arcs occur at the locations where gas streaming in from the leading edges of the bar meets that circulating in the x_2 orbits. It therefore seems reasonable to attribute the line broadening to shocks or cloud-cloud collisions resulting from the merger of these gas flows. The H I velocity field of Pedlar et al. (1992) lends further weight to this argument, as it exhibits significant departures from coplanar circular motion within the bar, which could be attributed to the inferred gas flows. The H I data also shows a bridge of material from the ring to the nucleus at an angle almost perpendicular to the main bar, and along the direction of the radiation cone. A similar inner gaseous structure in CO, is associated with the radiation cone of NGC 3227 (Mundell et al. 1995). As Fig. 2, which sketches the relationship between the inner structure of NGC 4151 and its radiation cones, illustrates there is an obvious resemblance of the the overall picture to that described above .

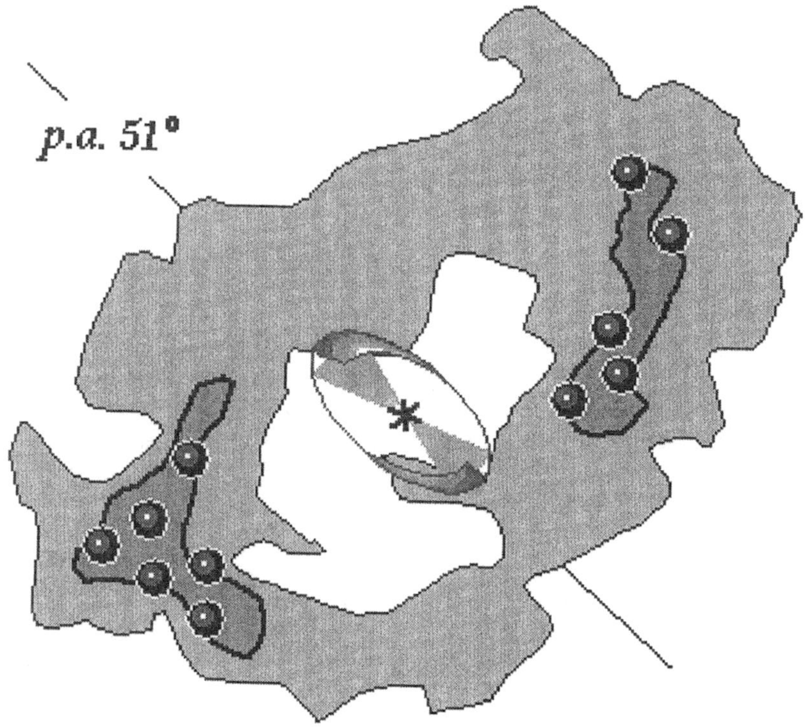

Fig. 2. Sketch showing the distribution of the main gaseous components of the galactic bar of NGC 4151. The large shaded region shows the H I distribution. The dark shading shows the H I arms, while the filled circles represent H II regions along the leading edges of the bar. The location of the Seyfert nucleus is identified with a cross. The ionization cone is directed along P.A. $\sim 50°$ perpendicular to the main bar, and is co-spatial with an inner H I bridge, which Pedlar et al. (1992) suggest might be an inner bar. The circumnuclear gas psuedo-ring, identified from CCD imaging, is represented by an ellipse, the shading marking the positions of its strong arm-like structures. In the model described in the paper the pseudo-ring marks the location of the outer ILR

4 The Formation of Molecular Tori

While the statistics are small, the striking spatial relationship between the radiation and the kiloparsec-scale inner bars and their associated circumnuclear gas rings which we have just described, so far account for around 40% of the Seyferts known to show radiation cones. As yet there is no known case, for which there is adequate data or which is not too highly inclined to properly resolve its structure, in which this is not true. We therefore propose that the *alignment is not a coincidence*, but rather, a direct consequence of the fact that the bar-driven gas flows provide the material for the opaque torus which collimates the AGN radiation field. To explain this relationship we propose that the presence and opening angle of such tori are directly related to the existence and strength of a second inner Lindblad resonance.

4.1 The Role of an Inner ILR

A galaxy with a high central mass concentration and a slowly rotating bar will have inner Lindblad resonances (ILR). The dominant stable orbit family interior to the outermost ILR (the x_2 family) is aligned perpendicular to the bar major axis and the longitudinal orbit family (x_1) which supports the bar (Contopoulos and Papayannopoulos 1980). ILRs will be present if the pattern speed of the bar, Ω_p, is smaller than the maximum of the curve $\Omega(r)-\kappa/2$, where κ is the epicyclic frequency and $\Omega(=v(r)/r)$ is the angular velocity (Binney and Tremaine 1987). Their locations are then given by the solutions to $\Omega(r)-\kappa/2=\Omega_\mathrm{p}$. As illustrated in Fig. 3 for the steeply rising solid body rotation curves, typical of AGN, $\Omega(r)-\kappa/2$ becomes a peaked curve with a maximum $(\Omega(r)-\kappa/2)_\mathrm{max}=\Omega_\mathrm{c}(1-1/\sqrt{2})$ at r_c and there will be *ILRs associated with both the rising and falling branches.* The steeper the rotation curve within the inner kiloparsec the closer the inner ILR will be to the nucleus.

In principle, it is possible to establish the existence and locations of the ILR from the rotation curve providing that the Ω_p is known. In practice, however, it is difficult to extract this information reliably from the observervations (cf. Heller and Shlosman 1994). Only an upper limit to the radius of the inner ILR can be derived. For typical observed rotation curves, it will be located only a few pc from the nucleus.

Since gas tends to follow the stable orbit families closely, just as circumnuclear rings can form around the x_2 orbits, gas falling in along the inner bar from the outer kiloparsec-scale ring accumulates in an inner circumnuclear ring at the inner ILR (Combes and Gerin 1985). The numerical simulations presented by Shlosman in these proceedings beautifully illustrate that rings can indeed form at the locations of both ILR.

The presence of an inner ILR provides a physical mechanism which enables the infalling gas to settle into stable orbits on scales of perhaps a few tens of parsecs. We propose that this gas ring forms the opaque torus enclosing the AGN continuum source and the broad emission line region (Fig. 4). The symmetry axis of the resulting radiation cone will coincide with the angular momentum vector of the collimating gas. As shown by Wada and Habe (1995) the infall of the gas can occur very rapidly, $\sim 10^7$ years, and thus an inner ring can be accumulated before the mass increase in the centre can lead to disruption of the bars (Hasan and Norman 1990).

4.2 The Orientation of the Tori Relative to the Gas Disk

In our model for the origin of the torus, the symmetry axis of the resulting radiation cone will coincide with the angular momentum vector of the collimating gas. In a number of Seyfert 1.5 (e.g. NGC 4151, Robinson et al. 1994), the

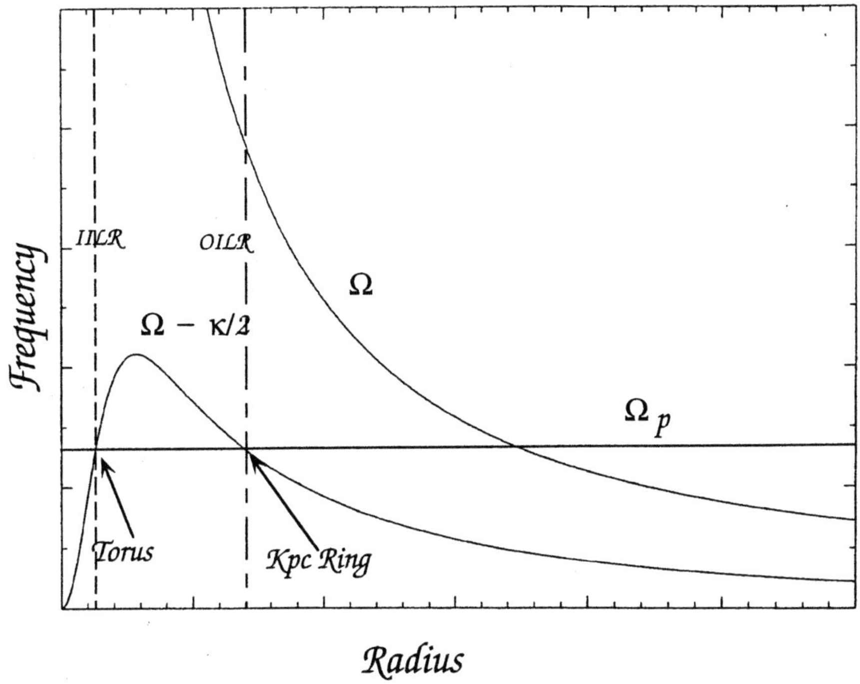

Fig. 3. Seyfert galaxies have steep central rotation curves. In this situation when the bar pattern speed is sufficiently slow, two Inner Lindblad Resonances exist as the $\Omega - \kappa/2$ curve is intersected twice, once on each of the rising and falling branches of the curve. For typical rotation curves the outer of these occurs on the kiloparsec scale, and results in the observed circumnuclear star formation pseudo rings and dust rings. An inner bar can form between these radii. The location of the inner resonance depends critically on how centrally condensed the mass distribution is, but for typical observed rotation curves, it will be located only a few pc from the nucleus. It is at this location that a second ring can be formed creating the torus of material responsible for collimating the nuclear radiation field

inferred radiation cone geometry requires that this common axis, and hence the inner torus, must be tilted relative to the rotation axis of the galactic disk. For the model to remain viable a mechanism of tilting the gas rings in the barred potential must be found. Given that the scale height of the gravitational potential will be much larger than the radius of the inner ILR, it seems plausible to suppose that the orbital plane of the gas within this resonance could be inclined relative to the galactic disk. The orbits can be tilted by the mechanism discussed by Tohline and Osterbrock (1982) or by vertical gravitational torques generated by the bar potential (Friedli and Benz 1993). Unfortunately, detailed

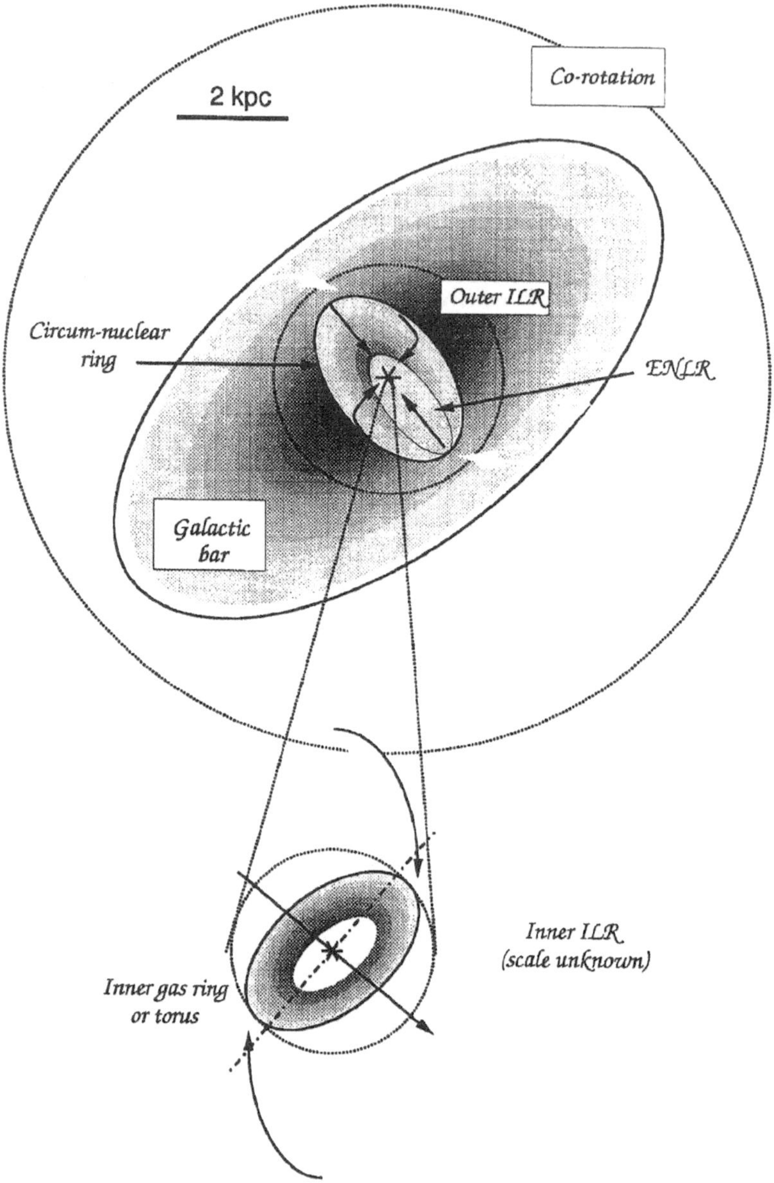

Fig. 4. Sketch illustrating how a double bar structure driven gas flow leads to the formation of a circumnuclear ring at the Outer Inner Lindblad Resonance and can also feed a molecular torus, responsible for collimating the nuclear radiation field at the Inner Inner Lindblad Resonance

3D N-body simulations of the behaviour with self-gravitating gas on these scales are currently not available to test either of these proposed mechanisms. Clearly this is an important area for modellers to address.

5 Conclusions

The results presented here show the existence of double bars in Seyfert and other AGN and point strongly to the idea that bars play a crucial role in fueling their nuclei. The spatial distribution of the gas and dust around the nuclei of these AGN closely matches the predictions of numerical simulations of gas flows induced by barred potentials in which the x_2 family of orbits is well developed, and in which the gas fraction is sufficiently high that its self-gravity is dynamically important.

The apparent connection between the inner bar axis and the direction of the radiation cones has led us to advance a model in which they are directly coupled. The collimating molecular torus is then formed by the bar–driven gas flow at an inner ILR. This correlation appears to hold true for around 40% of the Seyferts known to show radiation cones and, as yet, there are no adequately studied exceptions to this statement.

Almost exclusively, the known objects with inner bars are the nearest and thus the best resolved low inclination systems and all have linear radio sources. For the more distant Seyferts, HST-type resolutions are needed to see comparable structure in the inner kpc, particularly in the Seyfert type 1, in which the wings of the point spread function of the nucleus must be removed. The existing line-free imaging of Seyferts with HST is rather sparse, but several examples with inner bars have already been found using I band images (e.g. Mkn 573; see also the contribution to these proceedings by Nelson), and thus this kind of structure may be very common in the Seyfert population. Further HST IR and emission line imaging will be needed to test our picture of the relationship between anisotropy and bars in Seyferts as a whole.

References

Antonucci, R., Miller, J. (1985): ApJ **297**, 621

Athanassoula, E. (1992): MNRAS **259**, 345

Binney, J., Tremaine, S. (1987): Galactic Dynamics, Princeton University Press

Capetti, A., Axon, D.J., Macchetto, F., Sparks, W.B., Boksenberg, A. (1996): ApJ in press

Combes, F., Gerin, M. (1985): A&A **150**, 327

Contopoulos, G., Papayannopoulos, T. (1980): A&A **92**, 33

Friedli, D., Benz, W. (1993): A&A **268**, 65

Friedli, D., Martinet, L. (1993): A&A **277**, 27

Hasan, H., Norman, C. (1990): ApJ **361**, 69

Heller, C., Shlosman, I. (1994): ApJ **424**, 84

Krolik, J.H. (1992): in Testing the AGN paradigm, eds. S.S. Holt, S.G. Neff, Springer, Berlin, Heidelberg)
Macchetto, F., Capetti, A., Axon, D.J., Sparks, W.B., Boksenberg, A. (1995): ApJ **435**, L15
Mundell, C., Pedlar, A., Axon, D.J., Meaburn, J., Unger, S.W. (1995): MNRAS **277**, 641
Phinney, E.S. (1994): in Mass Transfer Induced Activity in Galaxies, ed. I. Shlosman, Cambridge University Press, Cambridge, p. 1
Pedlar, A., Howley, P., Axon, D.J., Unger, S.W. (1992): MNRAS **259**, 369
Robinson, A. et al. (1994): A&A **291**, 351
Sellwood, J.A., Wilkinson, A. (1993): Rep. Prog. Phys. **56**, 173
Shaw, M.A., Axon, D.J., Probst, R., Gatley, I. (1995): MNRAS **274**, 369
Shaw, M.A., Combes, F., Axon, D.J., Wright, G.S. (1993): A&A **273**, 31
Shlosman, I., Begelman, M.C., Frank, J. (1990): Nat **345**, 679
Shlosman, I., Frank, J., Begelman, M.C. (1989): Nat **338**, 45
Tohline, J.E., Osterbrock, D.E. (1982): ApJ **252**, L49
Unger, S.W., Lewis, J.R., Pedlar, A., Axon, D.J. (1992): MNRAS **258**, 371
Unger, S.W., Pedlar, A., Axon, D.J., Ward, M.J., Meurs, E.J.A., Whittle, D.M. (1987): MNRAS **228**, 671
Vila-Vilaro, B., Perez, E., Robinson, A., Axon, D.J., Baum, S., Gonzlalez-Delgado, R., Pedlar, A., Perez-Fournon, I., Perry, J.J., Tadhunter, C. (1995): A&A **302**, 58
Wada, K., Habe, A. (1995): MNRAS **277**, 433
Wilson, A.S., Bratz, J.A., Heckman, T.M., Krolik, J.H., Miley, G.K. (1993): ApJ **419**, L61
Wilson, A.S., Helfer, T.T., Haniff, C.A., Ward, M.J. (1991): ApJ **381**, 79
Woznaik, H., Friedli, D., Martinez, L., Martin, P., Bartschi, I. (1995): A&AS **111**, 115

Circumnuclear Starbursts in Barred Galaxies

Johan H. Knapen

Département de Physique, Université de Montréal, C.P. 6128, Succursale Centre-Ville, Montréal (Québec), H3C 3J7 Canada; and Observatoire du Mont Mégantic
Present address: Division of Physical Sciences, University of Hertfordshire, College Lane, Hatfield, Herts AL10 9AB, UK. E-mail knapen@star.herts.ac.uk

Abstract. I discuss new observations of circumnuclear regions in barred galaxies, both of their kinematics and their near-infrared (NIR) morphology. New kinematic observations of M 100 confirm that (1) the inner bar-like feature detected before in this galaxy is in fact a continuation of the large-scale stellar bar; and (2) the spiral arm-like sites of star formation in the circumnuclear zone are manifestations of density wave spiral structure in the core of this galaxy. From a NIR study of similar regions in other barred galaxies, I find that in many of these regions, sites of star formation are visible in the NIR images, implying that young stars are responsible for a non-negligible part of the NIR emission.

1 Introduction

Barred galaxies often experience starburst activity in their central regions, and in some cases the star formation (SF) is observed in more or less complete nuclear rings (see review by Kennicutt 1994). Imaging in the Hα emission line usually brings out these structures very well (e.g. Pogge 1989), but may be subject to dust extinction. This alone is sufficient reason to include near-infrared (NIR) imaging in the study of circumnuclear regions (CNRs) in barred galaxies.

In a recent study, we combined optical and NIR K-band imaging of the core of M 100 (=NGC 4321), a mildly barred galaxy, with dynamical modelling (Knapen et al. 1995a,b; Shlosman 1996). The SF in the CNR of this galaxy delineates two inner spiral arms, flanked by dust lanes, which connect outward through the bar and to the main spiral arms in the disc. The circumnuclear SF occurs between a pair of inner Lindblad resonances (ILRs), as indicated by the NIR morphology and by the modelling. The morphology in Hα and blue light is strikingly different from that at 2.2 μm (K-band; Knapen et al. 1995a). In K, the spiral armlets are hardly discernable: the CNR is mostly smooth where the SF occurs. Closer to the centre though, an inner bar-like feature is seen in the NIR, along with two small leading arms (Knapen et al. 1995b; Shlosman 1996).

In the present paper, I first present new velocity observations which confirm kinematically the inner bar-like structure and the spiral nature of the SF regions in the core of M 100. So far, these had only been inferred by us from optical and NIR imaging, and from dynamical modelling. I then discuss that the large difference between NIR and Hα core morphology as observed in M 100 seems to be rather unusual compared to similar systems. From new high-resolution

NIR imaging it is clear that in many, possibly most, cases of circumnuclear SF activity, the individual SF regions show up prominently in K (unlike in M 100).

2 Kinematics of the Core of M 100

2.1 CO

Recently, Rand (1995) observed parts of M 100, including the central region, in CO with the BIMA interferometer. We re-analyzed some of his data in order to specifically study the gas kinematics in the core region. We first removed noise peaks outside the area where CO emission is expected by setting pixel values at those positions to undefined. This was done interactively by inspecting the individual channel maps one by one, continually comparing with the same and adjacent channels in a smoothed data cube. The resulting data set was used to calculate the total intensity and velocity moment maps. The resulting moment images show the same structure as published by Rand (1995) but are somewhat more sensitive in the central region. We show the velocity field thus produced, overlaid on a grey-scale representation of the total CO intensity, in Fig. 1.

Density wave streaming motions are recognizable in the deviations from the regular shape of the velocity contours (isovels), especially toward the NE and SW of the nucleus near radii of some $7''$. These are due to the gas streaming near the spiral armlets (see also next section). In the central $\sim 5''$, the isovels do not run parallel to the minor axis but show a deviation characteristic of gas streaming along a bar. Such deviations were described before by Bosma (1981) in several barred galaxies. Knapen et al. (1993) confirmed the existence of a weak bar component in M 100 from the H I kinematics. Note, however, that the deviations seen here in CO occur on the much smaller scale of the inner bar-like feature as seen in the NIR (Knapen et al. 1995a), and confirm the gas streaming along this bar as seen in the numerical modelling (Knapen et al. 1995b; Shlosman 1996).

It is generally dangerous to use moment maps for interpretation of kinematical features in regions where profiles may deviate from a Gaussian shape, and/or be multiple-peaked. Since the inner region under consideration here is clearly such a region of enhanced risk, we produced a set of position-velocity diagrams along and parallel to the minor axis, shown in Fig. 2. It is immediately clear from this panel that multiple velocity components are present in the molecular gas, especially at the central position, and that the moment analysis may indeed not be valid there. But more interesting are 4 components which show up symmetrically in the position-velocity diagrams north and south of the minor axis. Two of these (labelled AN and AS in Fig. 2) are visible at RA offsets $\sim -8''$ (North) and $\sim +8''$ (S), with excess velocities with respect to v_{sys} of $\sim +25\,\mathrm{km\,s^{-1}}$ (N) and $\sim -25\,\mathrm{km\,s^{-1}}$ (S). We identify these components as the density wave streaming motions near the spiral armlets seen before in the velocity field. Their offsets, both in RA and in velocity, strongly support this interpretation, as well as their symmetric occurrence in the series of position-velocity diagrams.

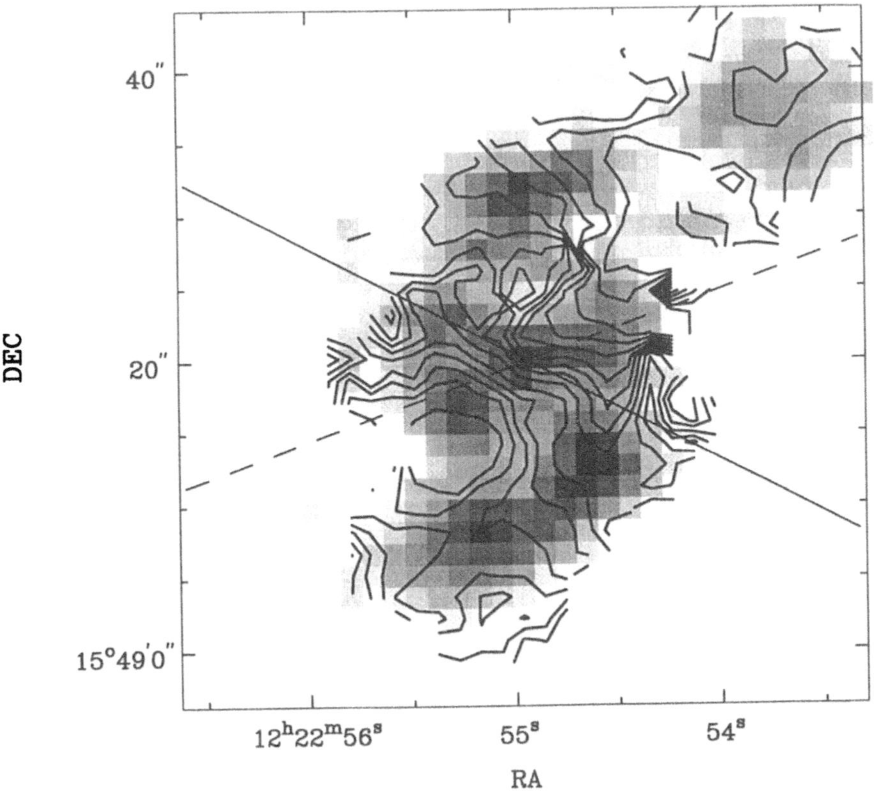

Fig. 1. CO velocity field of the central region of M 100 in contours, overlaid on the total intensity CO map. Drawn line indicates kinematic minor axis, dashed line is position angle of the large-scale bar. Contours are separated by 10 km s^{-1}. Epoch is J2000.0. CO data from Rand (1995)

The second set of components (labelled BN and BS in Fig. 2) has RA offsets of $\sim +3''$ (N) and $\sim -3''$ (S), and excess velocities of $\sim +70$ km s^{-1} (N) and ~ -70 km s^{-1} (S). This is the gas streaming along the inner bar-like feature, as again indicated strongly by the symmetric offsets in both position and velocity, and by qualitative comparison with the velocity field. This kinematically observed gas streaming confirms the existence of the inner barlike feature (continuation of the outer bar) so prominently seen in our NIR imaging and dynamical modelling (Knapen et al. 1995a,b).

It is interesting to compare Fig. 2 with Fig. 9 of Knapen et al. (1993). The latter are similar position-velocity diagrams along and parallel to the minor axis of M 100, but show the H I kinematics on the scale of the large bar, which extends some 4 kpc in radius along a position angle of $\sim 110°$. The similarity between the gas components labelled BN and BS here, and the H I components

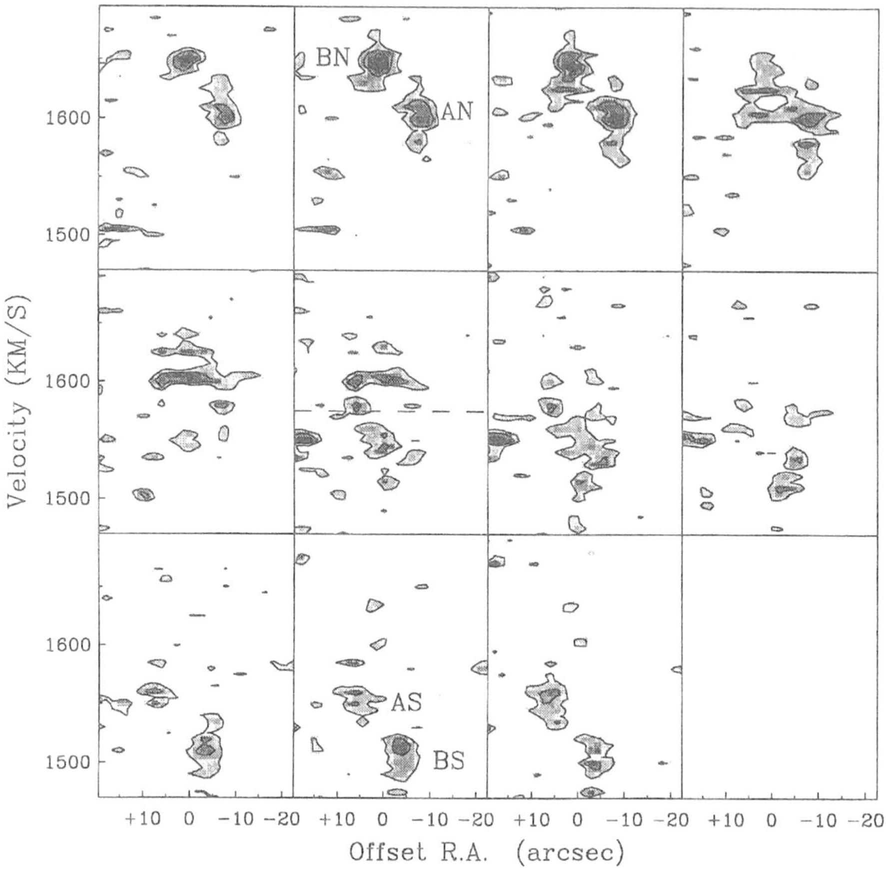

Fig. 2. Position-velocity diagrams in CO along (sixth panel) and parallel to the minor axis in the central region of M 100. Panels to the left of and above the middle panel are cuts North of the minor axis, panels to the right of and below are South. Cuts are separated by 1.″5 , or about half a beam. The systemic velocity of the galaxy is 1575 $\mathrm{km\,s^{-1}}$, indicated by the horizontal line in the minor axis (middle) panel. Named features are discussed in the text

with velocities symmetrically offset from v_{sys} is striking. The H I behaviour was interpreted by Knapen et al. (1993) as streaming along the large-scale bar of the galaxy. The CO velocities similarly represent the streaming along the inner bar. Both these kinematic observations indicate strongly that the gas is moving in the gravitational field of the same stellar bar.

2.2 Hα

Using the TAURUS instrument in Fabry-Perot (FP) mode on the 4.2-m WHT on La Palma, we obtained a data set giving full 2D velocity information in the Hα

emission line. We observed two different parts of the disc of M 100, and since the central region of the galaxy was present in both data sets, we combined these to obtain a core data cube with increased S/N. The data reduction procedure will be described in detail in a future publication. In the present paper, we present first results from a moment analysis in the central region of M 100: the total intensity map and velocity field, shown in Fig. 3. The individual Hα profiles over all but the very innermost ($< 1''$) region are well suited for a moment analysis, double-peaked profiles being practically absent.

The total intensity map as derived from the FP observations is comparable in quality to the image published by Knapen et al. (1995a). The resolution is around $0\farcs7$, indicating that the seeing was good and constant during the time of the FP observations (> 6 hrs). The velocity field of the central region shows rapidly increasing rotation velocities (the rotation curve as derived from the velocity field rises to more than $150\,\mathrm{km\,s^{-1}}$ in $\sim 2''$, or some 150 pc), but also important deviations from circular velocities. The first effect to note specifically is the S-shaped deviation of the isovels near the minor axis, very similar to the deviation seen in CO, above. Also in Hα, the streaming of gas along the inner part of the bar shows up clearly in the velocities, confirming the earlier findings.

Another important deviation in the velocity field is seen most clearly toward the NE and SW of the nucleus, at radii of some $9''$. This deviation is due to density wave streaming motions, strongest where we inferred the position of the inner spiral arms, just outside the well-defined dust lanes (Knapen et al. 1995a,b). Although the signature of the streaming motions is most obviously visible near the minor axis of the galaxy, they can in fact be recognized consistently up to some 60° on either side of the minor axis. As estimated from the velocity contours, the (projected) excess or streaming velocities are of the order of $40\,\mathrm{km\,s^{-1}}$. This kinematic detection confirms that indeed the inner spiral armlets are miniature density wave spiral arms, with a behaviour very similar to what one is used to see in discs of galaxies (e.g. Visser 1980).

3 NIR Imaging of CNRs in Barred Galaxies

The morphology of the central 2 kpc region in M 100 changes most strikingly from the blue to the NIR (Knapen et al. 1995a,b). Whereas in the blue, and also in Hα, the core region is dominated by a pair of miniature spiral arms which show strong SF activity, the K contours in the region are remarkably smooth. They hardly show any structure apart from two symmetrically placed "hot spots".

From a recent NIR imaging survey of a number of systems with M 100-like Hα morphology, it becomes clear that M 100 is more exception than rule. As an example, we show in Fig. 4 an Hα (left, taken with TAURUS on the WHT) and a K image (right, from MONICA on the CFHT), of the inner $\sim 20''$ region of the strongly barred Seyfert galaxy NGC 6951. Unlike in M 100, the K image of NGC 6951 shows a number of distinct sites of enhanced emission in the radial region where the intense SF occurs, at some $2'' - 4''$ from the nucleus. There is a general correspondence between the positions of the "hot spots" in Hα and in K, although some regions emit relatively more in Hα, others more in K. In some cases the Hα is seen slightly offset from a K-peak (e.g. the strongest K feature toward the SE, near which a strong dust lane can be seen in NIR colour index images).

From our mini-survey, we find that most of the studied objects are in fact similar to NGC 6951 in that the K images of the CNRs show a number of distinct emitting regions, often coinciding spatially with concentrations of Hα emission. Only a few of the objects studied show M 100-like smooth K emission in SF regions which are so prominent in Hα. Additional observational data is necessary to explain this difference between M 100 and other galaxies in our sample. We note, however, that one expects a region of massive SF to become prominent in K after some 10^7 yr, when the massive stars reach the supergiant phase in their evolution (e.g. Leitherer and Heckman 1995). This implies that the observations of NGC 6951 can be explained rather straightforwardly if one assumes that bursts of SF of relatively short duration (around 10^7 yr) occur randomly in the CNR. Then some regions emit more in Hα (starburst younger than 10^7 yr, no supergiants yet) and others more in K (slightly older, no more current SF). This interpretation implies that something prevents the SF regions from shining in K in M 100-like systems, unless the entire CNR is younger than 10^7 yr, which is unlikely. A more reasonable explanation is that the newly formed stars are dispersed throughout the CNR, so that by the time they become supergiants they are no longer as clustered as at birth (see Knapen et al. 1995a,b). This idea awaits further observational confirmation.

4 Discussion

The results of the K-band imaging (Sect. 3) seem to bring up more questions than they answer, and bring out the need for further detailed observations. Specifically, what needs to be addressed is what kinds of stars, and in which proportion, contribute to the NIR emission from the central regions of barred spirals. Because one often observes localized K-band emission from SF regions, at least in those cases an old stellar population cannot be entirely responsible for the K-emission in the central regions, and emission from young, massive stars must play a non-negligible role. There are several possible ways to quantify the relative contributions of old and young stars (basically: to disentangle the giant and supergiant contributions) to the NIR emission at the scales of interest here. One is detailed spectroscopy in the NIR (e.g. Goldader et al. 1995), another imaging in the NIR J, H and K bands and across the CO 2.3 μm absorption feature (e.g. Forster 1994; Armus et al. 1995; Haller et al. 1996).

Another open question is why in some galaxies the K light is smoothly distributed in the CNR, even though the SF as seen in Hα occurs in discrete regions (e.g. M 100: Knapen et al. 1995a), whereas in others the K follows the Hα emission, thus both trace the SF (e.g. NGC 6951: Fig. 4). The difference must be

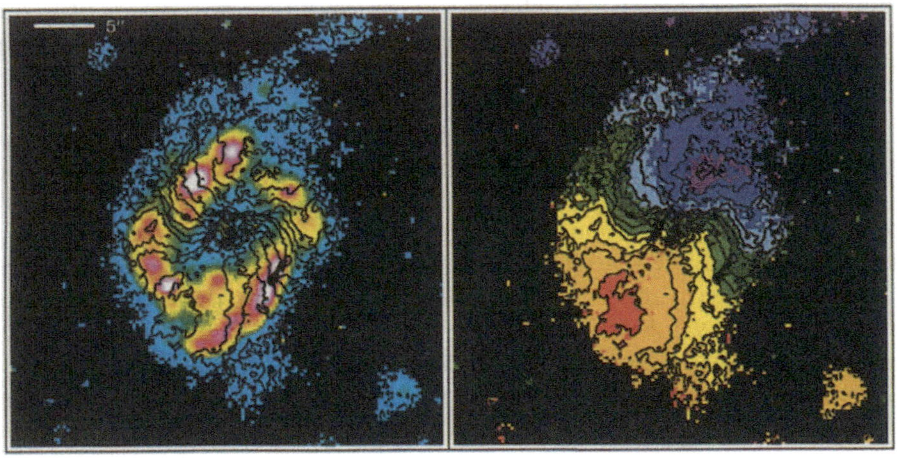

Fig. 3. Hα FP moment maps of the central region of M 100: the velocity field is shown in contours and overlaid on the total intensity Hα image (**left**), and on the velocity field itself (**right**). Contours are separated by $15\,\mathrm{km\,s^{-1}}$. N is up, E to the left

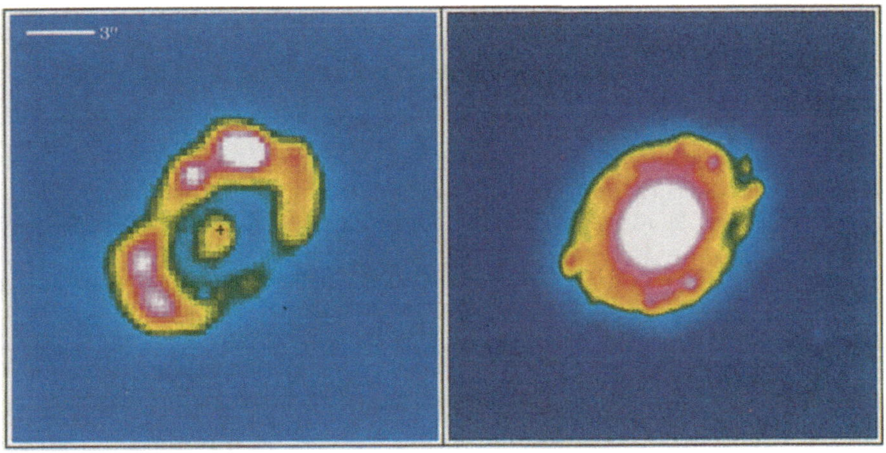

Fig. 4. Images of the core of NGC 6951 in Hα (**left**) and in the NIR K-band (**right**). The two images are at the same scale and orientation. N is up, E to the left. The bar runs almost horizontally

due to either different stellar populations in the two classes of galaxies, or to different dynamics. Two-dimensional kinematic observations can be used to study the latter possibility. It is interesting to note that in the case of NGC 6951 the velocity field as derived from Hα FP observations (Knapen, in preparation) is regular in the radial region where the massive SF is concentrated, and does not show density wave streaming motions, certainly not of the amplitude seen in M 100. Possibly the CNR of NGC 6951 is older and the mini-spiral is much more tightly wound, resembling a complete ring where the SF can occur (see e.g. Elmegreen 1994). In the latter case, stars will evolve more or less at the place where they formed, and Hα (tracing very young massive stars) and K (same stars after $\sim 10^7$ yr) emission spatially coincide. In the former case (M 100), stars may be displaced from their birthplace during the 10^7 yr of their evolution to supergiants, and dispersed enough to smooth their emission in the K-image, due to a sheared motion in the disc. Further observations will clarify this point.

Acknowledgements. I wish to thank the organizers of this symposium for making it possible for me to attend it. I am indebted to my collaborators on the various aspects of the work described in this paper: J.E. Beckman, R. Doyon, C.H. Heller, R.S. de Jong, D. Nadeau, R.F. Peletier, R.J. Rand, M. Rozas, and I. Shlosman. The WHT is operated on the island of La Palma by the RGO in the Spanish Observatorio del Roque de los Muchachos of the IAC. The CFHT is operated by the NRC of Canada, the CNRS of France and the Univ. of Hawaii.

References

Armus, L., Neugebauer, G., Soifer, B.T., Matthews, K. (1995): AJ **110**, 2610
Bosma, A. (1981): AJ **80**, 1825
Elmegreen, B.G. (1994): ApJ **425**, L73
Forster, N. (1994): Master's thesis, Univ. de Montréal
Goldader, J.D., Joseph, R.D., Doyon, R., Sanders, D.B. (1995): ApJ **444**, 97
Haller, J.W., Rieke, M.J., Rieke, G.H., Tamblyn, P., Close, L., Melia, F. (1996): ApJ **456**, 194
Kennicutt, R.C. (1994): in Proc. Conf. Mass-Transfer Induced Activity in Galaxies, ed. I. Shlosman, Cambridge Univ. Press, p. 131
Knapen, J.H., Beckman, J.E., Shlosman, I., Peletier, R.F., Heller, C.H., de Jong, R.S. (1995a): ApJ **443**, L73
Knapen, J.H., Beckman, J.E., Heller, C.H., Shlosman, I., de Jong, R.S. (1995b): ApJ **454**, 623
Knapen, J.H., Cepa, J., Beckman, J.E., del Rio, M.S., Pedlar, A. (1993): ApJ **416**, 563
Leitherer, C., Heckman, T.M. (1995): ApJS **96**, 9
Pogge, R.W. (1989): ApJS **71**, 433
Rand, R.J. (1995): AJ **109**, 2444
Shlosman, I. (1996): (this volume)
Visser, H.C.D. (1980): A&A **88**, 159

Circumnuclear Activity

Robert A. E. Fosbury[1,2]

[1] Space Telescope – European Coordinating Facility, D–85748 Garching bei München, Germany
[2] Affiliated to the Astrophysics Division, Space Science Department, European Space Agency

Abstract. Clues to the question of the origin of nuclear activity are provided by the response of the host galaxy and the surrounding intergalactic medium to the AGN power output. The nature of the rich variety of circumnuclear activity is discussed and particular emphasis is placed on attempts to understand these processes in the active galaxies, both quasars and powerful radio galaxies, which are being discovered at very high redshift. Although extreme in character, these objects give us one of the few direct views we have of galaxies and their formation processes at an early epoch.

1 Introduction

The phenomenon that we call 'nuclear activity' is most probably the consequence of events or processes which feed material into a high density region in the centre of a galaxy. The rapid evolution of this material, through various dissipative interactions, results in bursts of star formation, the formation of a massive collapsed object or both. The feeding of the nucleus can be the result of relatively quiescent gas flows such as those seen in bars — much discussed in this symposium — or catastrophic events associated with galaxy interactions or mergers. One of the key questions in the science of galaxies is the frequency of occurrence of currently unfed collapsed objects and the implications this carries for the evolution of the system.

The various interactions between an active nucleus and its host galaxy produce a rich variety of phenomena from which much can be learned about galactic structure and evolution. In the nearby Seyfert galaxies the axial symmetry of the AGN which produces the beautiful 'ionization cones' seen in highly excited emission lines appears to be independent of the symmetry of the generally spiral host galaxy (see particularly the contribution by Tsvetanov). The details of the relationship between the star formation and AGN processes can be well studied in these objects where high resolution imaging from the ground and from HST shows an apparently close coupling between circumnuclear rings and the nucleus. In the high redshift quasars and powerful radio galaxies — which we believe to host very luminous but partly hidden quasars — the AGN illuminates material within a huge volume. The evidence for this appears both in the very extended emission line regions seen around the objects and in the absence of absorption lines from low ionization species — notably Lyα — close to the emission line redshift of quasars: the 'proximity effect'. When we learn to read the signs written by ionization, fluorescence and scattering, we will witness aspects of the galaxy formation process which would be very hard or impossible to see in objects without an AGN at this early epoch.

From the optical observer's perspective, it appears that the dominant influence of the AGN on its host galaxy and extragalactic environment is via the photon radiation field. These photons ionize surrounding gas and scatter from dust and electrons. In the radio loud sources, however, there is much evidence for the interactions produced by the radio jets which extend up to tens or hundreds of kiloparsecs from the nucleus. The detailed physics of jet/cloud collisions is complex and difficult to unravel but may tell us much about the properties of the jets which feed the radio lobes and also something about the material on large spatial scales surrounding galaxies.

In this contribution, I want to place my emphasis on the prospects for the study of galaxies at the very highest redshifts. Here, the radio galaxies are currently our primary probes of the nature of galaxies at early epochs. This is not so much by design but is rather the result of the highly successful search techniques for the most distant objects based on radio properties (McCarthy 1993; Röttgering 1993). Indeed, one of our most important tasks is to recognise those aspects of the observed properties which are a direct result of the presence of a powerful AGN and to separate these from the stellar structures which may teach us about the formation of 'normal', or at least less extreme, galaxies.

Although I profess an interest in high redshifts, the diagnostic tools which help us have been developed by application to sources much closer-by. Consequently the Seyfert ionization cones, the radio jet/cloud interactions, the extended emission line regions around low redshift radio galaxies; indeed, all the trappings of extranuclear activity, will form part of the story which culminates at an epoch which is tantalisingly close to that where galaxies must be forming. I will start with a brief survey of the range of phenomena we observe around AGN and then go on to discuss some aspects in a little more detail.

2 The Phenomena

A substantial foundation for the subject was laid during the pioneering optical spectrophotometry of Seyfert nuclei carried out with the then-new electronic spectrum scanners principally by Osterbrock and his students and collaborators during the 70's. These early scanners were one-dimensional devices and so produced an integrated spectrum of the nuclear region of the galaxies which, for the nearby Seyferts, generally corresponds to the inner 100 pc or so. The slightly later advent of the two-dimensional detectors (Boksenberg's Image Photon Counting System (IPCS) and CCDs) led to the discovery that in some objects, most notably the radio galaxies, the high ionization emission lines which characterised an AGN were not necessarily restricted to the very nucleus but could be seen up to 100 kpc or so from the galaxy.

The astrophysical study of these extended emission line regions (EELR) led to the notion that they were excited by the AGN — most probably by the

EUV radiation field. This straightforward interpretation did, however, lead to a problem in some objects which became known as the 'photon deficit' (Robinson 1989). Using any reasonable extrapolation of the nuclear optical nonstellar continuum into the ionizing ultraviolet, there appeared to be insufficient photons to account for the line luminosity and the ionization state of the extended gas. There were two favoured explanations for the deficit problem (see Fig. 1). In the first, it was proposed that the continuum extrapolation could be wrong: instead of a power law extending into the Lyman continuum, there may be a hot ($1 - 2\times10^5$ K) 'thermal' ionizing spectrum which was essentially invisible in the optical. Distinguishing between power-law and hot black body ionizing spectra using the standard optical emission line plasma diagnostic techniques is not straightforward (Robinson et al. 1987) and it remains likely that 'EUV bumps' do play some role in exciting EELR. It was soon realised, however, that the observer and the extended ionized gas did not necessarily share the same view of the AGN. The resulting concept of an anisotropic radiation field has been remarkably successful, allowing us to understand a variety of AGN properties. In conjunction with a number of related contemporary studies it has led to the unification of apparently different classes of object based on the predominant importance of the viewing direction.

The recognition of the radiation anisotropy allowed the inference of nuclear properties even in objects whose AGN were not directly visible because of obscuration. Perhaps surprisingly, the diagnostic tools based on the optical emission line spectra — mostly ratios of forbidden and subordinate recombination lines — indicated some remarkably uniform properties for the EELR: both within an object and amongst different objects. Robinson et al. (1987, hereafter RBFT87) showed that the ionizing spectra could be characterised by a mean ionizing photon energy of 30 – 40 eV with rather little dispersion. Similarly, there could be little variation — within factors of two or three — in the chemical composition of the EELR in the sample of relatively low redshift radio galaxies studied. The dominant parameter representing the spread of spectra within the line ratio diagnostic diagrams is, in fact, the ionization parameter — the ratio of ionizing photon to gas density — which represents the degree of geometric dilution of the radiation field. This uniformity suggests to me that the EELR phenomenon, at least in these low redshift objects, is dominated by a single line excitation process, most probably nuclear photoionization. While there is strong evidence in some cases for local, extranuclear sources of ionizing radiation, e.g., in jet/cloud interactions, I do not believe that such processes contribute a very significant fraction of line emission in the global EELR phenomenon. This conclusion may, however, need to be revised at high redshifts where there is more evidence for disturbed kinematics and apparently a closer association between the line emission and the radio structures.

One of the tasks engaging us now is, indeed, the extension of these diagnostic techniques into the ultraviolet where we can use them to study the spectra of high redshift objects observed in the optical from the ground. Because the UV spectra are rich in resonance lines, geometric effects and the presence of dust

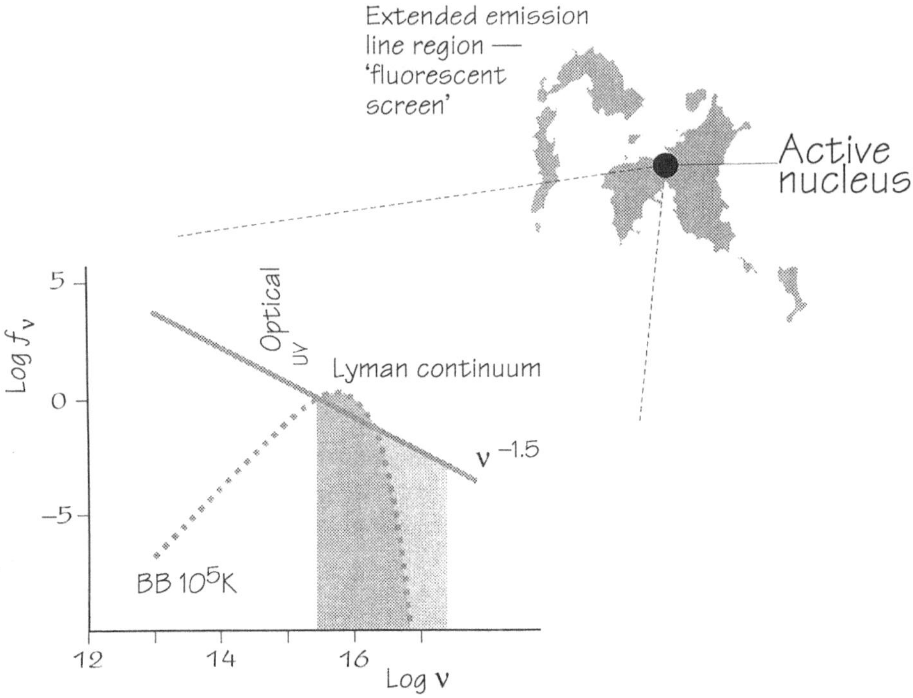

Fig. 1. An illustration of the photon deficit problem in AGN. The apparent absence of sufficient nuclear photons to ionize the extended nebulosity can be explained either as a result of radiation anisotropy or a spectrally 'hidden' ionizing source peaking in the EUV

become important factors in the development of tools. I will discuss some of this work later. The rewards are potentially great since we have the chance to study the properties and composition of gas lying 100 kpc from galaxies at redshifts up to 4 or so.

Another property of the gas which is readily studied from emission line spectroscopy is its kinematic state. If it can be shown to respond predominantly to the gravitation field, the gas provides an ideal tracer which extends to radii far beyond that seen in starlight. Although there are apparently chaotic gas motions in some objects (Baum et al. 1990; Baum et al. 1992), in many there is evidence of relatively quiescent rotational motion (Tadhunter, Fosbury and Quinn 1988; Hes 1995). An object at $z = 3.6$, studied by van Ojik et al. (1995) shows chaotic motion in the vicinity of the radio structure but smooth rotation in the outer parts of its 135 kpc diameter Lyman-α halo.

The idea of hidden sources led to the brilliant use of polarized light as a 'filter' to detect the small fraction of light scattered into our line of sight by dust or electrons close the nuclei of Seyfert 2 galaxies but outside of the regions which obscured our direct view. Antonucci and Miller (1985) were able to extract the broad line Seyfert 1 spectrum from the Seyfert 2 galaxy NGC 1068. Thenceforward, polarization became a powerful tool for unravelling the non-spherically symmetric geometry of AGN. In retrospect, I find it interesting to see how the recognition of polarization produced by scattering has produced such a rich harvest of understanding quite distinct from that arising from the various non-thermal processes which produce polarized light directly.

The discovery of co-axial radio/optical alignments in powerful radio galaxies beyond a redshift of about 0.7 (McCarthy et al. 1987; Chambers, Miley and van Breugel 1987) proved to be a landmark in the understanding of these distant galaxies. It had generally been thought that the measured blue colours were a result of the presence of a significant hot stellar population which might indicate the youth of the galaxy as a whole. The 'alignment effect' as it became known, spawned several explanatory theories including a proposal that the passage of the radio jets through an extended galactic 'atmosphere' would induce star formation which, for a short time ($\leq$ a dynamical timescale) would trace the jet passage with blue light (see McCarthy 1993 for a review and a complete set of references). The production of blue light by inverse Compton scattering within the region occupied by the relativistic electrons producing the radio emission (Daly 1992) was excluded in many cases because, while co-aligned, the optical and radio emission was generally not co-extensive (e.g., di Serego Alighieri, Cimatti and Fosbury 1993; Miley et al. 1992; Longair, Best and Röttgering 1995). Although the alignment effect is seen predominantly at high redshift — largely because of the shift of the observed waveband into the rest-frame ultraviolet, there are low redshift examples, e.g. in 3C 195 at $z = 0.1$ (Cimatti and di Serego Alighieri 1995).

It is important to understand whether the alignment is purely a K-correction effect or whether there are evolutionary factors involved. Answering this question is not as easy as it sounds because of the necessity to observe nearby objects in the ultraviolet. Here the HST, for all its fine resolution capabilities, is still a small telescope. There is also the extreme rarity in the present epoch of radio sources of comparable power to the objects being studied at high redshift. Cygnus A is the obvious example of a local, luminous radio source and its properties, in the context of the unified schemes, have been extensively studied by Tadhunter (1996) and Stockton and Ridgeway (1996). While there is compelling evidence for an anisotropic ionizing radiation field, the hidden AGN is surprisingly under-luminous. Alternatively it may be, in this case, that the radio source — because of the high density cluster environment — is *overluminous* and compact for its AGN power.

The growing realisation during this period that the powerful radio galaxies and the radio quasars could be considered the same class of objects differing principally in orientation (Orr and Browne 1982; Barthel 1989) led Tadhunter, Fosbury and di Serego Alighieri (1989) to suggest that the alignment effect may be caused simply by the scattered and fluorescently excited light from the hidden quasar which radiates photons mostly along the radio axis. This idea was already supported by the emission line spectroscopy which showed extended, high ionization lines which could not arise from ordinary stellar-photoionized H II regions but, instead, appeared similar in nature to the EELR seen at lower redshift. The unambiguous test of scattered light is its polarization state. It was at the limit of the capability of the 4-m class optical telescopes to demonstrate that the aligned radio galaxies were linearly polarized with an E-vector position angle fully consistent with the scattering hypothesis. Following the first measurements by di Serego Alighieri et al. (1989; see also Scarrott, Rolf and Tadhunter 1990), a significant body of polarization measurements was accumulated (see Cimatti et al. 1993) which showed that the strong polarization appeared when the observed passband moved below the rest-frame 4000Å break where dilution from the red stellar population disappeared. An objection to this interpretation which stood for some while was the apparent absence of scattered broad lines seen in the extended, aligned structures by analogy with the polarized broad line spectra seen in some Seyfert 2 galaxies. While it is true that the 4-m telescopes did manage to solve this problem we are now moving into the era of faint object spectropolarimetry with the Keck telescope which is giving a much clearer view of these phenomena. Before moving on to a more detailed discussion of the high redshift phenomena, I should like to dwell for a moment on what has been learned from studies of nearby objects.

3 Ionization Cones

The apparent biconical ionization cones seen most clearly in the Seyfert 2 galaxies will be discussed in detail by others at this symposium. I shall, nonetheless, review some of their properties as they bear on our observations of more distant and more powerful AGN. I will also note that their existence was presaged by much earlier observations of the distinctly different kinematic behaviour of the high- and low-ionization emission lines in long-slit spectra of Seyferts (e.g., in NGC 1365 by Phillips et al. 1983 and Jörsäter, Lindblad and Boksenberg 1984)

The hypothesis that the triangular regions of highly excited gas, with one apex at the galactic nucleus, are the projections on the sky of bi-conical zones of material illuminated and ionized by an equatorially obscured AGN (Fig. 2) appears to be well supported by observation. The sharp edges seen in some objects, notably in NGC 5252 (Tadhunter and Tsvetanov 1989), suggest that the shapes are limited by the radiation field rather than by the matter distribution — at least on scales up to 10 or 20 kpc. In the same object, Prieto and Freudling (1993) showed that there was apparent continuity, both in morphology and velocity, between the ionized and the neutral gas. The location of the source of radiation at the galactic nucleus is supported both by the cone morphology and by the absence of a sufficiently strong ultraviolet continuum within the cone which could represent a local source of ionization — although we must beware of the possible presence of 'spectrally hidden', hot thermal ionizing sources. The fact that many of the cones are apparently one-sided can be explained by the obscuration of the galactic disk. The general absence of clearly delineated cones in the Seyfert 1s is simply a result of projection: we do not recognise the cones when our line-of-sight lies inside them. What actually constitutes the edge of the shadowing body, however, is still not clear when nearby objects are examined in detail (e.g., NGC 4151, Evans et al. 1993; Pedlar et al. 1993; Robinson et al. 1994). The apparent opening angles of the cones can be influenced both by projection — the angle between the cone axis and our line of sight — and by the distribution of material, e.g., a disk with its plane lying within the cone. We do not yet, in my view, have a sufficient sample of appropriately observed Seyferts — both in terms of sample selection and total number — to rule out the hypothesis that all ionization cones have the same intrinsic opening angle (work in progress by Tsvetanov, Fosbury and Tadhunter).

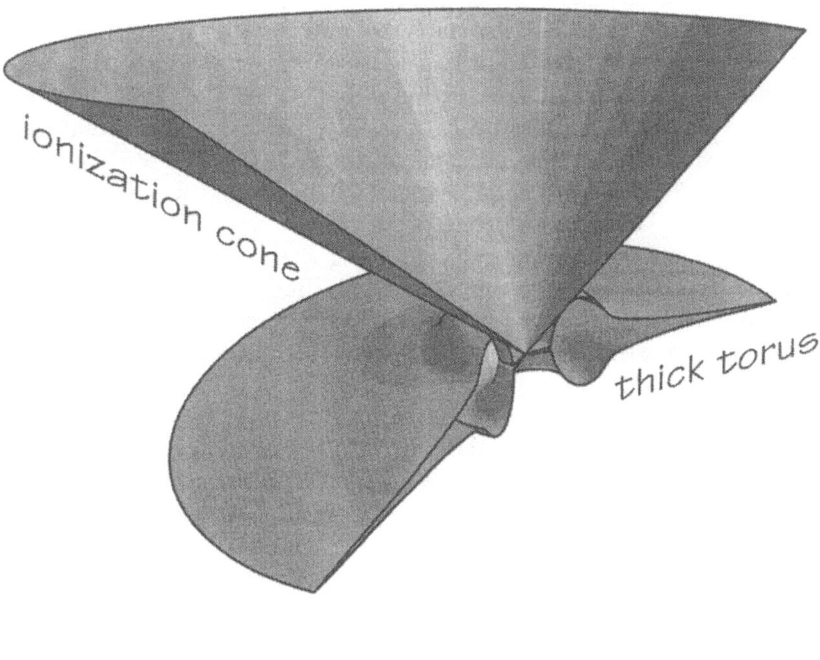

Fig. 2. A cartoon of the broad ionization cone produced by obscuration of an AGN by a thick torus

Amongst the known Seyfert 2s with detected ionization cones, there is a very close correspondence between the cone axis and the axis of symmetry of the nuclear radio source, especially when the latter is measured on the smallest spatial scales (Wilson and Tsvetanov 1994). This suggests that the collimation mechanisms of the radio plasma and the ionizing radiation are very closely related although the radio jets are clearly narrower than the cones. These authors also note, however, that there is no relation between the cone axis and that of the galactic disk except, perhaps, in late-type spirals where a weak trend could be produced by an observational selection effect. The apparently ubiquitous occurrence of the phenomenon and the coincidence of the radio and cone axes means that, in these studies, we have a wonderful example of what I call the 'shadow puppet' phenomenon: the use of observations on large spatial scales to map physical structures in galactic nuclei which are still well below direct resolution by telescopes. Since it is so important to be able to overlay accurately images at different frequencies, these studies will be greatly aided by the radical improvement in astrometric precision which will result in the near future from the application of the Hipparcos results to the HST Guide Star Catalog.

Even when seen in projection against the disk, the kinematic state of the cone material can usually be distinguished from it by its different ionization state — specifically by measuring the [O III] and the Hα lines respectively. Such studies are valuable for distinguishing between material which partakes of the general galactic rotation or flow within a bar or is in a radial flow, perhaps associated with radio jet plasma or, possibly, part of a concurrent starburst wind. Detailed studies of the gas kinematics in NGC 1365 are interpreted by Hjelm and Lindblad (1996) as an accelerated radial outflow with the velocities decreasing rapidly towards the cone edges. In the case of some powerful radio galaxies, there are velocity components along the radio axis with velocities as high as thousands of km s^{-1} which seem very likely to be due to entrainment of material with the radio jet (Tadhunter 1991). Such extreme velocities along the radio axis appear to be found more frequently at high redshift (McCarthy 1993).

In more distant radio galaxies there are, perhaps, no known cases of the well-delineated cones we see in the Seyferts. Is this simply a matter of lower intrinsic spatial resolution or are there differences in the collimating mechanism and/or matter distribution in the radio-loud objects? Amongst the high redshift radio galaxies (HzRG) which demonstrate the 'alignment effect', the rest-frame ultraviolet morphologies certainly do not resemble the Seyfert cones although, at least, the close axial alignment with the radio source is similar. We must remember, however, that the spatial scale of the extensions in the HzRG is about an order of magnitude larger than that in the nearby Seyferts: deep HST images will be important in resolving this issue and may already be doing so in a few cases.

4 Jet/Cloud Interactions

The particle beams which power the large-scale double radio lobes in the radio galaxies propagate through the host galaxy and its local environment. The beams are very efficient at transmitting power from the nucleus and radiate only very little of this to make themselves visible. Occasionally a jet will hit something more substantial than a diffuse interstellar/intergalactic medium and the resulting interaction produces enhanced radiation over a wide band of frequencies. This can tell a captivating story about both the jets and the galactic neighbourhood.

The detailed physics of such interactions is undoubtedly quite complicated (e.g., van Breugel et al. 1985) but one of the first tasks is to establish the influence of any photon beam generated at the base of the jet at the nucleus (Fig. 3). In the absence of any jet bending, the beam will travel along approximately the same path and can have a profound effect on the interaction site. Such photon beams are expected in both FR I and FR II radio sources (see the review by Urry and Padovanni 1995) and, when aimed at the observer, produce the BL Lac or Blazar phenomenon. One might expect to learn about these beams both from the statistics of completely identified samples of radio sources and from their effect on the excitation and ionization of the EELR. What evidence there is suggests that, while the photon density in the beams can be quite high — producing higher ionization parameters along the radio axis than in the diffuse EELR — the total fraction of the AGN luminosity carried by them is small.

Although there is a general tendency for the EELR excitation state to be higher along the radio axis, there are relatively few cases where a current jet/cloud interaction has been clearly recognised. These are presumably the most promising sites to examine for the presence of local sources of ionization associated with shocks but there may be, even here, a significant or even dominant contribution by ionizing photons from the AGN — especially if there is a beamed component. From the cases which have been studied, there appear to be two distinct kinds of behaviour (Clark and Tadhunter 1996). EELR associated with radio lobe hot-spots (e.g., PKS 2250-41, 3C 171, 4C 29.30 and Coma A) show evidence for shock excitation: highly disturbed kinematics and a relatively low ionization state. The EELR associated with the inner knots in the radio jet (e.g., PKS 2152-699 and the inner source in Coma A), in contrast, have a very high ionization state and a somewhat calmer velocity field. It may be that these latter objects are dominated by the effects of the beamed radiation from the AGN, a conclusion which is supported by the presence of a blue, polarized and presumably scattered continuum in PKS 2152-699 (Tadhunter et al. 1987; Tadhunter et al. 1988; di Serego Alighieri et al. 1988; Fosbury et al. 1990). The interaction site in this object radiates over a very wide band of frequencies, from the radio to the X-ray (Fosbury et al. in prep, see Fig. 4), and is in many ways similar in appearance to an AGN. This may be telling us that many of the observed properties of AGN could be due to jet interactions occurring close to the jet source.

Interesting and very nearby examples are the optical filaments along the radio axis of the FR I radio galaxy Centaurus A. It has been proposed that these are ionized by the BL Lac beam from the nucleus (Morganti et al. 1991, 1992). In this case, the required beam luminosity is very similar to that seen

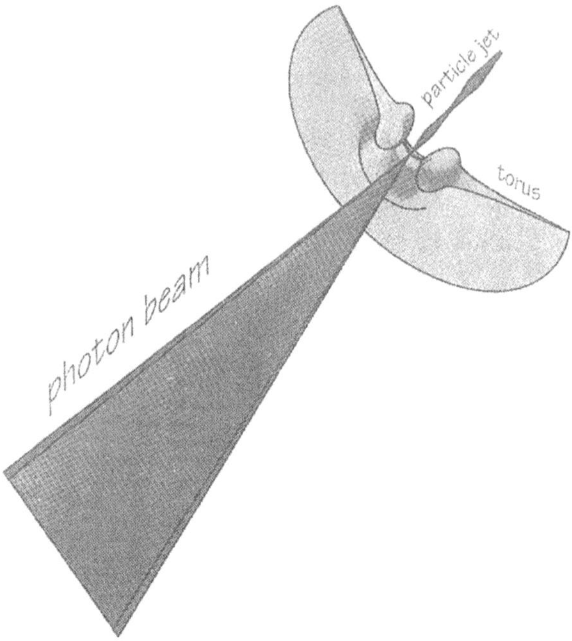

Fig. 3. In addition to the broad ionization cone there may be — in the radio-loud sources — a Doppler boosted photon beam which, initially at least, travels in the same direction as the particle jet which powers the outer radio lobes

directly in BL Lac itself. It would make Cen A into a 4th magnitude AGN if seen along the beam axis. An alternative explanation for the ionization of the filaments using the EUV cooling radiation from high speed shocks within the gas clouds (Sutherland, Bicknell and Dopita 1993), while attractive for some interactions, seems inappropriate here since there is no evidence from the radio observations of a jet currently propagating close to the excited gas — although the macroturbulent velocities within the filament structures are quite high.

5 Diagnostic Tools

In addition to the measurement of the structures of extended nebulosities which are assumed to be excited by the AGN radiation field, statistical methods can be used to infer the radiation pattern of the nucleus. To achieve this, complete samples of sources are needed which are selected on the basis of an isotropic property. The low-frequency radio emission is generally assumed to be the most reliable and the 3CR and 3CRR catalogues (McCarthy, van Breugel and Kapahi

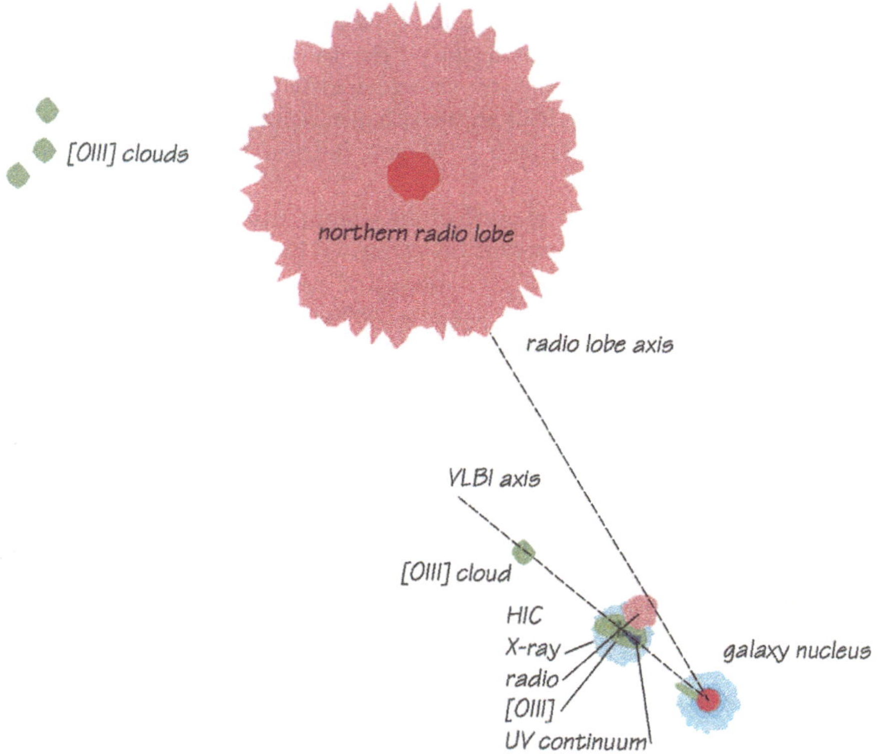

Fig. 4. A sketch of the jet/cloud interaction in PKS 2152-699. In this case, there is a common axis of apparently photon-induced activity which is traced from the VLBI scale out to at least 100 kpc. This axis is somewhat displaced from the current large-scale radio axis. The outer [O III]-emitting clouds are well outside the radio structure and are presumably photoionized

1991; Laing, Riley and Longair 1983) are valuable source lists for such work. Barthel (1989) was able to derive approximate parameters for the obscuration and beaming in these radio-loud objects by analysing the relative numbers of objects classified as blazars, quasars and radio galaxies. Considerable work has been done on the radio and X-ray beaming properties of both the FR I and FR II sources (see the review by Padovanni and Urry 1995) and Morganti at al. (1995) have used Monte-Carlo simulations to analyse the radio properties of their 2 Jy subsample in the south.

Such isotropic selection can also be used as a direct test of the unification hypothesis by examining the individual object classes on the basis of a second isotropic property. Jackson and Browne (1990) applied this test to the [O III] line emission of radio galaxies and quasars from the 3C but it failed for the reason, which we now understand, that much of the [O III] emission from these objects comes from a region which is little larger than the broad line region (BLR) and so is obscured in a similar way. This could be shown to be a plausible explanation if the [O III] line emission is significantly polarized in radio galaxies. By noting that the [O II] lines have a much lower critical density for collisional de-excitation, Hes, Barthel and Fosbury (1993) showed that carefully matched subsamples of quasars and radio galaxies, again from 3C, are indistinguishable from their [O II] luminosity.

A dominant characteristic of much of the circumnuclear activity we have been discussing is the notion of gas clouds being excited by an *external* radiation field. The observed properties of such configurations differ markedly from the classical internally stellar-photoionized H II regions, not only for their different ionizing spectra but also in the importance of the detailed geometry of the clouds, the position of the observer with respect to the illuminating source, and the effects of dust. This distinction becomes particularly apparent in the ultraviolet spectrum where several of the dominant lines are resonance transitions. There have been recent attempts, therefore, to develop photoionization models which take explicit account of the geometry and the physical effects of dust.

Optical studies of EELR in low redshift objects show, in general, little evidence for dust extinction. RBFT87 showed — for a sample of homogeneously observed radio galaxies — that, while the nuclear narrow line spectra exhibit small amounts of reddening, the EELR show none. This does not imply, however, that the EELR clouds are dust-free for it can be shown that reddening is a very insensitive diagnostic of dust in externally photoionized clouds. Because of its importance in understanding many aspects of objects at high redshift — both the stellar properties and the polarization of the aligned light — other methods have been developed to detect the presence of dust.

Measurements from the far-infrared to the mm of objects over a range of redshifts (e.g., Chini and Krügel 1994; Dunlop et al. 1994; Serjeant et al. 1995; Ivison 1995) have detected substantial quantities of dust (up to $10^8\,M_\odot$). Little is known, however, about the spatial distribution of the dust and it may be associated with the obscuring torus rather than the very extended extranuclear structures. Observations with ISO will be sensitive to thermal radiation from dust over the whole observed redshift range and, hopefully, the temperature distribution will tell us something about the spatial distribution.

A novel and very sensitive test for dust in emission nebulae is based on detecting the depletion of refractory elements from the gas phase onto dust grains. This was proposed by Ferland (1993) in order to understand the observed weakness of the forbidden ionized calcium doublet in the near-infrared spectra of LINERS. Standard photoionization models with solar gas-phase abundances predict the [Ca II]$\lambda\lambda$7291, 7324 lines to be amongst the strongest in the spectrum at the ionization level of these objects. Observations in the appropriate spectral region of a range of extended nebulosities around galaxies, including so-called 'cooling-flows' were made by Donahue and Voit (1993) who failed to detect the line and concluded the presence of dust. Villar-Martín and Binette (1996) have investigated the problem using the photoionization code MAPPINGS (Binette et al. 1993a,b), adapted to include several important aspects of dust physics, namely the effects of scattering on the radiation transfer and the contribution of dust to the thermal and ionization balance of the plasma. They examined in detail several alternative processes which could suppress the calcium lines but concluded that their absence over a certain range of the ionization parameter must imply depletion. Work in progress by Villar-Martín, Fosbury and Binette applies this modelling to observations of some low redshift radio galaxies and Seyferts with EELR.

The recognition of the importance of the rest-frame ultraviolet spectrum for the study of the nature of objects at the highest redshifts has led to studies of line formation in externally ionized clouds with particular emphasis on the geometrical aspects and the effects of dust on the resonance line transfer. This has many exciting applications for the physical study of objects at the intriguing early stages of galaxy formation and evolution. A particular attraction here for the objects with luminous AGN is the illumination of quiescent material at large radial distances which may only be in the very earliest stages of becoming part of a galaxy.

If we accept the unification of the powerful radio galaxies with the radio quasars — a picture which must be correct at some level (Antonucci 1993) — our studies of the HzRG are constrained by the knowledge of the general properties of the ionizing radiation field from the quasars. The obscuration of the quasar due to orientation does, however, leave us with a clear view of the much fainter EELR which is currently being exploited with the HST and large groundbased telescopes. A powerful technique which follows from the acceptance of unification is the use of the 'associated absorption' spectrum of radio quasars to give a complementary view of the extended structures we see in emission in the radio galaxies. This may help us quantify some of the geometrical factors which characterise the matter distribution and can be applied both with absorption lines and dust absorption/scattering.

The particular geometry of the radio galaxies, with the radio axis lying somewhere between the plane of the sky and about 45° to it, implies that we see clouds both from the front (illuminated) face and from the back. This can result in side-to-side asymmetries both in the line spectra and in the scattered light, especially if dust dominates over electron scattering. Such asymmetries should, in principle, give us a measure of orientation in 3-dimensions which can be compared with radio measurements of jet-sidedness and depolarization asymmetry (Laing 1988; Garrington 1988).

Villar-Martín, Binette and Fosbury (1996) have studied the particular case of the C IV/Lyα vs. C IV/C III] diagnostic diagram which, because of its inclusion of two resonance lines, is particularly sensitive to geometrical effects (Fig. 5). The computed diagrams are compared with a set of observations which has recently been very significantly enlarged by van Ojik's (1995) work on the ultra-steep radio spectrum selected high redshift objects. We find that the observed trends, in particular the large C IV/Lyα ratios, can be satisfactorily explained in almost all cases by the effects of seeing illuminated clouds from behind or in front. The only cases where dust is needed to quench Lyα are the highly obscured sources like F10214+4724 (Elston et al. 1994) and TX0211-122 (van Ojik et al. 1994) which are probably characterised by a more 'closed' geometry than the other radio galaxies. We also address the question of the neutral clouds — which are presumably outside the radiation cones and so do not see the AGN directly — which produce the spatially extended Lyα absorptions seen in objects like 0943-242 (Röttgering et al. 1995). These screens are clearly kinematically distinct from the line emitting clouds and so are presumably spatially separated from them. Such neutral clouds act as very efficient Lyα reflectors, even if they contain dust, and this process may be responsible for part of the extensive Lyα halos seen to enshroud objects like 1243+036, $z = 3.57$ (van Ojik et al. 1995). Such neutral reflectors could be recognised by the absence of other optical/UV emission lines and it is important to point out that pure Lyα emitters do not necessarily signify star-forming regions.

This particular diagnostic diagram is useful for objects observed from the ground with redshifts greater than about two. Similar tools need to be developed for the intermediate redshift sources.

The production of polarized light in the Seyferts and the radio galaxies by scattering of anisotropically emitted radiation from the AGN has led to an extension of the concept of a 'reflection nebula' to super-galactic scales. The quantitative understanding of these phenomena is hindered, particularly in the case of the HzRG, by the faintness of the sources, the complexity of the geometric structure and, not least, by our ignorance of the precise nature of the scatterers. Assuming the close resemblance of dust found at 100 kpc from a radio galaxy at a redshift of 4 to dust in the Solar neighbourhood today is a leap of faith of gigantic proportions. Nonetheless, models built on the basis of this assumption are quite successful in explaining at least the gross features of the observations (Manzini and di Serego Alighieri 1996). A purely empirical diagnosis of dust or electron scattering in any particular case is rather difficult. Thomson scattering by hot electrons can be distinguished by its thermal broadening of any scattered lines. Using the wavelength dependence of polarization can, however, be misleading because of the presence of one or more unpolarized diluting sources, possibly with different spectra. The effects of a complex geometry on the scattered spectrum and polarization can make it very hard to reach the underlying physical processes. I like to use the analogy of the twilight sky on Earth. This is dominated by Rayleigh scattering from molecules in the atmosphere and the process has a well understood wavelength dependence. The spectacular range of colours produced by a partially illuminated atmosphere viewed along different paths should, however, make us a little cautious when interpreting observations of the colour of galaxy-scale reflection nebulae. An investigation of the observed continua around a rest wavelength of 2200Å will, perhaps give us some direct evidence for the presence of dust and perhaps, even, its nature.

In some cases, particularly close to the AGN — as seen in some Seyfert 2s (e.g., Antonucci, Hurt and Miller 1995) — it is clear that there can be sufficient

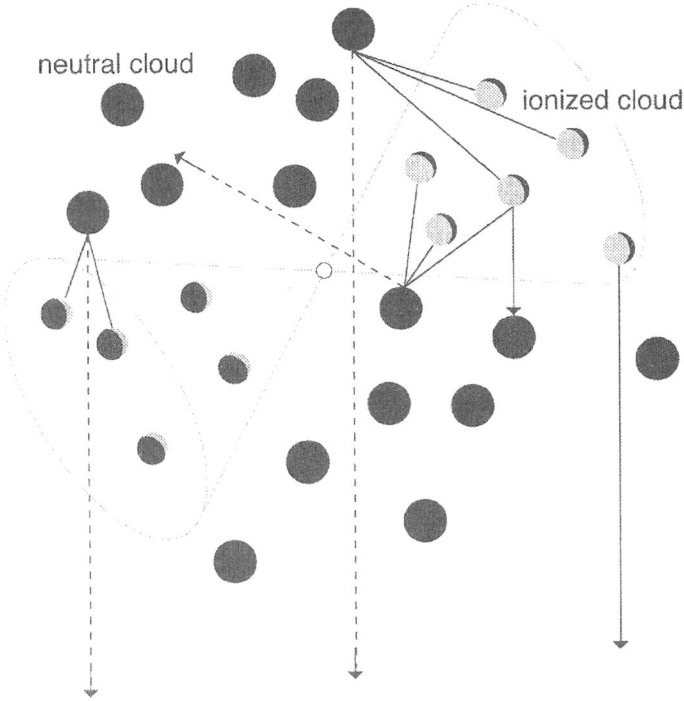

Fig. 5. An illustration of the viewing geometry which strongly affects the appearance of the UV emission spectrum of the EELR. The Lyα and C IV resonance lines escape preferentially from the front (illuminated) faces of the clouds. Lyα photons can also be reflected efficiently from neutral clouds which do not see the ionizing radiation from the AGN directly: such reflection may be responsible for part of the extended Lyα haloes seen in some high redshift radio galaxies

Thomson optical depth to produce diffuse electron scattered radiation. It seems, however, that on the larger scales, scattering by dust is much more plausible due to its considerably higher efficiency per unit mass. The recent discovery of polarized broad lines, similar in width to those seen in quasars, in some HzRG (di Serego Alighieri et al. 1996; Dey and Spinrad 1995) strongly favour dust scattering but does not entirely rule out scattering by a population of cool ($\leq 10^5$ K) electrons.

To finish this section, I should like to make a few remarks about the measurement of polarization in these faint sources. Although the polarimetric mode on the Low Resolution Imaging Spectrograph on the W M Keck 10-m telescope (LRIS, Oke et al. 1995) is now being used to make superbly high quality measurements, many of the observations which have established the importance of polarimetry for the study of these scattering phenomena have been made with

3 – 4-m-class telescopes. It is generally true that, for the faint radio galaxies, both imaging and spectropolarimetric measurements have an uncertainty which is dominated by photon statistics. Instrumental polarization and other systematic effects are calibrated as well as possible and are generally below the 1% fractional polarization which can be reached in reasonable integration times for many of these sources. Considerable care, however, is needed in the analysis of these statistical errors for any particular observation. This is particularly true for polarimetry (as is well known by the protagonists of high precision stellar polarimetry) since several of the quantities of interest have asymmetric distribution functions and results can be severely biased at low signal-to-noise ratios (Fosbury, Cimatti and di Serego Alighieri 1994 and references therein). We have found that the best way of calculating the error distributions, both as part of data analysis and in the design of an observational strategy, is to perform Monte-Carlo simulations with a relatively detailed stochastic model of the particular observational setup. Estimators for the Stokes q and u parameters are better behaved than those for fractional polarization and position angle.

6 AGN at Very High Redshift

A close examination of the superb images of some HzRG which have been obtained with the Hubble Space Telescope Wide Field and Planetary Camera 2 (e.g. Fig. 6 and Longair, Best and Röttgering 1995) will give us a good idea of the problems we face in interpreting the state of these distant galaxies. To my knowledge, there has so far been *no* confirmed identification of a young stellar population in any HzRG. The continuum spectra above a rest wavelength of ~4000Å appear to be dominated by an evolved, red stellar population and below 4000Å by the light responsible for the alignment effect (Rigler et al. 1992). The remarkably small dispersion of these objects in the K-band Hubble diagram (Lilly 1989, but see McCarthy 1993 for a recent review which includes more high redshift objects) also allows us to argue for an evolved stellar population although we must beware of strong emission lines moving into the photometric windows (Eales and Rawlings 1993).

A direct comparison of the K-band with the (rest-frame UV) HST images shows objects which are as different as chalk and cheese. This is beautifully consistent with the ensemble of polarization observations which show (Cimatti et al. 1993) the polarization 'switch on' as the observed passband (usually V or B) is moved by increasing redshift below the H and K break. In 3C 324 ($z = 1.2$, Longair, Best and Röttgering 1995) the position of the K-band nucleus occupies a central 'hole' in the very elongated HST image. It is clear from these pictures that, on the largest scale, these objects do not resemble the spectacular ionization cones seen in the Seyferts on the kpc scale. If the broad radiation cones are really there, they are not uniformly filled with matter which is, rather, distributed in a highly non-random manner. Although there is generally not a precise spatial coincidence between the radio source and the UV light (but

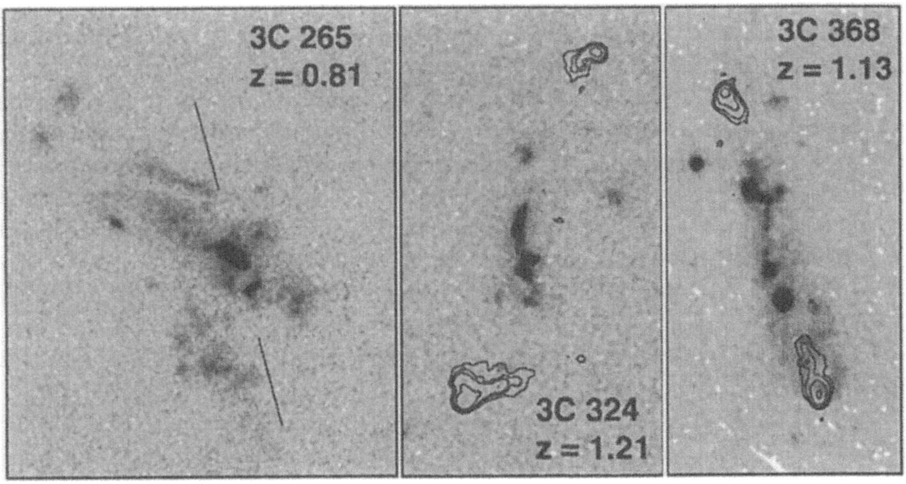

Fig. 6. HST images showing the rest-frame ultraviolet images of three $z \sim 1$ radio galaxies with overlayed radio contours (courtesy Malcolm Longair, NASA and NRAO)

see 4C 41.47, Miley et al. 1992), there are indications in some objects — and 3C 324 is a good example — that the light traces the path of the radio jet. This fact has often been used to argue the case for jet-induced star formation and, indeed, we cannot rule out the presence of blue stars in these regions which would dilute the fractional polarization of the scattered light. A salutary counter-example, however, is the $z = 0.811$ source 3C 265 which is misaligned with the radio source by about 35° and yet is highly polarized with an E-vector which is perpendicular to the UV extension (Jannuzi and Elston 1991; Dey and Spinrad 1995; di Serego Alighieri et al. 1996; Cohen et al. 1996) and *not* the radio axis. In some cases, emission beyond the radio lobes has been reported (van Breugel and McCarthy 1990; Eales and Rawlings 1993). Unless this material is the remnant of an earlier and now faded radio source, it must represent quiescent gas ionized by AGN photons.

A determination of the distribution of matter around these distant objects is crucial for coming to an understanding of their formation. The observations of the giant Lyα halo in the $z = 3.6$ radio galaxy 1243+036 (van Ojik et al. 1995) gives us a clue which I think helps us to draw some of the diverse threads together and begin to understand what is going on. The salient point here is the recognition of a component of the aligned EELR which is clearly associated with the (bent) radio structure and one which extends well beyond the radio source but is quite kinematically distinct. The gas contained within the region of the radio structure has a high velocity dispersion (~1 500 km s^{-1} FWHM) and, in particular, there is a blueshift at the point of the radio bend of 1100 km s^{-1} which coincides with a Lyα enhancement. The large scale (20″) halo, in contrast, has a velocity width of only 250 km s^{-1} and a global velocity gradient of 450 km s^{-1}. The faint extended UV continuum emission, while aligned with the principal (inner) radio axis, does not follow the bend in the structure.

I agree with van Ojik et al. that these observations argue strongly that the gaseous halo in 1243+036 must predate the AGN and its associated radio source. It also may go some way towards explaining the apparent dichotomy between those lower redshift EELR which appear to be smoothly rotating and those which appear chaotic: there may always be a component of the previously quiescent gas which has been disturbed after the birth of the radio source. The attraction of these quiescent halos is that they may represent material caught in the act of forming the galaxy. The presence of a powerful AGN illuminates the birth and give us opportunities for study which would be absent without the nuclear activity.

Having established the presence of a spatially extended scattered component which contributes to the alignment effect the polarization measurements are now being used as a major tool in attempts to separate out the different sources of radiation seen in these high redshift objects. Since the intrinsic fractional polarization of scattered radiation depends on the nature and on the geometrical configuration of the ensemble of scatterers, the measurement of the contribution of unpolarized diluting continuum is difficult. One technique, which has been applied to nearby Seyferts, is to look at changes in polarization across the scattered broad lines. An increase in polarization in the broad lines suggests a diluting unpolarized continuum called FC2 by Antonucci (1993, see also Binette, Fosbury and Parker 1993). This could be starlight or nebular continuum (Dickson et al. 1995).

The detection of broad, polarized lines in the aligned radio galaxies is now well established (di Serego Alighieri et al. 1996; Dey and Spinrad 1995) and in 3C 265, broad Mg II has even been seen in the extended component. They have also been seen in lower redshift objects like 3C 234 (Antonucci 1984; Tran, Cohen and Goodrich 1995) and Cygnus A (Antonucci, Hurt and Kinney 1994).

My conclusion is that, at these high redshifts, the nuclear radiation field is still playing a dominant role in the excitation of the extranuclear gas but that the spatial distribution of the material is influenced by the passage of the radio jet. I believe, however, that this may not constitute the whole story and CDM N-body simulations show us that the matter distribution over the scales of interest (100 kpc) was rather different at redshifts of a few from what it is now. Galaxies were forming within filamentary structures which, if illuminated by a powerful quasar, might look something like some of the aligned radio galaxies we see with HST rather than the filled ionization cones we see in the Seyferts on a much smaller scale.

Although I have been discussing the radio galaxies — objects with hidden quasars — there are some important implications for the work on quasar 'fuzz' which is directed towards the measurement of host galaxy properties. If scattering is redistributing a few percent of hidden quasar luminosity to make part or all of the continuum alignment effect in the galaxies, this scattered light will also appear in the quasar images. Indeed, if the scattering is from dust grains, the forward scattering which we would see when imaging a quasar could be a significantly larger fraction of the direct flux and so seriously contaminate any host galaxy starlight (Fig. 7). Without knowing the detailed distribution of the scatterers, this will be difficult to separate from the galaxy light and the colour information will be crucial.

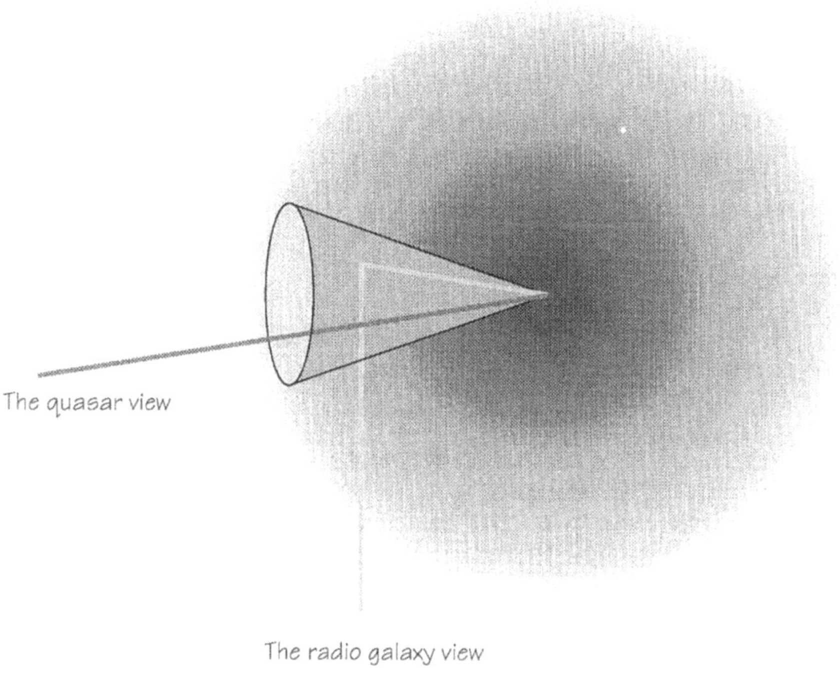

Fig. 7. The geometry which produces the scattered, aligned light in the radio galaxies can also produce forward-scattered haloes in quasars. In the higher redshift, luminous quasars where any stellar host galaxy may be apparently faint in the rest-frame ultraviolet part of the spectrum, such scattered haloes can produce pseudo-hosts which contribute as much as 10% of the total quasar light

7 Conclusions

The energy output of an active galactic nucleus has a profound effect on the state of its surroundings. The consequences of these interactions can be mapped on a scale of Mpc using the quasar absorption line 'proximity effect' and to a tenth of this in the radio galaxy extended emission lines. Studies of these phenomena have told us a great deal about the nature of the nuclei themselves and have allowed us to simplify our classification schemes for diverse AGN types. At the highest redshifts, AGN not only help us identify galaxies but they act as sources of illumination which enable us to study material in the process of forming galaxies.

Acknowledgements. This review is based partly on a long series of projects on various aspects of the subject over the years. From all of these, I should like to single out two of my longest-standing collaborators: Sperello di Serego Alighieri and Clive Tadhunter. Clive, in particular, provided me with a haven of tranquillity at the University of Sheffield where much of this was written. I should also like to mention Montse Villar-Martín and Luc Binette who allowed me to include some of the UV line modelling prior to publication.

References

Antonucci, R. R. J. (1984): ApJ **278**, 499
Antonucci, R. (1993): ARA&A **31**, 473
Antonucci, R., Hurt, T., Kinney, A. (1994): Nat **371**, 1994
Antonucci, R., Hurt, T., Miller, J. (1995): ApJ **430**, 210
Antonucci, R., Miller, J. (1985): ApJ **297**, 621
Barthel, P. D. (1989): ApJ **336**, 606
Baum, S. A., Heckman, T. M., van Breugel, W. (1990): ApJS **74**, 389
Baum, S. A., Heckman, T. M., van Breugel, W. (1992): ApJ **389**, 208
Binette, L., Fosbury, R. A. E., Parker, D. (1993): PASP **105**, 1150
Binette, L., Wang, J. C. L., Villar-Martín, M., Martin, P. G., Magris, C. G. (1993b): ApJ **414**, 535
Binette, L., Wang, J. C. L., Zuo, L., Magris, C. G. (1993a): AJ **105**, 797
Chambers, K. C., Miley, G. K., van Breugel, W. (1987): Nat **329**, 604
Chini, R., Krügel, E. (1994): A&A **288**, 33
Cimatti, A., di Serego Alighieri, S. (1995): MNRAS **273**, L7
Cimatti, A., di Serego Alighieri, S., Fosbury, R. A. E., Salvati, M., Taylor, D. (1993): MNRAS **264**, 421
Clark, N., Tadhunter, C. N. (1996): in Proc. NRAO Workshop – Cygnus A, NRAO, Greenbank, USA, 1–4 May 1995, eds C. Carilli, D. Harris, CUP, p. 15
Cohen, M. H., Tran, H. D., Ogle, P. M., Goodrich, R. W. (1996): in Proc. IAU Symp. 175 Extragalactic Radio Sources, Bologna, Italy, October 1995, in press
Daly, R. A. (1992): ApJ **386**, L9
Dey, A., Spinrad, H. (1995): ApJ in press
Dickson, R., Tadhunter, C. N., Shaw, M., Clark, N., Morganti, R. (1995): MNRAS **273**, L29
di Serego Alighieri, S., Binette, L., Courvoisier, T. J-L., Fosbury, R. A. E., Tadhunter, C. N. (1988): Nat **334**, 591
di Serego Alighieri, S., Cimatti, A., Fosbury, R. A. E. (1993): ApJ **404**, 584
di Serego Alighieri, S., Cimatti, A., Fosbury, R. A. E., Perez-Fournon, I. (1996): MNRAS submitted Nov. 1995

di Serego Alighieri, S., Fosbury, R. A. E., Quinn, P. J., Tadhunter, C. N., (1989): Nat **341**, 307
Donahue, M., Voit, G. M. (1993): ApJ **414**, L17
Dunlop, J. S., Hughes, D. H., Rawlings, S., Eales, S. A., Ward, M. J. (1994): Nat **370**, 347
Eales, S. A., Rawlings, S. (1993): ApJ **411**, 67
Elston, R., McCarthy, P. J., Eisenhardt, P., Dickinson, M., Spinrad, H., Januzzi, B. T., Maloney, P. (1994): AJ **107**, 910
Evans, I. N., Tsvetanov, Z., Kriss, G. A., Ford, H. C., Caganoff, S., Koratkar, A. P. (1993): ApJ **417**, 82
Ferland, G. J. (1993): in Proc. The Nearest Active Galaxies, Madrid, May 1992, eds Beckman, Colina and Netzer, p. 75
Fosbury, R. A. E., Cimatti, A., di Serego Alighieri, S. (1994): ESO Messenger **74**, 11
Fosbury, R. A. E., di Serego Alighieri, S., Courvoisier, T. J.-L., Snijders, M. A. J., Walsh, J., Wilson, W. (1990): in Evolution in Astrophysics, IUE astronomy in the era of new space missions, Toulouse, France, 29 May - 1 June 1990. ESA SP-310
Garrington,S. T. (1988): in Proc. NATO Advanced Research Workshop: Cooling Flows in Clusters and Galaxies, Cambridge, England, June 22–26, 1987, Kluwer, Dordrecht, p. 209
Hes, R. (1995): Orientation Effects in QSOs, Quasars and Radio Galaxies, Ph.D. Thesis, University of Groningen. Chap. 3
Hes, R., Barthel, P. D., Fosbury, R. A. E. (1993): Nat **362**, 326
Hjelm. M., Lindblad, P. O. (1996): A&A **305**, 727
Ivison, R. (1995): MNRAS **275**, L33
Jackson, N., Browne, I. W. A. (1990): Nat **343**, 43
Januzzi, B. T., Elston, R. (1991): ApJ **366**, L69
Jörsäter, S., Lindblad, P. O., Boksenberg, A. (1984): A&A **140**, 288
Laing, R. A. (1988): Nat **331**, 149
Laing, R. A., Riley, J. M., Longair, M. S. (1983): MNRAS **204**, 151
Lilly, S. J. (1989): ApJ **340**, 77
Longair, M. S., Best, P. N., Röttgering, H. J. A. (1995): MNRAS **275**, L47
Manzini, A., di Serego Alighieri, S. (1996): A&A in press
McCarthy, P. J. (1993): ARA&A **31**, 639
McCarthy, P. J., van Breugel, W., Kapahi, V. K. (1991): ApJ **371**, 478
McCarthy, P. J., van Breugel, W., Spinrad, H., Djorgovski, S. (1987): ApJ **321**, L29
Miley, G. K., Chambers, K. C., van Breugel, W., Macchetto, F. (1992): ApJ **401**, L69
Morganti, R., Fosbury, R. A. E., Hook, R. N., Robinson, A., Tsvetanov, Z. (1992): MNRAS **256**, 1p
Morganti, R., Osterloo, T. A., Fosbury, R. A. E., Tadhunter, C. N. (1995): MNRAS **274**, 393
Morganti, R., Robinson, A., Fosbury, R. A. E., di Serego Alighieri, S., Tadhunter, C. N., Malin, D. F. (1991): MNRAS **249**, 91
Oke, J. B. et al. (1995): PASP **107**, 375
Orr, M. J. L., Browne, I. W. A. (1982): MNRAS **200**, 1067
Pedlar, A., Kukula, M. J., Longley, D. P. T., Muxlow, T. W. B., Axon, D. J., Baum, S., O'Dea, C., Unger, S. W. (1993): MNRAS **263**, 471
Phillips, M. M., Edmunds, M. G., Pagel, B. E. J., Turtle, A. J. (1983): MNRAS **203**, 759
Prieto, M. A., Freudling, W. (1993): ApJ **418**, 668
Rigler, M. A., Stockton, A., Lilly, S. J., Hammer, F., Le Fèvre, O. (1992): ApJ **385**, 61

Röttgering, H. J. A. (1993): Ultra-Steep Spectrum Radio Sources: Tracers of Distant Galaxies, Ph.D. Thesis, University of Leiden.
Röttgering, H. J. A., Hunstead, R., Miley, G. K., van Ojik, R., Wieringa, M. H. (1995): MNRAS in press
Robinson, A. (1989): in ESO Workshop on Extranuclear Activity in Galaxies, ESO, Garching, 16–18 May 1989, eds. E.J.A. Meurs, R.A.E. Fosbury, p. 259
Robinson, A., Binette, L., Fosbury, R. A. E., Tadhunter, C. N. (1987): MNRAS **227**, 97
Robinson, A., Vila-Vilaro, B., Axon, D. J., Perez, E., Wagner, S. J., Baum, S. A., Boisson, C., Durret, F., Gonzalez-Delgado, R., Moles, M. (1994): A&A **291**, 351
Scarrott, S. M., Rolf, C. D., Tadhunter, C. N. (1990): MNRAS **243**, 5P
Serjeant, S., Lacy, M., Rawlings, S., King, L. J., Clements, D. L. (1995): MNRAS **276**, L31
Stockton, A., Ridgeway, S. (1996): in Proc. NRAO Workshop – Cygnus A, NRAO, Greenbank, USA, 1–4 May 1995, eds C. Carilli, D. Harris, CUP, p. 1
Sutherland, R. S., Bicknell, G. V., Dopita, M. A. D. (1993): ApJ **414**, 510
Tadhunter, C. N. (1991): MNRAS **251**, 46P
Tadhunter, C. N. (1996): in Proc. NRAO Workshop – Cygnus A, NRAO, Greenbank, USA, 1–4 May 1995, eds C. Carilli, D. Harris, CUP, p. 33
Tadhunter, C. N., Fosbury, R. A. E., Binette, L. A., Danziger, I. J., Robinson, A. (1987): Nat **325**, 504
Tadhunter, C. N., Fosbury, R. A. E., di Serego Alighieri, S. (1989): in Como Workshop on BL Lac Objects: 10 Years After, Como, Italy, September 1988, eds. Maraschi et al., Springer, Berlin
Tadhunter, C. N., Fosbury, R. A. E., di Serego Alighieri, S., Bland, J., Danziger, I. J., Goss, W. M., McAdam, B., Snijders, M. A. J. (1988): MNRAS **235**, 403
Tadhunter, C. N., Fosbury, R. A. E., Quinn, P. J. (1988): MNRAS **240**, 225
Tadhunter, C. N., Tsvetanov, Z. (1989): Nat **341**, 422
Tran, H. D., Cohen, M. H., Goodrich, R. W. (1995): AJ **110**, 2597
Urry, C. M., Padovanni, P. (1995): PASP **107**, 803
van Breugel, W., McCarthy, P. J. (1990): in Proc. The Hubble Centennial Symposium: The Evolution of Galaxies, ed. R.G. Kron, PASPC **10**, 359
van Breugel, W., Miley, G., Heckman, T., Butcher, H., Bridle, A. (1985): ApJ **290**, 496
van Ojik, R. (1985): Gas in Distant Radio Galaxies: Probing the Early Universe, Ph.D. Thesis, University of Leiden.
van Ojik, R., Röttgering, H. J. A., Carilli, C. L., Miley, G. K., Bremer, M. N., Macchetto, F. (1995): A&A in press
van Ojik, R., Röttgering, H. J. A., Miley, G. K., Bremer, M. N., Macchetto, F., Chambers, K. C. (1994): A&A **289**, 54
Villar-Martín, M., Binette, L. 1996): A&A in press
Villar-Martín, M., Binette, L., Fosbury, R. A. E. (1986): A&A submitted Nov. 1995
Wilson, A. S., Tsvetanov, Z. (1994): AJ **107**, 1227

Outflows from the Nearest Barred Galaxies

Timothy M. Heckman

Department of Physics & Astronomy, The Johns Hopkins University, Baltimore, MD 21218 USA

Abstract. I review the evidence for bars and outflows associated with three very nearby galaxies, beginning with the two nearest infrared starbursts (NGC 253 and M 82). These two starbursts and their outflows are rather similar to one-another, even though M 82 is a much less massive galaxy. The existence of a large (25 kpc) soft X-ray halo in NGC 253 and its lack in M 82 may be related to M 82's low mass. I also discuss the curious case of the circumnuclear region in M 31: are the goings-on there (including the nearest example of the LINER phenomenon) due to a bar, an outflow, or neither? Finally, I summarize recent empirical results concerning the link between galaxy disk dynamics and circumnuclear starbursts.

1 Introduction

The gas flows in and around typical starburst galaxies are clearly complex. This complexity arises both from the effects of a non axisymmetric gravitational potential (bars, oval distortions, and companion galaxies) and from the hydrodynamical consequences of the enormous rate of mechanical energy deposition inside the starburst. The former is intimately related to the causes of starbursts, while the latter is an effect of the starburst. Both processes causally connect the starburst with regions that are one or two orders-of-magnitude larger than the starburst itself. Understanding these gas flows is then central not only to understanding the starburst phenomenon, but for relating it to broader issues of cosmogony.

During this conference considerable attention has been focussed on the roles of gravity and bars in sending gas inward to fuel the starburst phenomenon. In this review my main aim is to give a brief review of the outflows that are one of the prime consequences of the starburst. I will also concentrate on the nearest examples of galaxies with bars and outflows, since these have the best and most complete sets of data.

In Sect. 2, I will outline the conceptual framework for the interpretation of observations of outflows from starburst galaxies. In Sects. 3 and 4, I will summarize the evidence for bars and outflows in the two nearest powerful starbursts: NGC 253 and M 82. A more 'general purpose' review of observations of starburst-driven outflows can be found in Heckman, Lehnert and Armus (1993).

I also want to stress the point that while bar-driven inflows and galactic winds are conceptually distinct, it is not always straightforward to tell them apart observationally. To illustrate this, in Sect. 5, I will discuss the central-most two kpc in M 31. Are the curious phenomena there (including the nearest-known LINER) related to a galactic wind or a bar or neither?

Finally, in Sect. 6, I will describe some results from a recent survey of edge-on starburst galaxies that illuminate the relationship between galaxy disk dynamics and the properties of circumnuclear starbursts.

2 Starburst-Driven Outflows: A Conceptual Overview

The engine that presumably drives the observed outflows in starburst galaxies is the mechanical energy supplied by massive stars in the form of supernovae and stellar winds (cf. Leitherer and Heckman 1995). For typical starburst parameters, the rate of supply of mechanical energy is of-order 1% of the bolometric luminosity of the starburst and typically 10 to 30% of the Lyman continuum luminosity. Some fraction of this mechanical energy may be radiated away by dense shock-heated material inside the starburst. However, observations of starburst galaxies imply that a significant fraction is available to drive an outflow (see below). Radiation pressure acting on dust grains may also play a role in driving the observed outflows, since the amount of momentum associated with photons is only a factor of several smaller than the equivalent momentum supplied by supernovae and stellar winds (cf. Leitherer and Heckman 1995).

The dynamical evolution of a starburst-driven outflow has been extensively discussed (cf. MacLow and McCray 1988; Tenario-Tagle and Bodenheimer 1988). Briefly, the deposition of mechanical energy by supernovae and stellar winds results in an over-pressured cavity of hot gas inside the starburst. This hot cavity will expand, sweep up ambient material and thus develope a bubble-like structure. If the ambient medium is stratified (like a disk), the bubble will expand most rapidly in the direction of the vertical pressure gradient. After the bubble size reaches several disk vertical scale heights, the expansion will accelerate, and it is believed that Raleigh-Taylor instabilities will then lead to the fragmentation of the bubble's outer wall. This allows the hot gas to 'blow out' of the disk and into the galactic halo in the form of a weakly collimated bipolar outflow (i.e. the flow makes a transition from a superbubble to a superwind). It seems clear that outflows driven by low-luminosity starbursts and outflows propagating into an ambient medium that is not strongly flattened will spend relatively longer in the superbubble phase.

The observational manifestations of superbubbles and superwinds are many and varied. The ambient gas (both the material in the outer superbubble wall and overtaken clouds inside the superbubble or superwind) can be photoionized by the starburst and shock-heated by the outflow. This material can produce soft X-rays and optical/ultraviolet emission and absorption lines. The hot gas that drives the expansion of the superbubble/superwind may itself be a detectable source of X-rays, especially if a significant amount of 'mass-loading' of the outflow occurs in or around the starburst (cf. Suchkov et al. 1996). Finally, cosmic ray electrons and magnetic field may be advected out of the starburst by the flow and produce a radio synchrotron halo and possibly an X-ray halo via inverse Compton scattering of soft photons from the starburst (cf. Seaquist and Odegard 1991).

Optical and X-ray data allow us to address the important question as to the fate of the gas flowing out of starbursts: will this gas escape the galaxian gravitational potential well and carry the newly synthesized metals into the intergalactic medium, or will the gas remain bound to the galaxy and perhaps cool and return as a galactic fountain?

Taking a spherically symmetric isothermal potential with an outer cut-off at r_{max} and a depth corresponding to the square of the circular orbital velocity (v_{circ}), then the escape velocity for material at a radius r is just

$$v_{esc}(r) = v_{circ}[2 + 2\ln(r_{max}/r)]^{1/2} \tag{1}$$

Following Wang et al. (1995), the 'escape temperature' for hot gas in a galaxy potential with an escape velocity v_{esc} is given by

$$T_{esc} = 2.3 \times 10^6 (v_{esc}/400\,\mathrm{km\,s^{-1}})^2\,\mathrm{K} \tag{2}$$

Thus, we can compare either the outflow speeds measured optically or the gas temperature derived from the X-ray data to the galaxy rotation curves to make inferences about the likely fate of the outflow.

3 NGC 253

The galaxy NGC 253 contains – along with M 82 – the nearest (3.4 Mpc) and best-studied example of a nuclear starburst. The starburst is quite compact, with a size of only a few hundred pc (cf. Pina et al. 1992; Forbes et al. 1993). Its luminosity is near the characteristic 'knee' in the IR galaxy luminosity function ($L_{bol} \sim 3\times10^{10}$ $L_{\odot}$; cf. Soifer, Houck and Neugebauer 1987).

NGC 253 itself is a typical late-type barred (SAB(s)c) galaxy with an absolute magnitude that is near the fiducial Schecter L_* value ($M_B = -20.3$) and a rotation speed $v_{rot} = 180$ km s^{-1}. These values make NGC 253 the most luminous and most massive member of the Sculptor group (Puche and Carignan 1991). Despite its group membership, it is not obviously undergoing any significant tidal interaction with other group members (cf. Puche, Carignan and van Gorkom 1991).

The nearly edge-on orientation of NGC 253 is ideal for studying its outflow, but made it difficult to discern the presence of a bar. The existence of a bar in NGC 253 was deduced by Pence (1980) based on a deprojected optical image, and he later observed large- scale gas streaming motions in its disk that he attributed to the barred potential (Pence 1981). Subsequently, near-IR imaging confirmed the presence of a stellar bar (Scoville et al. 1985), and a circumnuclear (0.6 × 0.2 kpc) molecular counterpart to the stellar bar has also been discovered (Canzian et al. 1988). Telesco, Dressel and Wolstencroft (1993) argue that the compactness of the starburst in NGC 253 is due to lack of an Inner Lindblad

Resonance, which allows the bar-driven inflow of gas to extend all the way into the nucleus.

As first seen by Fabbiano and Trinchieri (1984), there is a 'fan' of X-ray emission to the south-east of the starburst, along the galaxy minor axis (Fig. 1). The ROSAT PSPC data imply that this material has a temperature of about 7×10^6 K and an unabsorbed luminosity in the ROSAT band of about 4×10^{39} erg s^{-1} (Dahlem et al. 1996). This can be compared to the estimated mechanical luminosity of the starburst of $\sim 10^{42}$ erg s^{-1}.

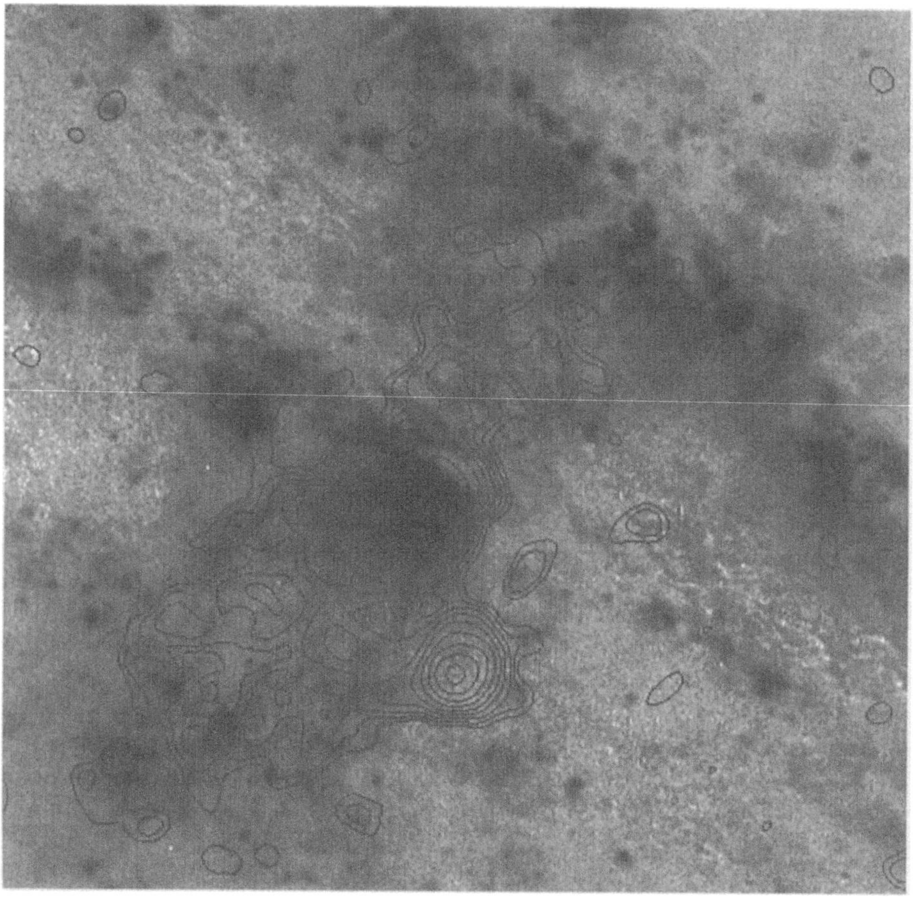

Fig. 1. An overlay of the ROSAT HRI X-ray image (contours) and an optical Hα image (greyscale) of the central 2!3 (2.3 kpc) of NGC 253

As noted by McCarthy, Heckman and van Breugel (1987), the minor-axis Hα filaments seem to lie along the sides of the X-ray source, suggesting that the optical emission arises in material that is being compressed and accelerated by the hot outflowing material visible in X-rays. This impression is confirmed by optical spectroscopy (Fig. 2), which shows that the emission-line gas here exhibits the 'line-splitting' characteristic of a hollow structure that is expanding or outflowing at a velocity of several hundred km s^{-1} (Ulrich 1978; Heckman, Armus and Miley 1990).

A counterpart of the outflow can be seen in the X-ray and Hα images on the opposite (north-west) side of the starburst. This north-west region is much fainter, presumably because it must be viewed through the obscuration of the galaxy disk. A molecular manifestation of this back-side outflow to the north-west was discovered by Turner (1985).

In addition to this kpc-scale 'nuclear' outflow, NGC 253 has a much larger X-ray halo (Fig. 3; see also Pietsch 1993). The X-ray halo is galaxy-sized (roughly 25 by 12 kpc), with its long axis aligned with the galaxy minor axis. Absorption of the soft X-ray emission by the near-side of the disk of NGC 253 can be seen on the north-west side. In maps made with the full angular resolution of the PSPC, the halo appears filamentary, with an overall 'X' shape (Pietsch 1993).

The X-ray halo is quite soft, and is much less conspicuous at energies above 0.5 keV (Pietsch 1993; Dahlem et al. 1996). The implied temperature is a few million K, and the unabsorbed luminosity of the halo component in the ROSAT band is about 10^{40} erg s^{-1}. Interestingly, the rotation speed for NGC 253 and equation (2) implies that the escape temperature from NGC 253 at radius of 10 kpc will be about 2.4 million K if it has an isothermal dark matter halo extending out to a radius of $\sim$50 kpc (cf. Puche and Carignan 1991). Thus, the observed hot halo gas is close to the escape temperature, and its fate is unclear.

Observations at 0.33 GHz by Carilli et al. (1992) show that NGC 253 is also one of the few known galaxies with a radio halo. The radio and X-ray halos are similar in size, but the radio halo is more nearly spherical and does not show the limb-brightened 'X' shape seen in the ROSAT image.

The relationship of the starburst and its kpc-scale outflow to the galaxy-scale X-ray and radio halos is a matter of debate. On morphological grounds, Carilli et al. (1992) suggest that the radio halo may be powered by an outflow from the inner part of the 'normal' star-forming disk of NGC 253, rather than from the nuclear starburst. Indeed – as Fig. 1 shows – the inner disk of NGC 253 is riddled with giant H II regions, and Carilli et al. note that a radio 'spur' seems to connect to a region of bright Hα emission located several kpc from the nucleus in the disk. ROSAT data also show that the inner disk (with a radius similar to the lateral extent of the X-ray halo) is a soft X-ray source. This could represent hot gas flowing out of the disk and into the halo.

Nevertheless, the rarity of X-ray halos in normal spiral galaxies on the one hand (cf. Vogler, Pietsch and Kahabka 1996) and their near-ubiquity in galaxies with powerful central starbursts (Dahlem et al. 1996) suggests to me at least that the X-ray halo in NGC 253 is primarily powered by the starburst. I would therefore propose the following possible interpretation, which is based in part on the numerical simulations presented by Suchkov et al. (1994, 1996).

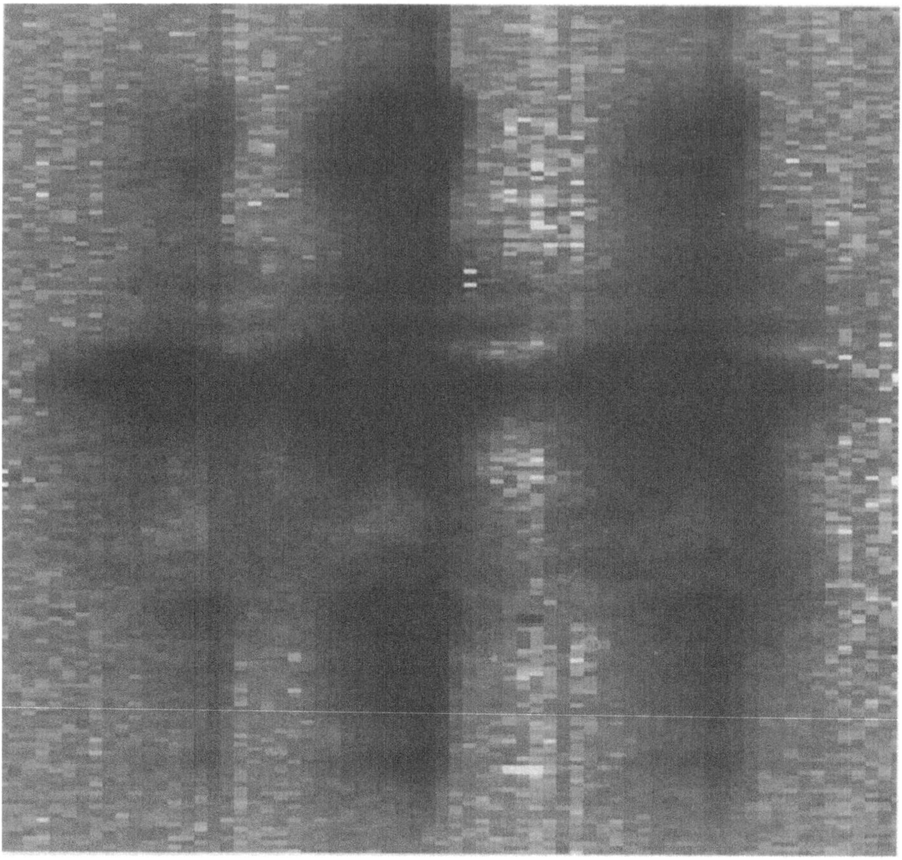

Fig. 2. A spectrum of (from left to right) the [N II]λ6548, Hα, and [N II]λ6584 emission-lines in a region 2.′5 (2.5 kpc) in total length along the minor axis of NGC 253. The spectrum is centered on the starburst, with northwest to the top. Note the complex kinematics throughout this region, and the double-peaked emission-lines with velocity separations of 300 to 600 km s^{-1} in the gas to the southeast of starburst (on the front side of the outflow)

Gas with a temperature of about 10^7 K is created in the starburst and flows out along the galaxy minor axis. The outflowing gas produces the kpc-scale X-ray source seen in Fig. 1 (cf. Bregman, Schulman and Tomisaka 1995; Suchkov et al. 1996), while swept-up, shocked-heated disk gas produces Hα emission along the walls of the outflow. As the outflow propogates into the halo, its X-ray emissivity drops dramatically due to the steep drop in its density and to adiabatic cooling. However, the outflow creates a high-pressure piston of gas that drives a shock into the pre-existing halo gas at a velocity of about 400 km s^{-1}, heating the halo gas to several million K, and producing a limb-brightened X-ray source (Suchkov et al. 1994). The fact that this halo source is ‘X’-shaped rather than ‘8’- shaped suggests that the outflow has recently ‘blown out’ of the galactic halo

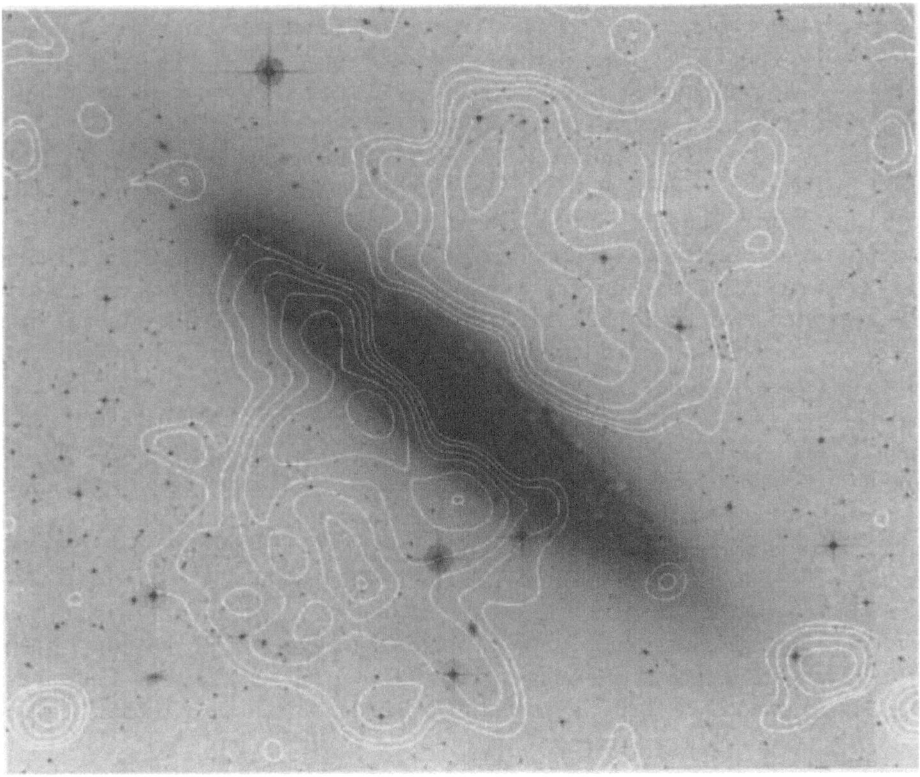

Fig. 3. An overlay of the ROSAT PSPC X-ray image of NGC 253 in the 0.25 kev band (in contour form) on an optical continuum image (greyscale). The displayed field size is 30′ (30 kpc) on a side. The ROSAT image has been processed by first digitally subtracting the point-like sources (including the nucleus) and then heavily smoothing the residual map to bring up diffuse, low-surface-brightness emission (see Dahlem et al. 1996)

(cf. MacLow and MacCray 1988). The size of the halo and the inferred shock speed would imply a dynamical age of a few$\times 10^7$ years – similar to the estimated starburst age.

4 M 82

The overall properties (size, luminosity, edge-on orientation) of the M 82 and NGC 253 starbursts are quite similar. However, they occur in very different galaxies. On the basis of its dust-riddled morphology, M 82 is classified as an Amorphous galaxy in the *Revised Shapley-Ames Catalog* and an I0 galaxy in the *Third Reference Catalog of Bright Galaxies.* It is a much smaller and less massive galaxy than NGC 253, with $M_B = -18.6$ and a rotation speed of only about 110 km s^{-1}. These properties are rather atypical of starbursts with comparable IR luminosities (Lehnert and Heckman 1996a).

M 82 is a member of the M 81 group. It is clearly interacting tidally with M 81, as is most evident in H I kinematic studies (cf. Yun, Ho and Lo 1993). M 82 apparently suffered a direct encounter with M 81 of-the-order 10^8 years ago, and its outer H I disk is being tidally disrupted as a result (Yun, Ho and Lo 1993). The latter authors argue that the gas in the starburst region has not been accreted from M 81, but is rather gas from M 82's pre-encounter H I disk that was driven inward by a bar that was in turn excited by the encounter with M 81.

However, the existence of a bar in M 82 is not as well-established observationally as in the case of NGC 253. Telesco et al. (1991) used near-IR images to argue on morphological grounds for the presence of a stellar bar with a length of about 1 kpc (several times larger than the starburst). Yun, Ho and Lo (1993) note several characteristics of the kinematics of the H I disk in M 82 that are consistent with (but do not require) the presence of a bar. Larkin et al. (1994) use near-IR imaging to identify a kpc- scale dust lane which apparently lies in front of the stellar population on one side of the nucleus and behind the stars on the other. They argue that this is suggestive of a stellar bar with leading dust lanes. Achtermann and Lacy (1995) have investigated the kinematics of the ionized gas at high spectral and spatial resolution using the 12.8μm line of [Ne II]. They find indirect morphological evidence for the presence of a bar (a nuclear ring and an emission-line counterpart to Larkin et al.'s dust lanes). However, they conclude that direct kinematic identification of a bar remains elusive (because the likely orientation of the bar is nearly in the sky-plane).

The presence of an outflow in M 82's famous Hα filament system was also debated for many years, and the long and complex history of this was nicely summarized by Bland and Tully (1988). While scattering of starburst light by dust in the halo of M 82 clearly plays an important role in what we see (cf. Scarrott, Eaton and Axon 1991), there is now no doubt that an outflow is also occurring. The kinematic signature of expanding superbubbles or a hollow biconic outflow was first noted by Axon and Taylor (1978). Bland and Tully (1988) later published optical Fabry-Perot data that clearly showed the presence of a large-scale bipolar outflow (superwind) along the optical minor axis of the galaxy (see also Heckman, Armus and Miley 1990). The outflow's kinematic signature is graphically illustrated in Fig. 4, which also shows how similar it is is to the one in NGC 253 (compare Figs. 2 and 4).

The outflow interpretation was also bolstered by X-ray observations by Watson, Stanger and Griffiths (1994), Fabbiano (1988), and Bregman, Schulman and Tomisaka (1995) showing diffuse X-ray emission co-extensive with the Hα emission-line filaments. The X-ray morphology of the M 82 outflow is rather

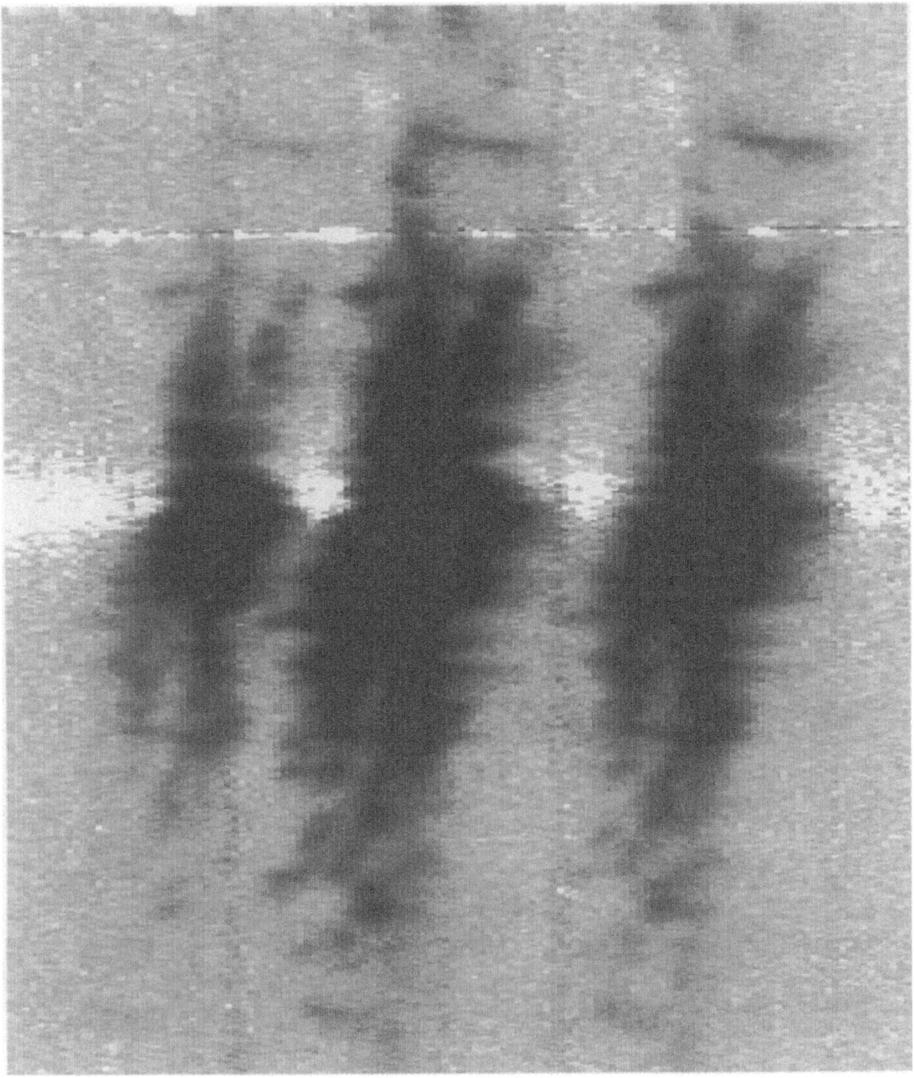

Fig. 4. A spectrum of (from left to right) the [N II]λ6548, Hα, and [N II]λ6584 emission-lines in a region 5.′1 (5.1 kpc) in total length along the minor axis of M 82. The spectrum is centered on the starburst, with northwest to the top. Note the complex kinematics and double-peaked emission-lines with velocity separations of 300 to 600 km s^{-1}. The gas to the southeast is on the front side of the outflow and is blueshifted relative to v_{sys}. Compare to Fig. 2 for NGC 253

different from that in NGC 253. In M 82, the entire 10-kpc-scale X-ray nebula resembles the small nuclear outflow in NGC 253. This can be seen by comparing Fig. 5 to Fig. 1.

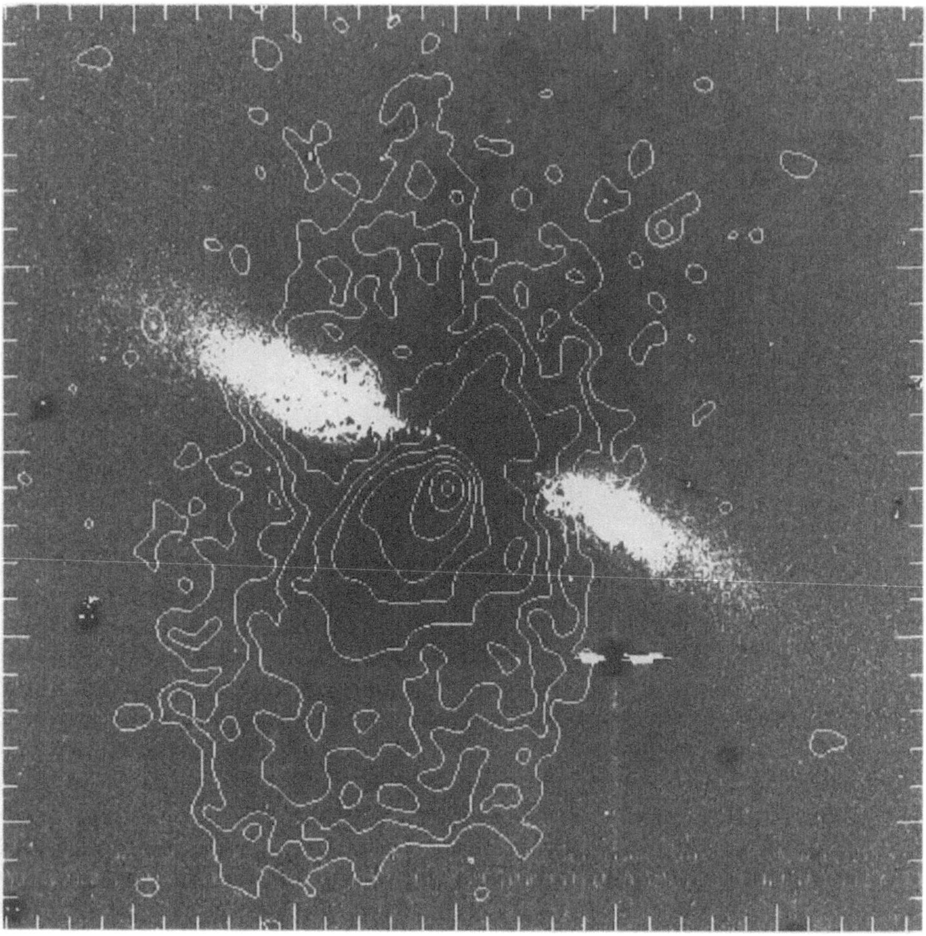

Fig. 5. An overlay of the archival ROSAT HRI X-ray image (contours) and an optical image (greyscale) of the Hα (dark) and red stellar continuum (light) of the central 5′ (5 kpc) of M 82. The optical image was kindly provided by Dr. Eric Smith

Note too the laterally limb-brightened morphology of the Hα filament system (particularly to the North of the galaxy). This is suggestive of a 'blow-out' of hot gas from M 82's disk into its halo, with the Hα emission produced by a sheath of shocked and entrained disk gas with a ruptured top (see Sect. 2 above). In support of this interpretation, the X-ray emission can be detected out to about a factor of two larger radius than the Hα emission ($\sim$5 kpc vs. $\sim$2.5 kpc).

Moran and Lehnert (1996) have analysed ROSAT PSPC and ASCA data that illustrate the spectral complexity of M 82 and its halo. They conclude that the hot gas in M 82 has a temperature of about 7×10^6 K in and near the starburst, declining to 4×10^6 K in the halo. The corresponding unabsorbed X-ray luminosities of these two spectral components are 10^{41} erg s^{-1} and 5×10^{39} erg s^{-1}. These can be compared to the estimated mechanical luminosity of the starburst ($\sim 10^{42}$ erg s^{-1}). The rotation speed for M 82 and Eq. (2) implies that the escape temperature from M 82 at a radius of 5 kpc will be about 1 million K, even if this small galaxy has an isothermal dark matter halo extending out to a radius of $\sim$50 kpc. Since the observed halo gas is about four times times hotter than this, it may well be escaping from M 82.

Like NGC 253, M 82 also has a radio-synchrotron halo extending out to a radius of at least 8 kpc (Seaquist and Odegard 1991). As in the case of NGC 253, this radio halo is more nearly spherical than the X-ray nebula. The spectrum of the radio emission in M 82 steepens markedly with increasing radius, which Seaquist and Odegard attribute to the more rapid cooling of the most energetic electrons by inverse Compton scattering of the starburst IR photons. They then derive an outflow speed for the relativistic plasma of about 2 000 km s^{-1} ($\gg v_{\rm esc}$). They also estimate that relativistic particles comprise only a few percent of the energy content of the M 82 outflow. They note that this ratio is similar to that observed in supernova remnants, as expected in the superwind model.

To summarize, the outflow in M 82 is similar to that in NGC 253 but differs in some important respects. In Sect. 3, I argued that the outer (25-kpc-scale) limb-brightened X-ray halo in NGC 253 is produced as the superwind drives a shock into a pre-existing gaseous halo surrounding NGC 253 (cf. Suchkov et al. 1994). Perhaps a dwarf galaxy like M 82 lacks such a halo (as Steidel 1993 suggests and may be generally true based on his identifications of galaxies responsible for Mg II absorption in quasar spectra). On the other hand, the X-ray outflow in M 82 is considerably brighter than the inner (kpc-scale) X-ray outflow in NGC 253. In the context of models by Bregman, Schulman and Tomisaka (1995) and by Suchkov et al. (1996), this may imply that there is a larger amount of mass-loading of the outflow in M 82 than in NGC 253. That is, Suchkov et al. (1996) show that for a given starburst energy injection rate, the X-ray luminosity of the outflow is proportional to roughly the cube of the mass outflow rate. Thus a factor of two increase in the mass outflow rate leads to nearly an order-of-magnitude increase in $L_{\rm X}$.

5 M 31

Let me begin this section by admitting that M 31 is a confusing case: it may or may not contain a bar, it may or may not have an outflow, and certainly does not contain a nuclear starburst. Why then am I discussing it here? First, I am intruiged by the mysterious goings-on in the circumnuclear region of M 31 (and their possible connections to bars, inflows, outflows, and the LINER phenomenon). Second, M 31's unmatched proximity offers us our best opportunity to study the circumnuclear region of a typical L_* spiral galaxy.

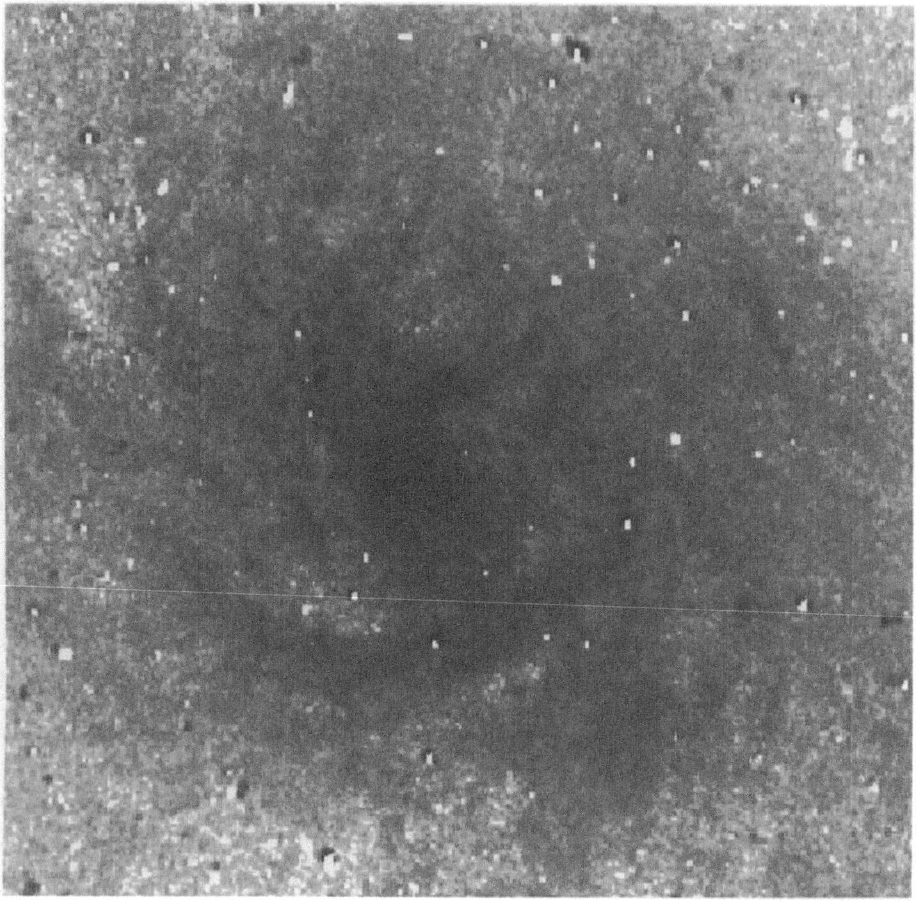

Fig. 6. An Hα+[N II] image of the innermost 9′ (1.9 kpc) of M 31, kindly provided by Dr. Nick Devereux

M 31 has long been known to have a kpc-scale circumnuclear region of low-surface-brightness optical emission-line gas (cf. Rubin and Ford 1971). This gas has been studied more recently in some detail by Jacoby, Ford and Ciardullo (1985), Ciardullo et al. (1988 – hereafter C88), and Devereux et al. (1994). The salient properties of the M 31 circumnuclear nebula shown in Fig. 6 are as follows:

1. It has an Hα luminosity of about 10^{39} erg s^{-1}, and a spatially-integrated [N II]λ6584 ([O II]λ3727) equivalent width of about 1Å (2.5Å). These parameters are typical of the weak emission-lines seen in the nuclei/inner bulges of early type galaxies (cf. Heckman 1980).

2. It has an overall size of about 2 kpc and a half-light radius of about 300 pc. It has a complex filamentary structure with a very different overall geometry than that of M 31's large-scale disk (the emission-line region is nearly round whereas M 31's disk is of course seen at high inclination).

3. The velocity field shows a globally-organized, large-scale shear with an amplitude of several hundred km s^{-1}, however the velocity field is considerably more complex than that of gas in simple circular orbits. The line-of-sight velocity dispersion of the gas (30 to 40 km s^{-1}) is much smaller than this shear, implying that the nebula must be geometrically thin along the line-of-sight.

Based on results 2 and 3 above, C88 concluded that the nebula is a warped, rotating disk that is highly inclined to M 31's galactic plane. They suggest that the complex kinematics results from a combination of rotation and outflow, with the latter induced by a galactic wind. The existence of a wind emanating from the otherwise totally quiescent nucleus of M 31 would be very important, since it would imply that galactic winds may in fact be ubiquitous. However, the nature of this wind's power source would be something of a mystery. Could it be the supermassive 'slumbering monster' in the nucleus (cf. Kormendy and Richstone 1995 and references therein)?

In contrast, Stark and Binney (1994) have argued that it is not necessary to invoke a galactic wind in M 31. They maintain that both the morphology and kinematics can be understood if we are observing gas moving in closed orbits *in* M 31's galactic plane under the influence of a weak barlike potential (with the bar being viewed nearly end-on). Stark and Binney contend that a bar (or a triaxial bulge) must be present in M 31 to account for the twisting of the isophotes in the bulge with respect to those of the outer disk (Lindblad 1956; Stark 1977). Others (McElroy 1983; C88) have not been persuaded that this is the case. It is clear that a kinematic test is required, which Stark and Binney say must involve a more complete map of the velocity field in the gas than has been constructed to date.

Motivated in part by these arguments, my JHU colleagues Gerhardt Meurer and Jonathan Gelbord and I are undertaking a detailed spectroscopic investigation of the kinematics and physical state of the ionized gas near the center of M 31. Our analysis builds on the extensive pioneering work by Rubin and Ford (1971) and C88. While our analysis effort is still in progress, we do have one interesting new result: the optical emission-line ratios in the nebula are typical of the class of Low Ionization Nuclear Emission-line Regions ('LINERs' – cf. Eracleous and Koratkar 1996). This should not be truly surprising: the centers of most early-type disk galaxies contain LINERs and M 31 is the nearest such galaxy. M 31's identity as a LINER was also foreshadowed by C88, who showed that [N II]λ6584 is brighter than Hα and that [S II]λ6724 is strong.

Since I cut my scientific teeth on LINERs, forgive me if I digress for a bit to consider the possible sources of ionization for the M 31 LINER. At present the upper limit on any nuclear UV point source rules out photoionization by an AGN, although a time-dependent model of the type discussed by Eracleous, Livio and Binette (1995) could be invoked. Likewise, there is no suitable population of young, high-mass stars and associated supernovae to photoionize or shock-heat the gas respectively (cf. C88). Stark and Binney suggest that the gas may be shock-heated as it makes a transition between the inner x_1 and outer x_2 orbits in their hypothesized barred potential. Jacoby, Ford and Ciardullo (1985) have concluded that the central stars of planetary nebulae in the bulge of M 31 (post-asymtotic giant branch stars, henceforth 'PAGB') may not be sufficient to photoionize the nebula.

However, I think this possibility of photoionization by PAGB stars is worth revisiting on account of two recent developments. First, Binette et al. (1994) have computed photoionization models utilizing such stars and have shown that a LINER-type spectrum can be readily produced. Second, we now know considerably more about the nature of the UV light in the central part of M 31.

I have compiled measurements of the vacuum-UV (1500Å) flux from M 31's bulge as measured through a variety of apertures ranging in effective radius from about 25 pc (IUE) to about 1 kpc (OAO). I then find that the growth curves (flux interior to r vs. r) for the Hα+[N II] emission and the UV continuum have the same shape. The fact that the UV light and the emission-line gas track one another in this overall sense is suggestive of an energetic link between them.

A Hopkins Ultraviolet Telescope spectrum of the central-most 30 by 370 pc of M 31's bulge was analysed by Ferguson and Davidsen (1993), who concluded that about 65% of the flux at 1500Å was produced by PAGB stars. Their stellar atmosphere model for the PAGB stars then provides a scaling factor between the flux at 1500Å and the flux of stellar Lyman continuum photons. If I make the huge assumption that the fraction of the UV light that is due to PAGB stars is the same throughout the inner bulge of M 31 as it is in the small region they measured, I then find that the predicted Lyman continuum flux is about twice as large as is required to explain the observed Hα emission. It is likely that a significant fraction of the UV light is also used to heat the dust grains responsible for the far-IR source that is roughly co-spatial with the emission-line nebula (Devereux et al. 1994).

Binette et al. (1994) have shown that photionization by PAGB stars leads to a LINER spectrum for an ionization parameter $\log U = -4$. They also use stellar evolutionary models for an old bulge population to self-consistently determine both the Lyman continuum (and hence Hα) luminosity and the luminosity of the red stellar continuum. This leads to a prediction that the Hα emission-line equivalent width should be about 1 Å (as is observed in M 31).

Thus, it would seem that there is no real problem with a very 'subdued' interpretation of the circumnuclear emission-line nebula in M 31: it represents diffuse gas in the disk of M 31 orbiting in a weakly-barred potential and irradiated by an ensemble of the central stars of planetary nebulae. It would then have a lot to do with bars, but nothing to do with galactic winds.

While this explanation may well be correct, it does not provide an obvious, natural explanation for the other phenomena occuring in the same region. First, there is a diffuse nonthermal radio source with a size and overall morphology that is very similar to the Hα nebula and far-IR source (Walterbos and Grave 1985). Curiously, this radio source and the co-spatial far-IR source fall *exactly* along the magical radio-FIR correlation obeyed by actively-star-forming and starburst galaxies (cf. Condon 1992)! Given the absence of a significant population of massive stars in this region, this is very hard to understand in the context of current models for the radio-FIR correlation. In the same region there is also a source of diffuse X-ray emission (presumably hot gas) with a luminosity of about 10^{39} erg s^{-1} (Primini, Forman and Jones 1993). How does this relate to the emission-line nebula, the far-IR source, the radio source, and the possible bar??

No matter what the final explanation, it seems clear that M 31 is a wonderful laboratory to study the dynamics and physical state of the interstellar medium in the bulges of spiral galaxies and the low-power end of the LINER phenomenon.

6 Disk Dynamics, Bars and the Starburst Phenomenon

Matt Lehnert and I (Lehnert and Heckman 1995, 1996a,b) have recently completed an optical spectroscopic and imaging survey of the ionized gas in a far-IR flux-limited sample of about 50 spiral galaxies whose disks were viewed from within 30° of edge-on. These galaxies were also selected to have a warm far-IR color temperature, to insure that the bulk of the far-IR emission is powered by a young stellar population. Most of the galaxies have IR luminosities in the range 10^{10} to $10^{11.5}$ $L_\odot$ and L_{IR}/L_{OPT} = 1 to 10 (i.e. they are moderately powerful starbursts). Our data consist of R and narrow-band Hα images and long-slit spectra along the major and minor axis.

In Lehnert and Heckman (1996b) we have discussed the implications of these data for the 'superwind' paradigm and concluded that such outflows are ubiquitous amongst the more 'IR active' half of the sample. In Lehnert and Heckman (1996a) we have used the rotation curves to study empirically the relationship between the disk kinematics and the size and strength of the starburst. I would like to highlight some results from this latter paper that I think are of particular relevance to this meeting:

1. The rotation curves of the starburst host galaxies generically have an inner solid body portion changing at the 'turnover radius' (r_{to}) to an outer flat portion. We find that r_{to} agrees very well with the starburst size as estimated by the Hα half-light radius. Thus, starbursts occur within the inner region of solid body rotation (Fig. 7). One possibility is that the competition between tidal shear and gravitational instability favors cloud growth in the region of solid body rotation (Kenney, Carlstrom and Young 1993). Models of bar-driven inflow (cf. Piner, Stone and Teuben 1995) also predict that rings of gas will accumulate near r_{to} (between two Inner Lindblad Resonances, if present).

2. The rather normal, symmetric rotation curves exhibited by most of our galaxies imply that fueling a circumnuclear starburst does *not* require severe dynamical disruption of the galactic disk, at least for the moderately luminous

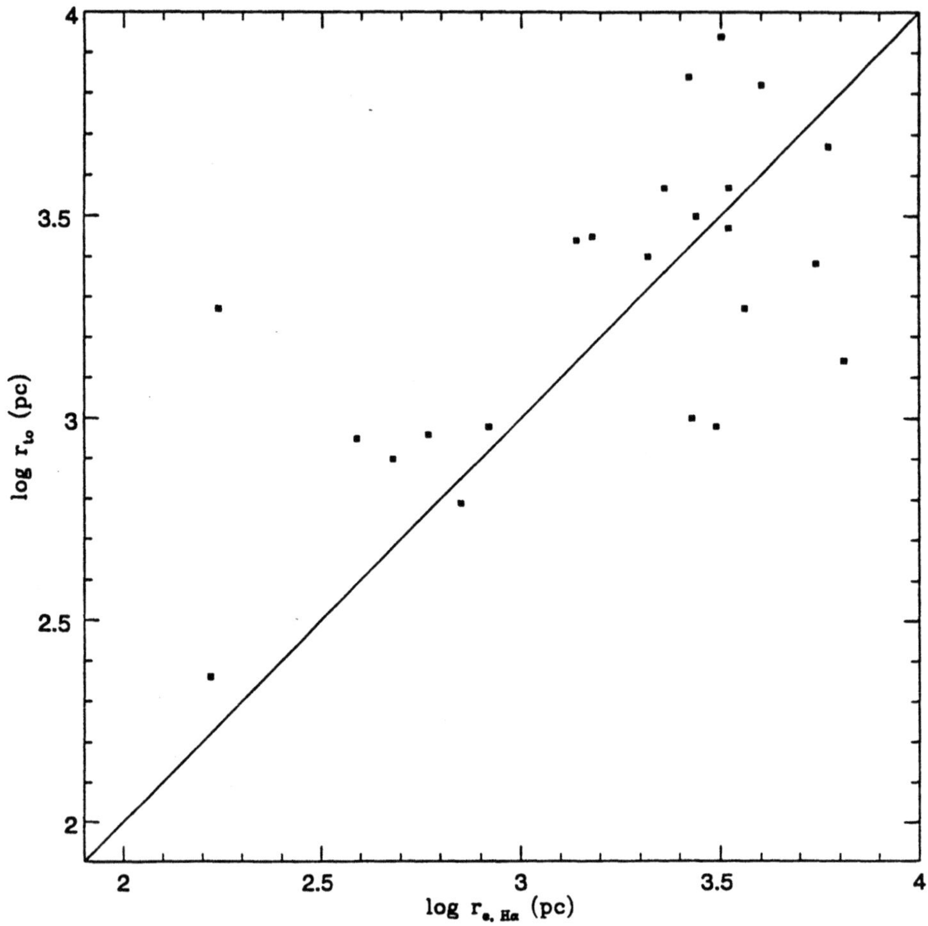

Fig. 7. The half-light radius of the Hα emitting region (the starburst) plotted versus the radius at which the galaxy rotation curve turns over from its inner solid-body to flat outer portion. The sample consists of far-IR-selected disk galaxies seen at moderate to high inclination. See Lehnert and Heckman (1996a) for details

starbursts in our sample (see above). This is a markedly different situation than for the rare but spectacular 'ultraluminous' IR galaxies in which major mergers are probably occuring (cf. Scoville and Soifer 1991; Mihos and Hernquist 1994a). Thus, the mechanism by which *typical* starbursts are triggered is not yet clear. Perhaps the capture of dwarf galaxies or mild grazing encounters with neighbors are responsible (cf. Mihos, Richstone and Bothun 1992; Mihos and Hernquist 1994b; Noguchi 1988). Secular evolution of the disk with a bar-driven inflow (possibly resulting from a mild tidal interaction) is suggested by the empirical links between bars and circum-nuclear starbursts (e.g. Devereux 1989; Ho, Filippenko and Sargent 1996).

3. Using the rotation speed as a surrogate for galaxy mass, the typical rotation speeds of 100 to 220 km s^{-1} in our sample imply that the starburst host galaxies have typical masses ranging from about 5% to 100% of the mass (M_*) of a fiducial L_* spiral galaxy. We find very little correlation between the host galaxy rotation speed and the IR ($\sim$ bolometric) luminosity of the starburst. Even quite low mass galaxies (10% M_*) can apparently 'host' quite powerful starbursts ($L \sim 10^{11}$ $L_\odot$).

4. Weedman (1983) pointed out that the emission-line widths in starburst nuclei were surprisingly small compared to expectations for gas in virial equilibrium in the bulge of a typical disk galaxy. He proposed that the starburst was therefore not in dynamical equlibrium, and would rapidly evolve into a highly compact configuration (as a precursor to the formation of a Seyfert nucleus). We confirm Weedman's result (by comparing $v_{\rm rot}$ to the width of the nuclear emission-line profiles). However, the narrowness of the nuclear lines arises in our sample because the gas in the nucleus (few arcsec) does not sample the full range of the rotation curve (which is solid body in the central region).

5. The maximum star-formation rate allowed by causality is one in which all the gas in a starburst is converted into stars in one dynamical time-scale. This leads to a prediction that the maximum starburst luminosity will scale like the cube of the starburst velocity dispersion (Heckman 1993). Comparing the far-IR luminosities and rotation speeds implies that most of the starbursts in our sample are forming stars at a rate that is typically an order-of-magnitude below this limit (even assuming a normal Initial Mass Function and a normal gas fraction). However, several extreme objects have such high luminosities for their rotation speeds that they violate this limit unless they are forming only massive stars and/or gas dominates the mass of the starburst. Both of these hypotheses have been offered on other grounds for starbursts (cf. Scoville and Soifer 1991; Rieke et al. 1993).

In the long run, these survey results can hopefully be combined with detailed investigations of a small sample of the nearest and brightest starburst galaxies to shed some light on the 'care and feeding' of starbursts.

Acknowledgments. Many people have worked with me over the past several years on various aspects of the work described above. I am particularly grateful to M. Dahlem, J. Gelbord, M. Lehnert, G. Meurer, A. Suchkov, and K. Weaver who are leading the efforts in the on-going projects discussed in this paper. I would also like to thank N. Devereux and G. Jacoby for each sending me a digital copy of his Hα+[N II] image of M 31, and likewise E. Smith in the case of M 82. H. Ferguson helped by providing me with the details of his models for the UV light in M 31. My research on galactic outflows has been supported in large part by NASA grants NAGW-4025 and NAG5-1991, and by NSF grant AST-90-20381.

References

Achtermann, J., Lacy, J. (1995): AJ **439**, 163

Binette, L., Magris, C., Stasinska, G., Bruzual, G. (1994): A&A **292**, 13

Bregman, J., Schulman, E., and Tomisaka, K. (1995): ApJ **439**, 155

Canzian, B., Mundy, L., & Scoville, N. (1988): ApJ **333**, 157

Carilli, C., Holdaway, M., Ho, P., De Pree, C. (1992): ApJ **399**, L59

Ciardullo, R., Rubin, V., Jacoby, G., Ford, H., Ford, W.K. (1988): AJ **95**, 438

Condon, J. (1992): ARA&A **30**, 575

Dahlem, M., Weaver, K., Heckman, T., Wang, J. (1996): in preparation

Devereux, N. (1989): ApJ **346**, 126

Devereux, N., Price, R., Wells, L., Duric, N. (1994): AJ **108**, 1667

Eracleous, M., Koratkar, A. (1996): The Physics of LINERs in View of New Observations, San Francisco, ASP, in press

Eracleous, M., Livio, M., Binette, L. (1995): ApJ **445**, L1

Fabbiano, G. (1988): ApJ **330**, 672

Fabbiano, G., Trinchieri, G. (1984): ApJ **286**, 491

Ferguson, H., Davidsen, A. (1993): ApJ **408**, 92

Forbes, D., Ward, M., Rotaciuc, V., Blietz, M., Genzel, R., Drapatz, S., van der Werf, P., Krabbe, A. (1993): ApJ **406**, L11

Heckman, T. (1980): A&A **87**, 152

Heckman, T., Armus, L., Miley, G. (1990): ApJS **74**, 833

Heckman, T., Lehnert, M., Armus, L. (1993): in The Evolution of Galaxies and their Environments, eds. S.M. Shull, H. Thronson, Kluwer, Dordrecht, p. 455

Ho, L., Filippenko, A., Sargent, W. (1996): in IAU Colloq. 157 Barred Galaxies, ed. R. Buta, B. Elmegreen, D. Crocker (San Francisco: ASP), in press

Jacoby, G., Ford, H., Ciardullo, R. (1985): ApJ **290**, 136

Kenney, J., Carlstrom, J., Young, J. (1993): ApJ **418**, 687

Kormendy, J., Richstone, D. (1995): ARA&A **33**, 581

Larkin, J., Graham, J., Matthews, K., Soifer, B.T., Beckwith, S., Herbst, T., Quillen, A. (1994): ApJ **420**, 159

Lehnert, M., Heckman, T. (1995): ApJS **97**, 89

Lehnert, M., Heckman, T. (1996a): ApJ submitted

Lehnert, M., Heckman, T. (1996b): ApJ in press

Leitherer, C. Heckman, T. (1995): ApJS **96**, 9

Lindblad, B. (1956): Stockholm Obs. Ann. **19**, No. 2

MacLow, M., McCray, R. (1988): ApJ **324**, 776

McCarthy, P., Heckman, T., van Breugel, W. (1987): AJ **93**, 264

McElroy, D. (1983): ApJ **270**, 485

Mihos, J.C., Hernquist, L. (1994a): ApJ **425**, L13

Mihos, J.C., Hernquist, L. (1994b): ApJ **431**, L9

Mihos, J.C., Richstone, D., Bothun, G. (1992): ApJ **400**, 153

Noguchi, M. (1988): A&A **203**, 259

Pence, W. (1980): ApJ **239**, 54

Pence, W. (1981): ApJ **247**, 473

Pietsch, W. (1993): in Panchromatic View of Galaxies - Their Evolutionary Puzzle, eds. G. Hensler et al., Editions Frontieres, Gif-Sur-Yvette, p. 137

Pina, R., Jones, B., Puetter, R., Stein, W. (1992): ApJ **401**, L75

Piner, B., Stone, J., Teuben, P. (1995): ApJ **449**, 508

Primini, F., Forman, W., Jones, C. (1993): ApJ **410**, 615
Puche, D., Carignan, C. (1991): ApJ **378**, 487
Puche, D., Carignan, C., van Gorkom, J. (1991): AJ **101**, 456
Rieke, G., Loken, L., Rieke, M., Tamblyn, P. (1993): ApJ **412**, 99
Rubin, V., Ford, W.K. (1971): ApJ **170**, 25
Scoville, N., Soifer, B.T., (1991): in Massive Stars in Starburst Galaxies, eds. C. Leitherer, N. Walborn, T. Heckman, C. Norman, Cambridge University Press, p. 233
Scoville, N., Soifer, B.T., Neugebauer, G., Young, J., Matthews, K., Jerka, J. (1985): ApJ **289**, 129
Seaquist, E., Odegard, N. (1991): ApJ **369**, 320
Soifer, B.T., Houck, J., Neugebauer, G. (1987): ARA&A **25**, 187
Stark, A. (1977): ApJ **213**, 368
Stark, A., Binney, J. (1994): ApJ **426**, L31
Suchkov, A., Balsara, D., Heckman, T., Leitherer, C. (1994): ApJ **430**, 511
Suchkov, A. Berman, V., Heckman, T., Balsara, D. (1996): ApJ submitted
Telesco, C., Campins, H., Joy, M., Dietz, K., Decher, R. (1991): ApJ **369**, 135
Telesco, C., Dressel, L., Wolstencroft, R. (1993): ApJ **414**, 120
Tenario-Tagle, G., Bodenheimer, P. (1988): ARA&A **26**, 145
Turner, B. (1985): ApJ **299**, 312
Ulrich, M.-H. (1978): ApJ **219**, 424
Vogler, A., Pietsch, W., Kahabka, P. (1996): A&A **305**, 74
Walterbos, R., Grave, R. (1985): A&A **150**, L1
Wang, D., Walterbos, R., Steakley, M., Norman, C., Braun, R. (1995): ApJ **439**, 176
Weedman, D. (1983): ApJ **266**, 479
Yun, M., Ho, P., Lo, K.Y. (1993): ApJ **411**, L17

The Nuclear High Excitation Outflow Cone in NGC 1365

Per Olof Lindblad, Maja Hjelm, Steven Jörsäter and Helmuth Kristen

Stockholm Observatory, S-13336 Saltsjöbaden, Sweden

Abstract. The morphology and kinematics of the high excitation outflow cone in the nuclear region of the Seyfert 1.5 galaxy NGC 1365 is investigated. An empirical model based on ground-based [O III] emission line data consists of a somewhat hollow double cone with its apex at the Seyfert nucleus. The cone axis is well aligned in space with the normal to the symmetry plane of the galaxy and the position angle of its projection on the sky coincides closely with that of a jet-like radio feature. The opening angle of the cone is 100° and the orientation such that the line of sight to the Seyfert 1.5 nucleus falls inside the cone. The outflow velocities within the cone are accelerated and fall off towards the edge.

An HST FOC narrow band exposure of the nuclear region in [O III] emission shows the cone to be resolved into a large number of discrete clouds. A number of star forming regions surround the nucleus, of which the brightest seem to have absolute magnitudes in B of about -17^{m}.

1 Introduction

Emission line regions around active galactic nuclei (AGN) often are eccentric and lack circular symmetry around the nucleus. Regions of high excitation, characterized by strong [O III] line radiation, may tend to axial symmetry, sometimes in the shape of a cone or double cone with its apex at the nucleus. Wilson and Tsvetanov (1994) list 11 cases of such so called ionization cones. These ionization cones generally seem to be aligned with linear radio sources or radio jets, when present, while their orientation with respect to the symmetry plane of the galaxy may be dependent of the galaxy type (Wilson and Tsvetanov 1994). It is generally assumed that ionizing radiation from the active nucleus is confined, possibly by an obscuring torus, to this cone.

When the ionization cone has an opening angle and orientation such that it intersects with the galactic plane, it may create an extended narrow line region (ENLR) in the plane that need not be aligned with the axis of the actual ionization cone (Pedlar et al. 1993). The motions within this ENLR will agree with the motions of galactic rotation.

Where the ionization cone does not interfere with matter in the plane, we may find that the gas of high excitation within the cone shows a peculiar kinematic behaviour and in some cases clear indications of outflow from the nucleus. It is important to observe the detailed kinematics of the ionization cone, as well as the morphology and the relation of the cone structure to the radio structure, in order to understand these phenomena.

An illuminating case is the galaxy NGC 1365, where the difference in distribution between high and low excitation gas is demonstrated in Fig. 2. In the present paper we will discuss this case in more detail.

2 NGC 1365

NGC 1365 is a supergiant galaxy in the Fornax cluster. An optical image of the galaxy is shown as Fig. 1a. It contains a strong bar that should cause inflow of interstellar matter to the nuclear region (P.A.B. Lindblad et al. these proceedings). In Fig. 1b we show a CCD image of the nuclear region of the galaxy obtained with the ESO NTT at very good seeing. In this image we can see a large number of H II regions resolved in the surrounding of the nucleus.

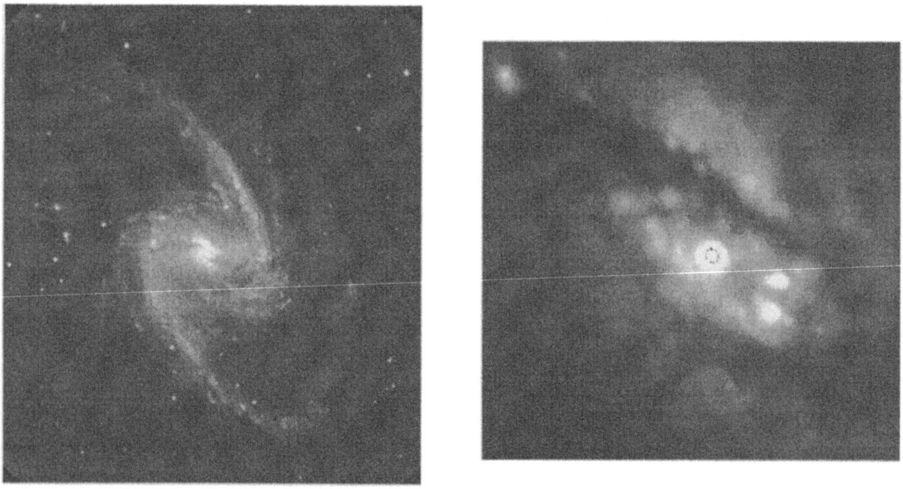

Fig. 1. a NGC 1365 photographed at the prime focus of the ESO 3.6 m telescope on a IIIa-J plate. The width of the frame is 11′. **b** The nuclear region of NGC 1365 obtained with CCD in the R band at the ESO NTT (courtesy M. Tarenghi). The width of the frame is 36″

Note in particular the two strong star forming regions to the right and slightly below the nucleus, which we will return to later. The very nucleus itself is a Seyfert 1.5 which is strong in the red and infrared and saturated in this image. Its spectrum shows both broad and narrow Hα components, the broad component having a velocity width of 1800 km s^{-1}. We also see the strong dust lanes coming in from the front side of the bar, and the Seyfert nucleus is found to be absorbed by several magnitudes in V.

As reported by Sandqvist (these proceedings) the nucleus is surrounded by a number of resolved and unresolved ($0\farcs10 \times 0\farcs25$ FWHM) continuum radio sources as well as a molecular ring. In particular, there is a 5″ long jet-like radio source extending from the nucleus in position angle 125°.

Observing this nuclear region with narrow band filters, the difference in distribution of Hα and [O III] becomes obvious as seen in Fig. 2. Hα is dominant in the star forming regions while [O III] shows a one sided distribution. Considering that NW is the near side, we may indeed get the impression of a cone-like feature emerging from the plane and being projected against the far side of the disk, while simultaneously we have a system of star forming regions, presumably confined to a circumnuclear disk in the symmetry plane of the galaxy.

3 Kinematics

We placed a number of spectrograph slits across the nuclear region. The wavelength region covered Hβ and [O III] λ5007. Velocity fields of the nuclear region of this galaxy have previously been derived by Jörsäter (1984) and Teuben et al. (1986) for the low excitation lines, by Edmunds et al. (1988) for the high excitation lines and by Jörsäter and Lindblad (1989) for the two sets of lines.

As first shown by Phillips et al. (1983), the [O III] lines turn out to be double in some areas. To see the character of these velocities let us look at the velocities along a selected slit, presented in Fig. 3, passing over the cone. We see here the [O III] velocities as a function of position along the slit, and we see the lines to be split at the two ends of the slit. The Hβ velocities share the velocities of that [O III] component which runs from positive velocities over a sharp drop to negative. These velocities make up what we call the 'disk' velocity field. The other [O III] component runs from negative to positive velocities, also via very sharp gradients, and makes up what we will call the 'cone' velocity field. In the middle these two fields cannot be separated.

We combine all these measurements along the slits to give the velocity fields in Fig. 4. Left we see the normal velocity field as given by measurements mainly of Hβ, and right the cone velocity field given by [O III].

The disk velocity field has been interpreted by model simulations, as described by P.A.B. Lindblad et al. in these proceedings, and the strong gradients and twists seen in Fig. 4a seem to be due to the twisting of closed gaseous orbits in connection with an inner resonance, as was also concluded by Teuben et al.

The cone field (Fig. 4b) shows large negative velocities, while the disk field has large positive velocities and vice versa. There also seems to be a discontinuity in the field that makes it difficult to close the contours. If we think in terms of a cone emerging from the plane, the negative velocities would mean outflow motion. Positive velocities could refer to a counter cone seen through the plane of the galaxy.

4 The Cone Model

As a further development of suggestions by Edmunds et al. and by Jörsäter and Lindblad we have designed a model with the following characteristics, involving 13 independent parameters (Hjelm and Lindblad 1996):

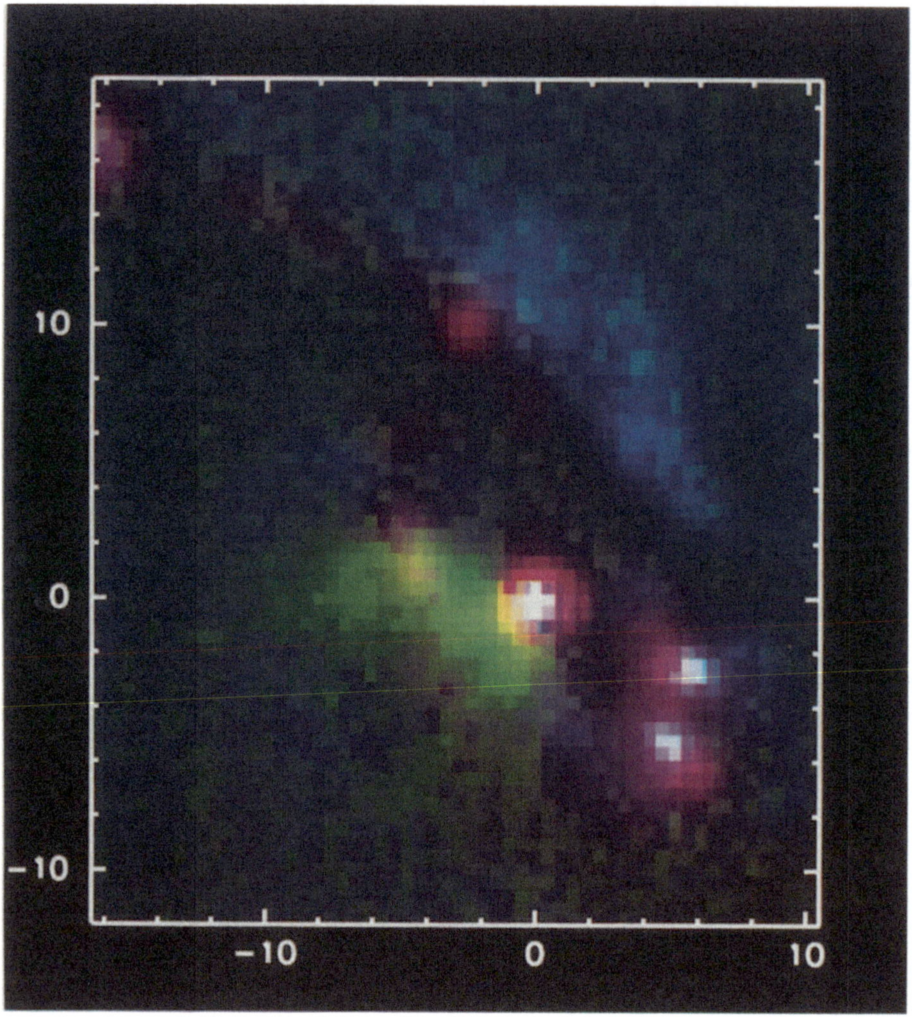

Fig. 2. Combined colour picture of the nuclear region of NGC 1365, where the continuum is blue, Hα is red and [O III] λ5007 is yellow. Danish/ESO 1.5 m telescope. The scale indicates offsets from the nucleus in arcseconds

The model is an outflow cone with its apex at the centre of the galaxy and the counter cone, partly absorbed, behind the plane of the galaxy. The cone is somewhat hollow and the outflow velocity, as well as the intensity of line emission, can vary with distance from the centre and distance from the cone axis.

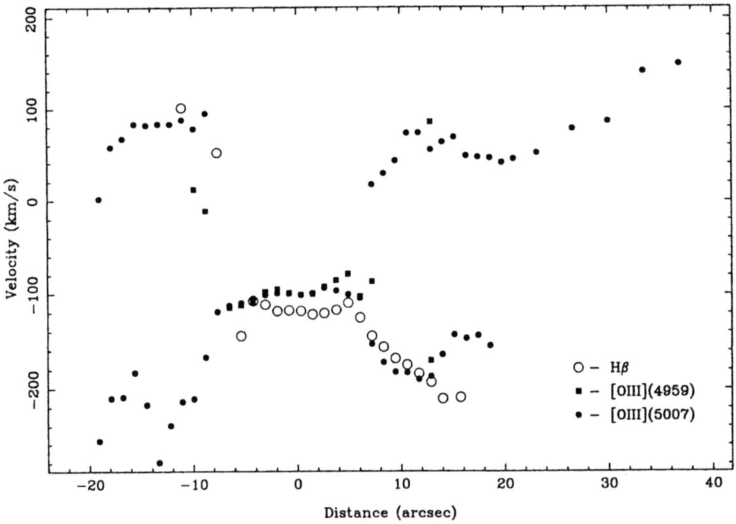

Fig. 3. Radial velocities along a slit passing over the cone 5″ East of the nucleus with position angle 332° (cf. Fig. 2). The zero point of the horizontal scale is the H II region 4″ E and 2″ N of the nucleus, and the distance scale is increasing in the direction of PA = 332°. The zero point of velocity is +1630 km s^{-1}

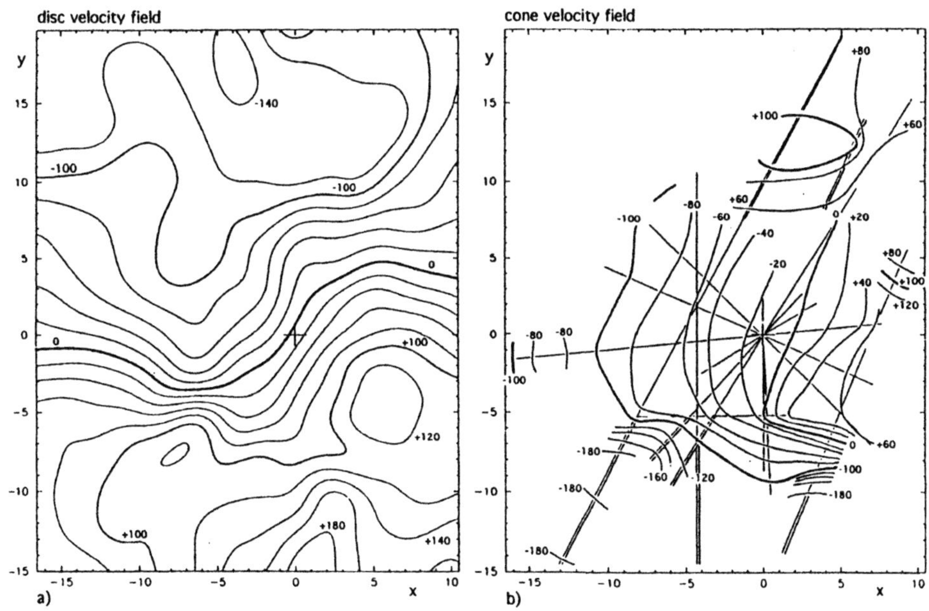

Fig. 4. Nuclear velocity fields. **a** The disk velocity field as given by the Hβ emission lines. **b** The cone velocity field as given by [O III] emission. The spectral slit positions are shown and the double lines indicate where double line profiles are measured. The units on the axes are arcseconds offset from the centre

The [O III] lines are assumed to contain emission from:

– The near half of the double cone seen in front of the plane of the galaxy.
– The counter cone, symmetric with the cone but absorbed in the plane. The absorption is derived from an extinction map, based on three colours.
– A component in the plane with the same distribution and velocities as the hydrogen lines.

The best fitting model to the velocities measured and to the morphology is shown in Fig. 5. It has the following characteristics:

– The cone opening angle is 100°.
– The cone axis deviates 5° from the normal to the symmetry plane of the galaxy as derived from H I 21-cm line observations (Jörsäter and van Moorsel 1995).
– The position angle of the cone axis, as projected on the sky, deviates 4° from the minor axis.
– The position angle of the cone axis lies within 1° from the position angle of the radio jet.
– The radial outflow velocity is accelerated and increases with the square of the distance from the centre. The velocity is largest at the axis of the cone and decreases to zero at the edge of the cone.

With these characteristics the cone fits well with the extreme end of the relation of Wilson and Tsvetanov. The cone is wide enough that the line of sight falls inside the cone which permits us to see the Seyfert nucleus within the cone and thus also the broad line region.

The close agreement between the position angle of the cone axis and the radio jet indicates a physical relationship, and that the jet is actually oriented along the cone axis. With knowledge of the orientation of the jet relative to the line of sight, of the surface brightness of the jet relative to the noise, and assuming that the disk is optically thin to radio radiation, we can deduce a value for the jet velocity $> 0.7c$ in order for a counter-jet not to be seen. However, it cannot be excluded that the disk is optically thick in the radio domain.

Meaningful line ratios in the presence of such a cone can only be made where the contributions from the cone and the disk can be separated, i.e. where the spectral lines are double. In the light from the cone Hβ is too weak to be accurately measured. [N II]/Hα ratios are possible to estimate in the counter cone, where lines are redshifted with respect to the disk line components. In the diagnostic diagram this line ratio lies in the region occupied by AGNs.

5 HST Observations

We have obtained pre-refurbishment observations with the Faint Object Camera (FOC) on the Hubble Space Telescope (HST) of the nuclear region of NGC 1365

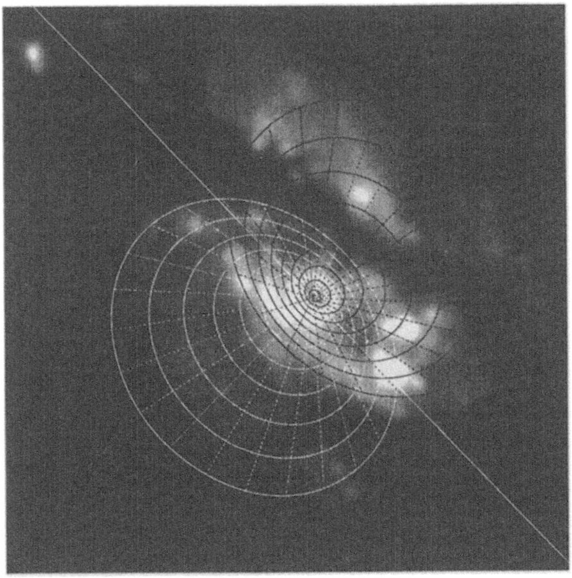

Fig. 5. Model ionization cone. The NW (upper right) side of the galaxy is the near one. The cone axis falls closely along the normal to the plane of the galaxy and the white cone component is the near one. The counter cone (black) is seen through the galactic plane. The straight line marks the line of nodes

(Kristen et al. 1996). The FOC was used in the F/96 mode which gives an $11'' \times 11''$ field which was positioned to include the nucleus and the two bright nuclear hot spots mentioned above. Two narrow band filters were used to extract the [O III] λ5007 line and the continuum on the blue side of the line.

The continuum-subtracted [O III] image is reproduced in Fig. 6. We see only the strongest [O III] emission in the bottom of the cone about $2''$ SE of the nucleus (Fig. 2). The image has been deconvolved with the Richardson-Lucy algorithm.

The cone is resolved into a large number of discrete clouds, which was also the case for the [O III] cones of NGC 1068 (Macchetto et al. 1994) and NGC 4151 (Boksenberg et al. 1995). We also see an intriguing streak directed (almost) towards the nucleus. This is not seen in the point spread function and thus should be real. The nuclear [O III] source has a FWHM of 2.8 pixels, corresponding to $0\overset{''}{.}06$. This suggests an unresolved central source with an upper limit for its radius of 3 pc.

The blue continuum, obtained with the HST FOC and deconvolved with the Richardson-Lucy algorithm is shown in Fig. 7. We see a large number of bright spots tending to fall on an elongated ring around the nucleus. This phenomenon may be related to the molecular ring described by Sandqvist in these proceedings.

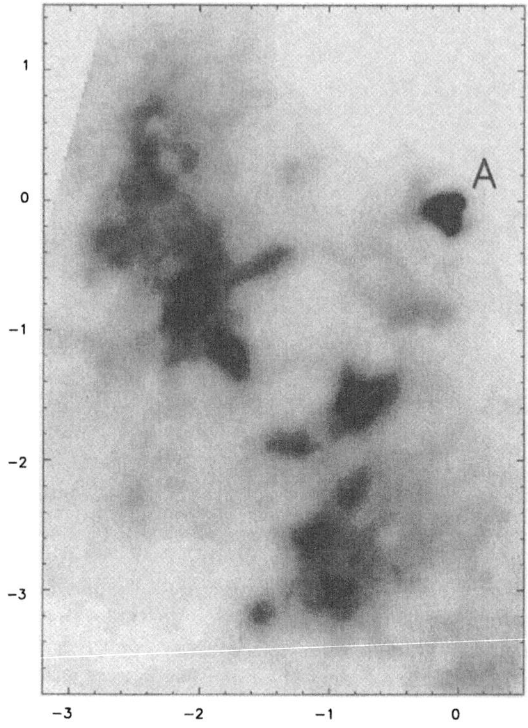

Fig. 6. Inner part of the ionization cone obtained with the HST FOC in the light of [O III] λ5007. The frame is continuum-subtracted and deconvolved with the Richardson-Lucy algorithm. The units on the axes are arcseconds offset from the centre. The unresolved nucleus is indicated (A)

The two brightest star forming regions, designated L2a and L3a, have B magnitudes of 17^{m}. With reddening estimated from the Balmer lines we get absolute B magnitudes of -17^{m}. Their radii are < 5 pc. These star forming regions might be of a similar nature as the super star clusters found by Barth et al. (1995) in nuclear rings of barred galaxies with M_{V} of -14^{m} to -15^{m}, and with radii < 4 pc. Barth et al. suggest these luminous regions to be young globular clusters. The relation of these super star clusters to the continuum radio sources, however, remains obscure.

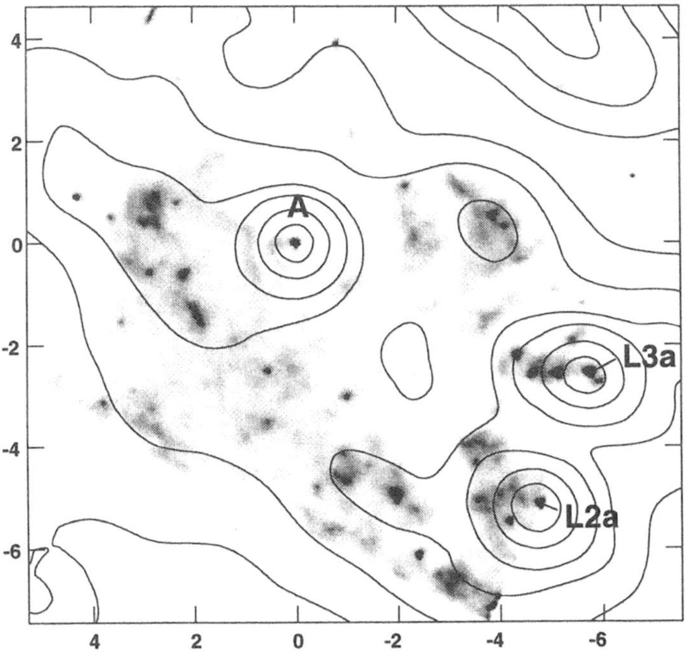

Fig. 7. The blue continuum, obtained with the HST FOC and deconvolved. Overlay isophotes represent a ground based B image. The unresolved Seyfert nucleus (A) and the brightest hot spots (L2a and L3a) are indicated. The units on the axes are arcseconds offset from the centre

References

Barth, A.J., Ho, L.C., Filippenko, A.V., Sargent, W.L.W. (1995): AJ **110**, 1009
Boksenberg, A., Catchpole, R.M., Macchetto, F. et al. (1995): ApJ **440**, 151
Edmunds, M.G., Taylor, K., Turtle, A.J. (1988): MNRAS **234**, 155
Hjelm, M., Lindblad, P.O. (1996): A&A **305**, 727
Jörsäter, S. (1984): Thesis, Stockholm University
Jörsäter, S., Lindblad, P.O. (1989): in ESO Workshop on Extra Nuclear Activity in Galaxies, eds E.J.A. Meurs and R.A.E. Fosbury, ESO Conf. and Workshop Proc. **32**, p. 39.
Jörsäter, S., van Moorsel, G.A. (1995): AJ **110**, 2037
Kristen, H., Jörsäter, S., Lindblad, P.O., Boksenberg, A. (1996): A&A submitted
Macchetto, F., Capetti, A., Sparks, W.B., Axon, D.J., Boksenberg, A. (1994): ApJ **435**, L15
Pedlar, A., Kukula, M.J., Longley, D.P.T., Muxlow, T.W.B., Axon, D.J., Baum, S., O'Dea, C., and Unger, S.W. (1993): MNRAS **263**, 471
Phillips, M.M., Turtle, A.J., Edmunds, M.G., Pagel, B.E.J., (1983): MNRAS **203**, 759
Teuben, P.J., Sanders, R.H., Atherton P.D., van Albada, G.D. (1986): MNRAS **221**, 1
Wilson, A.S., Tsvetanov, Z.I. (1994): AJ **107**, 1227

Hubble Space Telescope Observations of the Centers of Elliptical Galaxies

H. Ford[1,2], L. Ferrarese[1], G. Hartig[2], W. Jaffe[3], Z. Tsvetanov[1] and F. van den Bosch[3]

[1] Department of Physics and Astronomy, Johns Hopkins University, 3701 San Martin Drive, Baltimore, Maryland 21218 USA
[2] Space Telescope Science Institute, 3700 San Martin Drive, Baltimore, Maryland 21218 USA
[3] Huygens Laboratory, Neils Bohr Weg. 2, Leiden, Netherlands

Abstract. Hubble Space Telescope observations of the centers of elliptical galaxies reveal that there are two types of ellipticals, divided by luminosity and morphology. The first type, denoted Type I (or core galaxies), have luminosity profiles which can be fitted accurately with a double power law for radii smaller than ~ 1 kpc, and are characterized by high luminosity, relatively low central surface brightness, and low ellipticity. Type II galaxies (power law galaxies) have power law brightness profiles that rise steeply into the very center, and are characterized by high central surface brightness, high ellipticity, and relatively low luminosity. The Type II galaxies often have a distinct, bright, thin ($r < 25$ pc) nuclear disk. There are no elliptical galaxies with isothermal cores. Dust is the rule rather than the exception in Virgo ellipticals. Elliptical galaxies often have gaseous disks, dusty disks being the most common. Recent COSTAR plus FOS observations show that the small nuclear disks in two luminous radio ellipticals, M 87 and NGC 4261, are rotating around central masses greater than $\sim 10^9$ $M_\odot$. The respective mass to light ratios are $M/L_V \sim 3000$ and 5000, leading to the conclusion that these galaxies host massive black holes.

1 Introduction

A great deal of study in the decade preceding the Hubble Space Telescope (HST) revealed that elliptical galaxies are not the ancient, homogeneous, and nearly pure stellar systems envisioned by Hubble and Baade. Instead, they often are found to have a significant interstellar medium, multiple, discrete dynamical components, and stellar populations with a wide range of ages. These facts have forced a revision of our ideas about the origin and evolution of elliptical galaxies. Before we can develop a detailed understanding how ellipticals evolve, there are many additional fundamental issues which must be addressed. What are the stellar light distributions and deprojected spatial densities of stars in ellipticals? For example, are the centers isothermal as inferred by ground based observations? What are the ages and metallicities of the nuclear stars? What are the types and distributions of orbits, and are cores rotationally supported? How do these properties vary as a function of luminosity and environment? Perhaps an unlikely question a few years ago, are massive black holes (MBHs) present in the centers of some or all early type galaxies, and does their presence account for any of the properties we observe?

Although all the facts about elliptical galaxies are still far from being determined, e.g., we know little about the distribution of stellar orbits or ages, we have learned a great deal from HST, and found real surprises. In this review we will concentrate on two topics, HST observations of dust and morphology in the centers of early type galaxies, and HST observations of MBHs in the centers of the elliptical galaxies M 87 and NGC 4261.

2 Dust and Morphology in the Centers of Elliptical Galaxies

Two groups, one led by Walter Jaffe and the other by Sandra Faber, used the WFPC-1 (pre-refurbishment) to observe the centers of early-type (E, E/S0, and S0) galaxies. Because the two groups used different galaxy samples and analysis procedures, we will discuss the two programs separately. Although only the first in a series of papers from the Faber program has been published, it is reassuring that the two groups come to similar conclusions.

2.1 HST Observations of a Complete Sample of Elliptical Galaxies in the Virgo Cluster.

Jaffe's group selected a sample of elliptical galaxies unbiased by any criteria other than membership in the Virgo cluster as defined by Huchra (1984), classification as E or E/S0 in the RSA (Revised Shapley-Ames Catalog: Sandage and Tammann 1981), and brighter than $B_{\mathrm{T}} = 13.45$. This statistically complete, luminosity-limited sample consists of the 12 brightest E and E/S0 galaxies in the Virgo cluster, plus M 87 and NGC 4472, which were added to the sample from the HST archives. Because the Virgo cluster is relatively near ($D \sim 15$ Mpc; Jacoby et al. 1990), the galaxies can be observed and compared at the same high spatial resolution without differences and uncertainties introduced by different distances.

The galaxies were observed for 700 s + 1700 s with the WFPC-1 Planetary Camera through the F555W filter (Johnson V). All images were re-reduced with the most appropriate WFPC-1 calibration files. The images were deconvolved with the Richardson-Lucy algorithm and with a Fourier filtering algorithm. Both methods gave comparable results. Details of the observations, reductions, analysis, and conclusions given below can be found in Jaffe et al. (1994, hereafter J94), Ferrarese et al. (1994, hereafter F94), and van den Bosch et al. (1994).

The Dimorphism of Early Type Galaxies. A rank ordering of the galaxies in the Jaffe sample shows an interesting effect; the most luminous galaxies have low ellipticity, whereas the flattened E and E/S0 galaxies are almost exclusively

1 to 2 magnitudes fainter. This simple observation presages the discovery that shape and the central luminosity profile divide elliptical galaxies into two types.

The first type, denoted Type I, have luminosity profiles which can be fitted accurately with a double power law for radii smaller than 15″ (~ 1 kpc at Virgo), as shown in Fig. 1. The cores have shallow power laws which continue into the innermost measurable radius (~ 0.″2 = 14 pc). The Type I ellipticals are characterized by high luminosity ($B_T < 11$, which implies $M_B < -20$), relatively low central surface brightness (typically $17 > \mu_V > 16$ mag arcsec^{-2}), low ellipticity (< 0.4), and boxy isophotes.

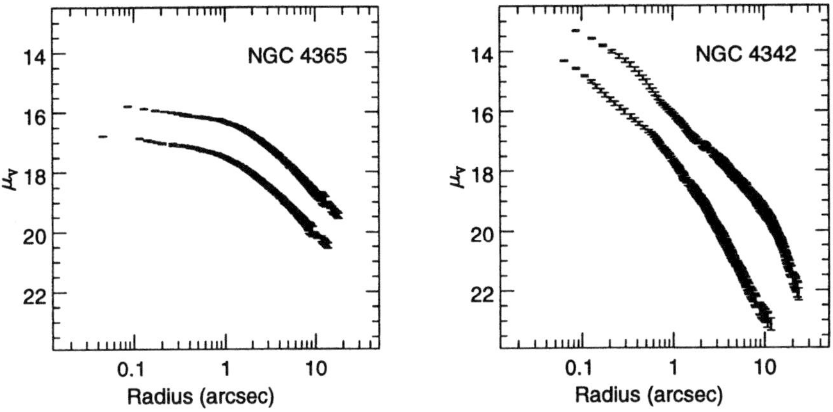

Fig. 1. The brightness profiles for a Type I galaxy (NGC 4365) and a Type II galaxy (NGC 4342), derived from the Richardson-Lucy deconvolved PC images by finding the average surface brightness along the isophotes. The upper curves show the profiles along the major axis of the galaxy, while the lower curves show the profiles along the minor axis

All active galaxies in the Jaffe sample are Type I, consistent with the correlation between radio and X-ray brightness and galaxy luminosity. None of the six Type I galaxies (nor any of the Type IIs) have isothermal cores.

Type II galaxies have power-law brightness profiles that rise steeply into the very center, as shown in Fig. 1. Put another way, Type II galaxies do not have cores. Type II galaxies are characterized by high central surface brightness (a consequence of the steep power law into the center), relatively low luminosity ($B_T > 11; M_B > -20$), flattening, and strongly varying ellipticity caused by a disk. As noted by Lauer et al. (1995), the high central surface brightnesses imply luminosity densities as high as $10^3\,L_\odot$ pc^{-3} to $10^4\,L_\odot$ pc^{-3}. If the power laws persist at scales smaller than HST resolution, the luminosity densities will be even higher.

Another surprise was the discovery that Type II galaxies often have a distinct, bright, thin ($r < 25$ pc) nuclear disk. Figure 2 shows the stellar nuclear disk found by van den Bosch et al. (1994) in NGC 4342.

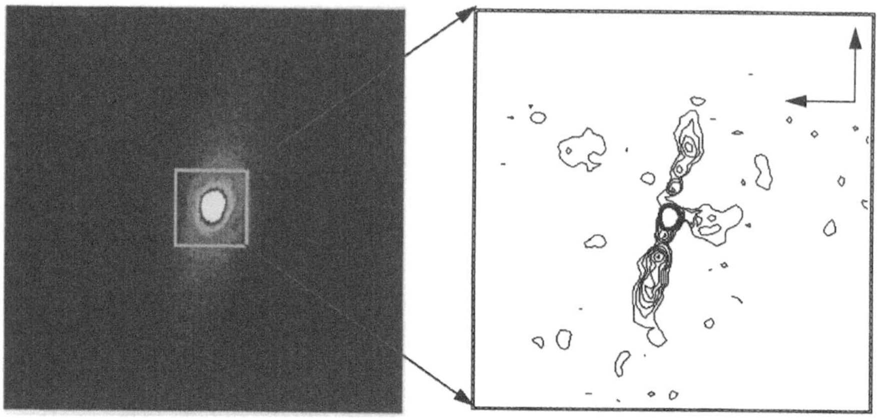

Fig. 2. The HST image of NGC 4342 and the residual map which shows the stellar nuclear disk in this Type II galaxy. The map was constructed by subtracting a model galaxy from the deconvolved image. North is up and East is to the left. The inset and right hand panel are $2''.6 \times 2''.6$

Type II ellipticals appear to be composite galaxies with a bulge, kpc-scale disk on scales of 100 – 1000 pc, in addition to the nuclear disks. The dichotomy seen on small scales is well established on large scales by authors such as Capaccioli et al. (1993) in a survey of 1500 galaxies. They refer to the two classes as "bright" and "ordinary," and find the two classes separate at $M_B \sim -19.3$ ($B_T \sim 12$ at Virgo).

Finally, in the Jaffe sample of Type I galaxies there is no correlation between core size or central surface brightness and galaxy luminosity.

Dust in Early-type Galaxies. Dust was found near the nucleus in 11 of the 14 galaxies. Including the ground-based detection of dust in M 87 (Sparks et al. 1993), 12 of the 14 have nuclear dust.

The three active (radio) galaxies in the Jaffe sample, M 87, NGC 4261, and NGC 4374, have small ($r \sim 100$ pc) gaseous nuclear disks or disk-like structures. Figure 3 shows the WFPC-1 F555W images of the latter two galaxies. The "striations" in NGC 4374 are morphologically similar to the "dust lanes" in Centaurus A, which are due to a highly inclined, rotating, warped disk settling into one of the principal planes in Cen A's triaxial potential. In all three galaxies the major axes of the disks (or long direction of the dust lanes in NGC 4374) are approximately perpendicular to the axis of the radio jets. This alignment suggests that the angular momentum in the gaseous disks may determine the

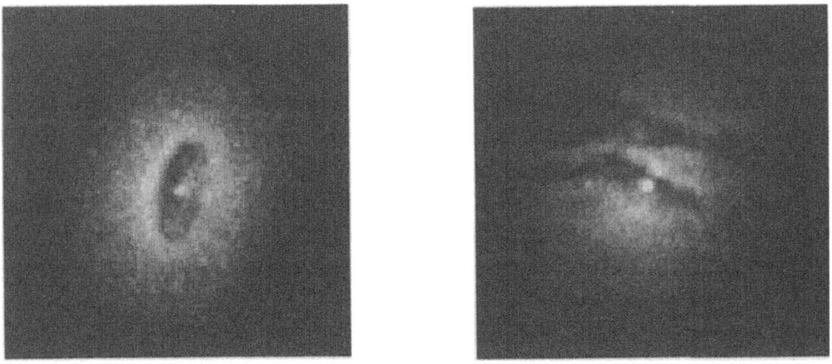

Fig. 3. The **left** panel (5.″8 × 5.″8) shows the deconvolved image of the dusty disk in the nucleus of NGC 4261 and the **right** panel (11.″2 × 11.″2) shows the disk-like striations in NGC 4374. North is up and East is left

The disk in M 87 (Ford et al. 1994, hereafter F94) is largely ionized with a mass of $\sim 3.9 \pm 1.3 \times 10^3$ $M_\odot$. The inclined disk in NGC 4261 (Ferrarese et al. 1996) is very dusty with an optical depth $\tau_V \sim 1$. Assuming a galactic extinction law and using the Bohlin et al. (1978) calibration of $E(B-V)$ versus surface density of atomic plus molecular hydrogen, the mass of neutral and molecular hydrogen is $5.4 \pm 1.8 \times 10^4$ $M_\odot$. Jaffe and McNamara (1994) used the VLA and JCMT to detect H I and CO absorption in NGC 4261 against the radio emission from the nucleus. The absorption velocities and velocity dispersions are respectively 2237 ± 44 and 2235 ± 15 km s^{-1} and 44 and 30 km s^{-1}. The RC3 systemic velocity of the galaxy is 2210 ± 14. The surface density of H I from the VLA observations is 7.2×10^{18} T_S cm^{-2}, and the inferred surface density of H_2 from the CO observations is 2×10^{20} T_S cm^{-2}, where the spin temperatures T_S are respectively likely to be ~ 100 K and ~ 30 K.

Several facts suggest that the dusty disk in NGC 4261 resulted from capturing a smaller galaxy. The strongest argument for capture is the fact that the axis of the disk is perpendicular to the angular momentum of the stars in NGC 4261's center as measured by Davies and Birkinshaw (1986). The geometrical center of the disk is displaced by 90 ± 27 mas from the isophotal center of the galaxy and 67 mas from the nucleus. Perhaps even more surprising, the nucleus is displaced by 23 ± 5 mas, which is 3.3 ± 0.7 pc, assuming NGC 4261 is behind the Virgo cluster at a distance of 30 Mpc (Nolthenius 1993). The displacement of the disk relative to the galaxy and nucleus suggest that it has been captured and has not yet reached dynamical equilibrium.

Although the dust in Type I galaxies such as NGC 4374 and NGC 4261 is most likely the result of capturing a smaller late-type galaxy, the cold stellar disks in the Type II galaxies would be disrupted by collision with a larger late-type galaxy. The dust in Type II galaxies either originates from the capture of a galaxy with $M_{\rm captured} \leq 10^9 M_\odot$, or comes from stars in the parent galaxy.

Conclusions and Inferences from the Jaffe Sample

1. The dichotomy between luminous, round, pressure-supported ellipticals and the fainter, flattened, rotationally supported ellipticals suggests that these galaxies formed in fundamentally different ways. Understanding the origin of Type I and II galaxies will undoubtedly tell us a great deal about the origin and evolution of elliptical galaxies. One obvious hypothesis is that the brighter galaxies formed from mergers of the fainter galaxies, or mergers of the fainter protogalaxies.
2. The dynamically cold stellar disks and highly flattened structures of Type II ellipticals suggest that these galaxies formed during conditions when gaseous dissipation was important and specific angular momentum was high.
3. The persistence of the extremely thin, cold disks with scale heights less than 25 pc suggests these galaxies have not undergone major dynamical disturbances since their formation. Alternatively, the galaxies have not undergone major dynamical disturbances since the disks formed.
4. The high surface brightness of the Type IIs is caused by the nuclear disks seen near to edge-on. As a corollary, we provisionally conclude that essentially all fainter early type galaxies ($B_{\rm T} > 11$ at Virgo) are of Type II. This is a statistical inference based on the fact that the faint round galaxies in Virgo can be accounted for completely by projection effects given the number of flattened faint galaxies.
5. Dust is the rule rather than the exception in the nuclei of early-type galaxies. The detection of gaseous disks in the three radio galaxies suggests that for the first time we are seeing the immediate fuel supply for the engines in these active galaxies. The angular momentum in the gaseous disks appears to be causally connected to the direction of the radio jets.

2.2 HST Observations of the Centers of a Large and Heterogeneous Sample of Early-Type Galaxies

Sample and Observations. Lauer et al. (1995, hereafter L95) observed a heterogeneous sample of 45 galaxies with a selection emphasis designed to fill out the core parameter relations of Lauer (1985) and Kormendy (1985). The sample includes Local Group Es, high luminosity galaxies that appeared to have small cores, galaxies like NGC 3115 that are candidates to host MBHs, four M 32-like dwarfs in Virgo, and "a few galaxies with unusual central morphologies and kinematics".

The data were WFPC-1 PC F555W images with exposure times which varied from 200 s to 5100 s and were calculated to obtain 10 000 photons/pixel in the galaxy center. Spherical aberration was removed from the images with 80 iterations of the R-L algorithm, using three composite stellar PSFs.

Results and Conclusions. Although the authors will publish detailed conclusions in a series of papers, their first paper has the following conclusions:

1. There are two types of elliptical galaxies, those with cores (fit by a double power law), and (steep) power-law galaxies (no cores). These two types correspond closely to the J94 Type I and II galaxies. With the exception of the companion to M 87, NGC 4486 B, which may be a tidally stripped galaxy, all of the "Core" galaxies are high luminosity. If $I(r) \sim r^{-\gamma}$, then in "Core" galaxies $\gamma \sim 1$ for $r > r_{\rm b}$, and $0 < \gamma < 0.3$ for $r < r_{\rm b}$. In the power-law galaxies γ is greater than 0.5. They do not find any galaxies with $0.3 < \gamma < 0.5$; the distribution is bimodal as in the Jaffe sample. Figure 4 shows the observed frequency of the two types as a function of absolute V magnitude. Although L95 did not observe a statistically complete sample of galaxies, the figure shows a clear division into two types by luminosity, as found in the Virgo sample.

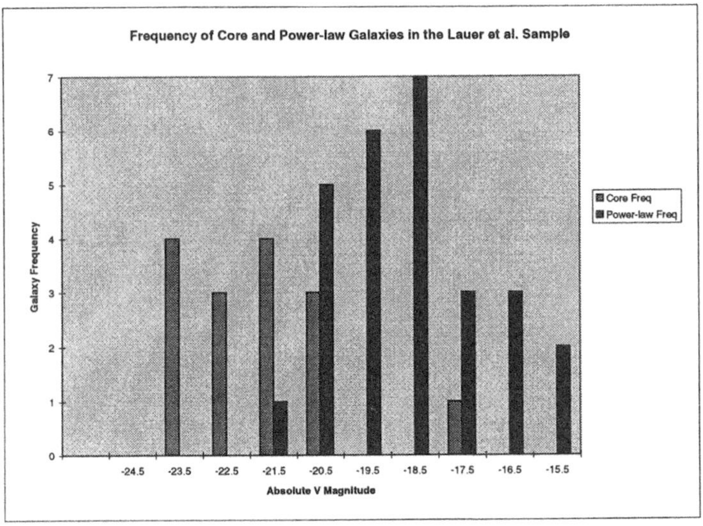

Fig. 4. The observed frequency of core and power-law galaxies in the Lauer sample as a function of absolute V magnitude. Although the sample is not statistically complete, the figure shows a clear division into two types by luminosity, as found in the Virgo sample

Like J94, they do not find any constant surface brightness cores, (as in globular clusters). King profiles do not work, and the cores are not isothermal. The very galaxies that might be expected to have captured nuclei or bright steep cusps due to a black hole do not.

2. L95 found stellar disks in five power-law and one unclassified galaxy. Because they failed to find stellar disks in many highly inclined power-law galaxies, they conclude that the disks are sometimes but not always present in power-law galaxies, contrary the hypothesis by J94 that stellar nuclear disks are always present in Type II galaxies. Unlike J94, they find abundant power-law galaxies with low central ellipticity. Although L95's failure to detect disks possibly could be due to distance or some artifact of the reduction, their identification of power-law galaxies with low central ellipticity argues against the J94 hypothesis that edge-on disks are the origin of steep power-law profiles.
3. Luminosity densities are $\sim 10^3\,L_\odot$ to $10^4\,L_\odot$ for many galaxies at the limiting HST resolution ($r \sim 0\farcs1 - 0\farcs2$). Some galaxies appear to contain nuclear star clusters, with luminosity densities $\sim 10^7\,L_\odot$. These appear exclusively in the less-luminous power-law galaxies. *Core galaxies, precisely the galaxies that might be expected to have remnants of accreted less luminous galaxies, do not have nuclei at limits* $\sim 10^5\,L_\odot$.
4. A few galaxies have non-concentric isophotes in the central regions, i.e., displaced centers.
5. L95 found spiral patterns and dust in a number of galaxies. Specifically, they found dust in 13 of the 45 galaxies. The lower frequency of nuclear dust in the Lauer sample relative to the Jaffe sample could be due to more conservative criteria for dust detection, lower spatial resolution in galaxies at distances greater than Virgo, or could be a real difference. If the latter, it appears that a cluster environment is very favorable for dust to fall into galaxies.

   Three of the L95 galaxies (NGC 4697, NGC 5845 and NGC 5322) have NGC 4261-type nuclear dust disks. The beautiful disk in NGC 4697 ($r \sim 3''$) is very symmetrical and, like the disk in NGC 4261, has sharp edges.

2.3 Conclusions About Morphology and Dust in Early-type Galaxies

The dimorphism of elliptical galaxies appears to be well established by J94 and L95. Type II galaxies often have cold stellar nuclear disks. Dust is common in elliptical galaxies, and in Virgo ellipticals is the rule rather than the exception. Gaseous nuclear disks, especially dusty disks, occur in 10% to 20% of ellipticals. These disks may be a prerequisite for fueling the central engines of radio ellipticals.

3 Massive Black Holes in Elliptical Galaxies

Massive black holes fueled by accretion disks are plausible candidates for the engines in active galaxies. Using stellar velocity dispersions to measure the central masses in ellipticals has been difficult from the ground because seeing prevents observing close enough to the center to rule out models with changing velocity anisotropies and no central dark mass (cf. Dressler and Richstone 1990). Stellar velocity dispersion measurements with the HST are difficult in luminous ellipticals because of HST's small aperture and relatively low throughput, and the fact that luminous ellipticals (Type Is) have low central surface brightnesses.

The discovery of small gaseous nuclear disks in NGC 4261 (Jaffe et al. 1993) and M 87 (F94) provided the possibility of measuring central masses from the circular motion of ionized gas in the disks. The motion of course may not be circular, but if the gas has a disk-like morphology, it is reasonable to make the observations and then test the observed velocities against the simplest possible assumption, Keplerian motion around a point mass. Here we discuss recent observations of the disks in M 87 and NGC 4261.

3.1 HST FOS/COSTAR Observations of the Ionized Disk in M 87

F94 used HST WFPC-2 narrow-band images of M 87 to find a 100 pc scale disk of ionized gas with a major axis approximately perpendicular to the jet. Harms et al. (1994, hereafter H94) used COSTAR and the FOS to measure the velocity at five positions in the disk found by F94, and concluded that there was Keplerian rotation around a mass of 2.4×10^9 $M_\odot$. The large values of the mass-to-light ratio, $(M/L)_I = 170$ and $(M/L)_V \sim 500$, led them to conclude there is a MBH in the center of M 87. If this is indeed true, the velocities should rise parabolically toward the center. To further investigate the dynamics in the disk and the mass of the BH, we reobserved M 87 with the apertures and positions shown in Fig. 5. In addition to the small $0\farcs086$ aperture positions along the apparent major axis, we used the $0\farcs26$ aperture to measure the velocities at two positions along the apparent minor axis of the disk, the direction of the jet.

Representative spectra at these new positions are shown in Fig. 5. We interpret the multiple components as major outflows along the axis defined by the jet. The gas everywhere is very turbulent with a typical FWHM ~ 600 km s^{-1}.

The conclusion that we are observing Keplerian motion around a MBH is supported and strengthened by the new observations. Fig. 6 shows a preliminary fit of the H94 [O III] velocities and the new [O III] velocities to the projection of a Keplerian disk onto the line of sight (the equation for ${V_K}^2$ in H94). The slope of the fit gives the Keplerian mass. The plot would be a straight line for a perfect fit, and a scatter diagram in the absence of a central mass. Because at least three parameters can be varied (central mass, inclination of the disk, and position angle of the line of nodes), the fit is not unique. However, all combinations of the parameters which minimize the rms residuals require a central mass $\sim 2 \times 10^9$

$M_\odot$. The best fit kinematical inclination $i = 36°$ is close to the $42° \pm 5°$ geometrical inclination of the disk (F94), and the best fit kinematical line of nodes $\theta = 6°$ agrees acceptably well with $11° < \theta < 22°$ determined geometrically. The kinematical line of nodes is only 6° from the direction perpendicular to the jet, again suggesting a causal relationship between the angular momentum in the disk and the direction of the jet.

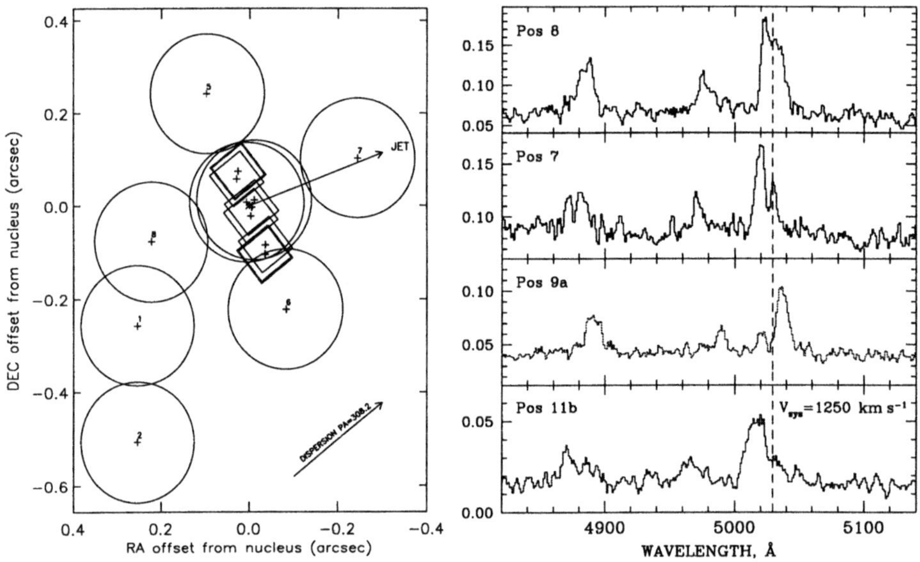

Fig. 5. The **left** panel shows a schematic of the FOS aperture positions relative to the nucleus and jet in M 87. The small 0.″086 square aperture observations were made at two different epochs at each position, accounting for the small differences in centering. The **right** panel shows representative COSTAR plus FOS spectra at four positions in the center of M 87. The ordinate is observed flux in counts per second

The dashed lines in Fig. 6 are the extremes which bound the data, and show that the fit requires a mass between 1 and 3.5×10^9 $M_\odot$. Several factors suggest that the dynamics in the disk are more complicated than simple Keplerian motion. The first is the turbulence in the disk, which shows that the motion cannot be perfectly Keplerian. The apparent spiral structure noted by F94 suggests shocks, which can accelerate gas. Finally, the disk in NGC 4261 is eccentric relative to the nucleus, and shows signs of mild warps. There may be similar effects in the M 87 disk. In view of these, it is remarkable that the Keplerian model works as well as it does. Though we likely could improve the fit to the data by adding more parameters such as eccentricity, we do not think this is justified at this time.

The mean radius of the four points closest to the nucleus is 0$''$.087 = 6 pc. Using the Lauer et al. (1992) WFPC-1 observations of M 87 to estimate the luminosity (excluding light from the non-thermal point source), we provisionally get $(M/L)_\mathrm{I} \sim 1000$ and $(M/L)_\mathrm{V} \sim 3000$! In view of the good fit of the data to the model, and the fact that M/L continues to rise and reaches very high values as we approach the center, we conclude there is a MBH with a mass $\sim 2 \times 10^9$ $M_\odot$ in the center of M 87.

The mass of the black hole and M/L are likely underestimated because we have not accounted for the kinetic energy in the turbulent motion. Two important unanswered questions are, what drives the turbulence in the disk, and how do we kinematically and physically account for the outflow (large non-circular velocities)?

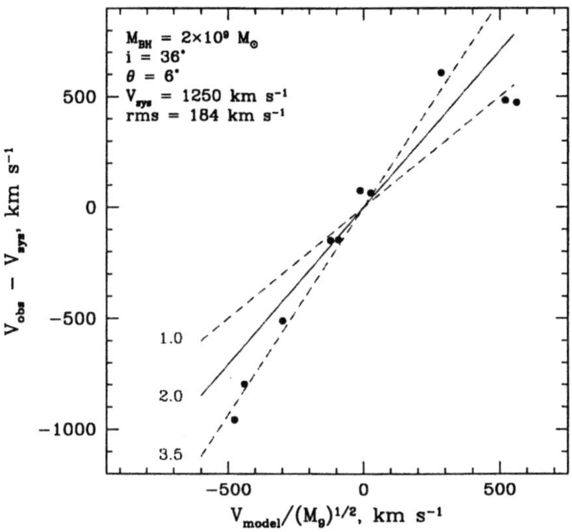

Fig. 6. A preliminary fit of the H94 [O III] M 87 velocities and the new [O III] velocities to the projection of a Keplerian disk onto the line of sight. The slope of the fit gives the Keplerian mass in units of 10^9 $M_\odot$

3.2 HST WFPC-2 and FOS/COSTAR Observations of NGC 4261

Ferrarese et al. (1996, hereafter F96) use the Planetary Camera in the WFPC-2 to take F547M (V), F675W(R), and F791W (I) images of NGC 4261. F96 used the images to derive the optical depth in the disk and construct an Hα+[N II] image of the nucleus. Figure 7 shows a true color composite image of the galaxy. The new images are notable in showing "spiral-like" structure in the disk, and a modest warp on the NE side of the disk. The color of the bright striations in the disk shows that these are regions of low opacity rather than star-forming regions. The nucleus is unresolved in the V and I images, but is extended in the R image and in the constructed Hα+[N II] image. The mass of ionized gas is not very large, a few tens to a few hundred solar masses. As noted earlier, the nucleus is not in the center of the disk.

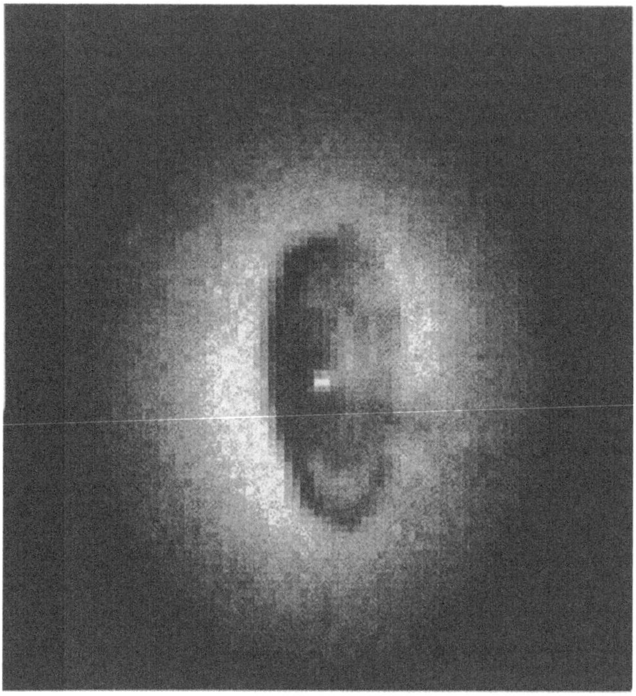

Fig. 7. A true color composite image of the V, R, and I images of NGC 4261. The major axis of the disk is approximately 1.″7 long. North is approximately at the top, and East to the left. The reddening of background starlight seen through the spiral-like gaps in the disk is readily evident. A linear dust feature can be seen extending beyond the disk on both sides of the disk. The projected disk appears widest on the north side

F96 used COSTAR and the FOS 0.″086 aperture to take G570H spectra at 5 positions along the major axis and 2 positions on either side of the nucleus along the minor axis. The center of the cross was on the nucleus, and 4 additional positions filled in the corners of the cross. The positions were spaced by 0.″086 in

x and y. Figure 8 shows the best fit of the data to the projection of a Keplerian disk onto the line of sight. The data shows the clear signature of rotation. The kinematical inclination of the disk, $i = 69°$, is in good agreement with the 63° geometrical inclination of the disk. The position angle of the kinematical line of nodes (19°) rotates 36° from the geometrical major axis, suggesting that the disk twists toward the center.

The central mass from the best fit to the Keplerian model is $1.2 \pm 0.4 \times 10^9$ $M_\odot$. Using the observed light distribution corrected for the extinction in the disk, F96 obtain $(M/L)_V \sim 5200$! This very large M/L suggests that there is a MBH in the center of NGC 4261 which is fueled by the gas and dust in the disk.

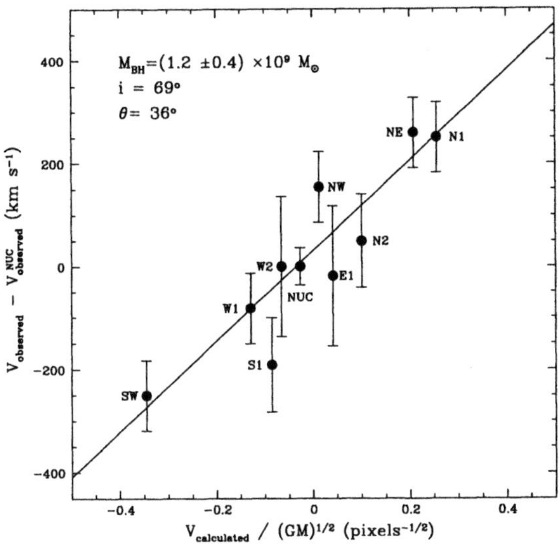

Fig. 8. The observed velocities in NGC 4261 fit to the projection of a Keplerian disk

3.3 Conclusions

There is now strong evidence that there are MBHs with masses $\sim 10^9$ $M_\odot$ in two luminous radio elliptical galaxies. Miyoshi et al. (1995) and Greenhill et al. (1995) used VLBI synthesis maps of the luminous H_2O maser in the nucleus of the Seyfert galaxy NGC 4258 to map a beautiful Keplerian velocity curve in an edge-on molecular ring at a radius ~ 0.1 pc. The central mass is 2.1×10^7 $M_\odot$, and the mass density exceeds 3.5×10^9 $M_\odot$ pc^{-3}. This provides very strong evidence for a MBH in NGC 4258. Confirmation of the paradigm that active galaxies are powered by MBHs in these three galaxies lends considerable confidence to the conclusion that the paradigm is true in all active galaxies.

Research on MBHs appears poised to move from seeking to establish the existence of MBHs to a study of black hole demographics. Are there MBHs in all ellipticals, and in all galaxies? How do the masses depend on parent mass?

When and how did MBHs become so massive, and what role do they play in the evolution of galaxies? Seeking the answers to these and many other questions will profitably occupy astronomers for the next decade.

Acknowledgements. This research was supported by NASA grants NAS 5-29293, NAG5-1630, and NAS/STScI grant GO-2607.01-87A. H. Ford and Z. Tsvetanov thank the Nobel Foundation for supporting the conference, and especially thank Drs. Steven Jörsäter, Aage Sandqvist and Per Olof Lindblad for their warm hospitality.

References

Capaccioli, M., Caon, N., D'Onofrio, M. (1993): in ESO Conference Proc. 45 Structure, Dynamics and Chemical Evolution of Elliptical Galaxies, eds. I.J. Danziger, W.W. Zeilinger, K. Kjär, p. 43
Davies, R.L., Birkinshaw, M. (1986): ApJ **345**, L45
Dressler, A., Richstone, D. (1990): ApJ **348**, 120
Ferrarese, L., Ford, H.C., Jaffe, W., O'Connell, R.W.O. (1996): ApJ in press
Ferrarese, L., van den Bosch, F.C., Ford, H.C., Jaffe, W., O'Connell, R.W. (1994): AJ **108**, 1598
Ford, H.C., Harms, R.J., Tsvetanov, Z.I., Hartig, G.F., Dressel, L.L., Kriss, G.A., Davidsen, A.F., Bohlin, R.A., Margon, B. (1994): ApJ **435**, L27
Greenhill, L.D., Jiang, D.R., Moran, J.M., Reid, M.J., Lo, K.Y., Claussen, M.J. (1995): ApJ **440**, 619
Harms, R.J., Ford, H.C., Tsvetanov, Z.I., Hartig, G.F., Dressel, L.L., Kriss, G.A., Bohlin, R.A., Davidsen, A.F., Margon, B., Kochhar, A.K. (1994): ApJ **435**, L35
Huchra, J.P. (1984): in ESO Conference Proc. 2 The Virgo Clusters of Galaxies, eds. O.-G. Richert, B. Binggeli, p. 181
Jacoby, G., Ciardullo, R., Ford, H.C. (1990): Bull. CFH **22**, 15
Jaffe, W., Ford, H.C., Ferrarese, L., van den Bosch, F., O'Connell, R.W.O. (1993): Nat **364**, 213
Jaffe, W., Ford. H.C., O'Connell, R.W.O., van den Bosch, F.C., Ferrarese, L. (1994): AJ **108**, 1567
Jaffe, W., McNamara, B.R. (1994): ApJ **434**, 110
Kormendy, J. (1985): ApJ **295**, 73
Lauer, T.R. (1985): ApJ **292**, 104
Lauer, T.R., Faber, S.M., Lynds, C.R., Baum, W.A., Ewald, S.P., Groth, E.J., Hester, J.J., Holtzman, J.A., Kristian, J., Light, R.M. (1992): AJ **103**, 703
Miyoshi, M., Moran, J., Herrnstein, J., Nakai, N., Diamond, P., Inoue, J. (1995): Nat **373**, 127
Nolthenius, R. (1993): ApJS **85**, 1
Sandage, A., Tammann, G.A. (1981): A Revised Shapley-Ames Catalog of Bright Galaxies, Carnegie Inst. of Washington, Washington, DC
van den Bosch, F.C., Ferrarese, L., Jaffe, W., Ford, H.C., O'Connell, R.W.O. (1994): AJ **108**, 1579

Lecture Notes in Physics

For information about Vols. 1–439
please contact your bookseller or Springer-Verlag

Vol. 440: H. Latal, W. Schweiger (Eds.), Matter Under Extreme Conditions. Proceedings, 1994. IX, 243 pages. 1994.

Vol. 441: J. M. Arias, M. I. Gallardo, M. Lozano (Eds.), Response of the Nuclear System to External Forces. Proceedings, 1994, VIII. 293 pages. 1995.

Vol. 442: P. A. Bois, E. Dériat, R. Gatignol, A. Rigolot (Eds.), Asymptotic Modelling in Fluid Mechanics. Proceedings, 1994. XII, 307 pages. 1995.

Vol. 443: D. Koester, K. Werner (Eds.), White Dwarfs. Proceedings, 1994. XII, 348 pages. 1995.

Vol. 444: A. O. Benz, A. Krüger (Eds.), Coronal Magnetic Energy Releases. Proceedings, 1994. X, 293 pages. 1995.

Vol. 445: J. Brey, J. Marro, J. M. Rubí, M. San Miguel (Eds.), 25 Years of Non-Equilibrium Statistical Mechanics. Proceedings, 1994. XVII, 387 pages. 1995.

Vol. 446: V. Rivasseau (Ed.), Constructive Physics. Results in Field Theory, Statistical Mechanics and Condensed Matter Physics. Proceedings, 1994. X, 337 pages. 1995.

Vol. 447: G. Aktaş, C. Saçlıoğlu, M. Serdaroğlu (Eds.), Strings and Symmetries. Proceedings, 1994. XIV, 389 pages. 1995.

Vol. 448: P. L. Garrido, J. Marro (Eds.), Third Granada Lectures in Computational Physics. Proceedings, 1994. XIV, 346 pages. 1995.

Vol. 449: J. Buckmaster, T. Takeno (Eds.), Modeling in Combustion Science. Proceedings, 1994. X, 369 pages. 1995.

Vol. 450: M. F. Shlesinger, G. M. Zaslavsky, U. Frisch (Eds.), Lévy Flights and Related Topics in Physics. Proceedigs, 1994. XIV, 347 pages. 1995.

Vol. 451: P. Krée, W. Wedig (Eds.), Probabilistic Methods in Applied Physics. IX, 393 pages. 1995.

Vol. 452: A. M. Bernstein, B. R. Holstein (Eds.), Chiral Dynamics: Theory and Experiment. Proceedings, 1994. VIII, 351 pages. 1995.

Vol. 453: S. M. Deshpande, S. S. Desai, R. Narasimha (Eds.), Fourteenth International Conference on Numerical Methods in Fluid Dynamics. Proceedings, 1994. XIII, 589 pages. 1995.

Vol. 454: J. Greiner, H. W. Duerbeck, R. E. Gershberg (Eds.), Flares and Flashes, Germany 1994. XXII, 477 pages. 1995.

Vol. 455: F. Occhionero (Ed.), Birth of the Universe and Fundamental Physics. Proceedings, 1994. XV, 387 pages. 1995.

Vol. 456: H. B. Geyer (Ed.), Field Theory, Topology and Condensed Matter Physics. Proceedings, 1994. XII, 206 pages. 1995.

Vol. 457: P. Garbaczewski, M. Wolf, A. Weron (Eds.), Chaos – The Interplay Between Stochastic and Deterministic Behaviour. Proceedings, 1995. XII, 573 pages. 1995.

Vol. 458: I. W. Roxburgh, J.-L. Masnou (Eds.), Physical Processes in Astrophysics. Proceedings, 1993. XII, 249 pages. 1995.

Vol. 459: G. Winnewisser, G. C. Pelz (Eds.), The Physics and Chemistry of Interstellar Molecular Clouds. Proceedings, 1993. XV, 393 pages. 1995.

Vol. 460: S. Cotsakis, G. W. Gibbons (Eds.), Global Structure and Evolution in General Relativity. Proceedings, 1994. IX, 173 pages. 1996.

Vol. 461: R. López-Peña, R. Capovilla, R. García-Pelayo, H. Waelbroeck, F. Zertuche (Eds.), Complex Systems and Binary Networks. Lectures, México 1995. X, 223 pages. 1995.

Vol. 462: M. Meneguzzi, A. Pouquet, P.-L. Sulem (Eds.), Small-Scale Structures in Three-Dimensional Hydrodynamic and Magnetohydrodynamic Turbulence. Proceedings, 1995. IX, 421 pages. 1995.

Vol. 463: H. Hippelein, K. Meisenheimer, H.-J. Röser (Eds.), Galaxies in the Young Universe. Proceedings, 1994. XV, 314 pages. 1995.

Vol. 464: L. Ratke, H. U. Walter, B. Feuerbach (Eds.), Materials and Fluids Under Low Gravity. Proceedings, 1994. XVIII, 424 pages, 1996.

Vol. 465: S. Beckwith, J. Staude, A. Quetz, A. Natta (Eds.), Disks and Outflows Around Young Stars. Proceedings, 1994. XII, 361 pages, 1996.

Vol. 466: H. Ebert, G. Schütz (Eds.), Spin – Orbit-Influenced Spectroscopies of Magnetic Solids. Proceedings, 1995. VII, 287 pages, 1996.

Vol. 467: A. Steinchen (Ed.), Dynamics of Multiphase Flows Across Interfaces. 1994/1995. XII, 267 pages. 1996.

Vol. 468: C. Chiuderi, G. Einaudi (Eds.), Plasma Astrophysics. 1994. VII, 326 pages. 1996.

Vol. 469: H. Grosse, L. Pittner (Eds.), Low-Dimensional Models in Statistical Physics and Quantum Field Theory. Proceedings, 1995. XVII, 339 pages. 1996.

Vol. 470: E. Martínez-González, J. L. Sanz (Eds.), The Universe at High-z, Large-Scale Structure and the Cosmic Microwave Background. Proceedings, 1995. VIII, 254 pages. 1996.

Vol. 471: W. Kundt (Ed.), Jets from Stars and Galactic Nuclei. Proceedings, 1995. X, 290 pages. 1996.

Vol. 472: J. Greiner (Ed.), Supersoft X-Ray Sources. Proceedings, 1996. XIII, 350 pages. 1996.

Vol. 474: Aa. Sandqvist, P. O. Lindblad (Eds.), Barred Galaxies and Circumnuclear Activity. Proceedings of the Nobel Symposium 98, 1995. XI, 306 pages. 1996.

New Series m: Monographs

Vol. m 1: H. Hora, Plasmas at High Temperature and Density. VIII, 442 pages. 1991.

Vol. m 2: P. Busch, P. J. Lahti, P. Mittelstaedt, The Quantum Theory of Measurement. XIII, 165 pages. 1991. Second Revised Edition: XIII, 181 pages. 1996.

Vol. m 3: A. Heck, J. M. Perdang (Eds.), Applying Fractals in Astronomy. IX, 210 pages. 1991.

Vol. m 4: R. K. Zeytounian, Mécanique des fluides fondamentale. XV, 615 pages, 1991.

Vol. m 5: R. K. Zeytounian, Meteorological Fluid Dynamics. XI, 346 pages. 1991.

Vol. m 6: N. M. J. Woodhouse, Special Relativity. VIII, 86 pages. 1992.

Vol. m 7: G. Morandi, The Role of Topology in Classical and Quantum Physics. XIII, 239 pages. 1992.

Vol. m 8: D. Funaro, Polynomial Approximation of Differential Equations. X, 305 pages. 1992.

Vol. m 9: M. Namiki, Stochastic Quantization. X, 217 pages. 1992.

Vol. m 10: J. Hoppe, Lectures on Integrable Systems. VII, 111 pages. 1992.

Vol. m 11: A. D. Yaghjian, Relativistic Dynamics of a Charged Sphere. XII, 115 pages. 1992.

Vol. m 12: G. Esposito, Quantum Gravity, Quantum Cosmology and Lorentzian Geometries. Second Corrected and Enlarged Edition. XVIII, 349 pages. 1994.

Vol. m 13: M. Klein, A. Knauf, Classical Planar Scattering by Coulombic Potentials. V, 142 pages. 1992.

Vol. m 14: A. Lerda, Anyons. XI, 138 pages. 1992.

Vol. m 15: N. Peters, B. Rogg (Eds.), Reduced Kinetic Mechanisms for Applications in Combustion Systems. X, 360 pages. 1993.

Vol. m 16: P. Christe, M. Henkel, Introduction to Conformal Invariance and Its Applications to Critical Phenomena. XV, 260 pages. 1993.

Vol. m 17: M. Schoen, Computer Simulation of Condensed Phases in Complex Geometries. X, 136 pages. 1993.

Vol. m 18: H. Carmichael, An Open Systems Approach to Quantum Optics. X, 179 pages. 1993.

Vol. m 19: S. D. Bogan, M. K. Hinders, Interface Effects in Elastic Wave Scattering. XII, 182 pages. 1994.

Vol. m 20: E. Abdalla, M. C. B. Abdalla, D. Dalmazi, A. Zadra, 2D-Gravity in Non-Critical Strings. IX, 319 pages. 1994.

Vol. m 21: G. P. Berman, E. N. Bulgakov, D. D. Holm, Crossover-Time in Quantum Boson and Spin Systems. XI, 268 pages. 1994.

Vol. m 22: M.-O. Hongler, Chaotic and Stochastic Behaviour in Automatic Production Lines. V, 85 pages. 1994.

Vol. m 23: V. S. Viswanath, G. Müller, The Recursion Method. X, 259 pages. 1994.

Vol. m 24: A. Ern, V. Giovangigli, Multicomponent Transport Algorithms. XIV, 427 pages. 1994.

Vol. m 25: A. V. Bogdanov, G. V. Dubrovskiy, M. P. Krutikov, D. V. Kulginov, V. M. Strelchenya, Interaction of Gases with Surfaces. XIV, 132 pages. 1995.

Vol. m 26: M. Dineykhan, G. V. Efimov, G. Ganbold, S. N. Nedelko, Oscillator Representation in Quantum Physics. IX, 279 pages. 1995.

Vol. m 27: J. T. Ottesen, Infinite Dimensional Groups and Algebras in Quantum Physics. IX, 218 pages. 1995.

Vol. m 28: O. Piguet, S. P. Sorella, Algebraic Renormalization. IX, 134 pages. 1995.

Vol. m 29: C. Bendjaballah, Introduction to Photon Communication. VII, 193 pages. 1995.

Vol. m 30: A. J. Greer, W. J. Kossler, Low Magnetic Fields in Anisotropic Superconductors. VII, 161 pages. 1995.

Vol. m 31: P. Busch, M. Grabowski, P. J. Lahti, Operational Quantum Physics. XI, 230 pages. 1995.

Vol. m 32: L. de Broglie, Diverses questions de mécanique et de thermodynamique classiques et relativistes. XII, 198 pages. 1995.

Vol. m 33: R. Alkofer, H. Reinhardt, Chiral Quark Dynamics. VIII, 115 pages. 1995.

Vol. m 34: R. Jost, Das Märchen vom Elfenbeinernen Turm. VIII, 286 pages. 1995.

Vol. m 35: E. Elizalde, Ten Physical Applications of Spectral Zeta Functions. XIV, 228 pages. 1995.

Vol. m 36: G. Dunne, Self-Dual Chern-Simons Theories. X, 217 pages. 1995.

Vol. m 37: S. Childress, A.D. Gilbert, Stretch, Twist, Fold: The Fast Dynamo. XI, 410 pages. 1995.

Vol. m 38: J. González, M. A. Martín-Delgado, G. Sierra, A. H. Vozmediano, Quantum Electron Liquids and High-T_c Superconductivity. X, 299 pages. 1995.

Vol. m 39: L. Pittner, Algebraic Foundations of Non-Commutative Differential Geometry and Quantum Groups. XII, 469 pages. 1996.

Vol. m 40: H.-J. Borchers, Translation Group and Particle Representations in Quantum Field Theory. VII, 131 pages. 1996.

Vol. m 41: B. K. Chakrabarti, A. Dutta, P. Sen, Quantum Ising Phases and Transitions in Transverse Ising Models. X, 204 pages. 1996.

Vol. m 42: P. Bouwknegt, J. McCarthy, K. Pilch, The $\mathcal{W}_3$ Algebra. Modules, Semi-infinite Cohomology and BV Algebras. XI, 204 pages. 1996.